OIL AND GAS ASSESSMENT
- Methods and Applications -

AAPG Studies in Geology #21

OIL AND GAS ASSESSMENT
- Methods and Applications -

edited by
Dudley D. Rice

Published by
The American Association of Petroleum Geologists
Tulsa, Oklahoma 74101, U.S.A.

Library of Congress Cataloging-in-Publication Data

Oil and gas assessment.

    (AAPG studies in geology ; #21)
    Bibliography: p.
    Includes index.
    1. Petroleum--Reserves.   2. Gas, Natural--Reserves.
I. Rice, Dudley D.   II. American Association of Petroleum
Geologists.   III. Series: AAPG studies in geology ;
no. 21.
TN871.0352 1986   622'.1828   86-14158

ISBN: 0-89181-027-7

Association Editor: James Helwig
Science Director: Ronald L. Hart
Special Projects Manager: Victor V. Van Beuren
Project Editor: Anne H. Thomas
Special Editor & Production Supervisor: Kathy A. Walker
Design: S. Wally Powell
Typographers: Eula Matheny, Elaine DiLoreto, and Phyllis Kenney

PREFACE

There is a continual need to update estimates of oil and gas resources remaining to be discovered, and also to refine the methodologies for making these assessments. In 1974, AAPG sponsored a research conference dealing with the above topics. Many of the papers presented at that conference were published in 1975 in AAPG Studies in Geology No. 1, entitled "Methods of estimating the volume of undiscovered oil and gas resources." As a follow-up to that conference and volume, a USGS Workshop on "Oil and gas resource appraisal methodology for the future" was convened in 1983, and a special session for the 1984 National AAPG meeting was held on "Assessment of oil and gas resources." The purpose was to bring together the experts from government, academia, and private industry who were working on assessment of oil and gas resources. Written versions of many of the talks presented at these sessions, in addition to several other papers, are presented in this volume. The papers have been grouped into two types: those describing methodologies for evaluating resources and those presenting assessments of both conventional and unconventional resources.

This volume is the work of many dedicated authors, who were willing to meet deadlines and to make changes suggested by the editors. We thank the technical reviewers for their important contribution to the volume.

Dudley D. Rice
Denver, Colorado
May 1986

TABLE OF CONTENTS

Resource Appraisal Methods: Choice and Outcome

Betty M. Miller
U.S. Geological Survey
Reston, Virginia

An overview and critique are presented for five basic categories of resource appraisal methods used to generate petroleum resource estimates during the last three decades: (1) areal and volumetric yield techniques, in combination with geologic analogy; (2) Delphi or subjective consensus assessments; (3) historical performance or behavioristic extrapolations; (4) geochemical material balance techniques; and (5) combination methods using geologic and statistical models (e.g., exploration–play analysis). The results of selected resource estimates are compared from several geographic areas in which different methods have been used. Major issues fundamental to resource assessment discussed are (1) the basic requirements and criteria essential in a systematic approach to the selection of the resource appraisal methods to be used in assessing petroleum resources; (2) a critique of the strengths and weaknesses of the basic resource appraisal methods; (3) the differences among resource estimates that result from different methods being used in the assessments; and (4) the status of resource assessment methodology for the last decade and goals for the future.

A comprehensive analysis is provided for three case studies: nationwide resource assessments for both Canada and the United States and regional resource assessments for the Permian basin of western Texas and southeastern New Mexico. Evidence is presented that suggests (1) that a significant number of the differences among resource estimates result from the use of different appraisal methods, (2) that consistent patterns do occur in the magnitude of the estimates, and (3) that the relative magnitude of the estimates is predictable depending upon the specific method of assessment used.

INTRODUCTION

At the First IIASA Conference on Energy Resources, M. F. Searl, in his referral to the "methodology for assessing resources at the aggregate or macro level," stated: "The methodology of resource assessment is a much neglected topic of research in the United States." He continued: "There have been few systematic efforts at establishing resource assessment methodologies" (Searl, 1975, p. 71). My paper addresses the issue of resource appraisal methodology and whether the situation has shown improvement during the succeeding decade.

There are many methods now for estimating petroleum resource potential and innumerable variations on the basic techniques, each supported by its respective proponents. Each basic method requires a different level of geologic knowledge or degree of available information. For the application of these different appraisal methods, it is also important to consider the needs of the users and the purposes for which the assessments may be used.

To date, all the published methods have recognized limitations and have both strengths and weaknesses. Major problems, however, have arisen because of misinterpretation and misuse of the resulting estimates and because of the lack of recognition by the users (deliberate or not) of the limitations of the methods employed, the assumptions made, or the data used. No single method has universal application or acceptance, and until some petroleum province has been completely explored and completely produced, we will not know definitively the reliability of any of these methods for estimating oil and gas resources.

The purpose of this paper is (1) to compare the results of selected resource assessments from several geographic areas in which different methods have been used, and (2) to answer certain fundamental questions raised. What should an estimator consider in making the selection of a method or combination of methods to conduct an assessment of the petroleum resources of a specified area? What should the user of the resource estimates be cognizant of regarding the inherent limitations of the estimates that result from the methods used in making the assessments? Is there a significant difference in the outcomes of the assessments relative to the selection of the methodology? Has any progress been made since Searl's observations in 1975 toward developing a systematic approach to a total resource appraisal methodology that addresses all basic issues relative to the petroleum potential of a designated area?

First, the basic types of resource appraisal methods referred to in this paper are reviewed, as well as their general data requirements. Second, the considerations that the estimator must analyze and resolve when selecting a resource appraisal method or combination of methods are summarized. Several case studies are analyzed and discussed demonstrating the outcomes of petroleum resource assessments that are significantly dependent on the choice of the resource appraisal method used. I will deal mainly with ideas and approaches to resource assessment problems based on my personal experience while in private industry and with the U.S. Geological Survey (USGS). No attempt is made here at complete coverage of any of the methods.

REVIEW OF BASIC ASSESSMENT METHODS

The many estimates of petroleum resources that have been made in the last three decades fall into five basic categories of methodology. These categories, simplified for classification purposes, are (1) areal and volumetric yield, in combination with geologic analogy; (2) Delphi or subjective consensus assessment; (3) performance or behavioristic extrapolation based on historical data; (4) geochemical material balance; and (5) combinations of geologic and statistical models. The optimum application of each category of methods is shown in Figure 1.

Resource estimates can be made on any level of basic geologic data. The amount of data available, however, can determine the quality of the estimate and should be the primary factor for determining the method(s) that might be used for the appraisal. The chosen methods or procedures may be altered as the amount and nature of the available data change for a specific basin, or if the purpose for the assessment should change.

The selection of resource appraisal methods should be dependent on the degree of geologic assurance at a given stage of exploration within a petroleum province. In the early frontier stages of exploration, when scant information exists concerning the gross interpretation of the basin geology and when the amount of data are minimal, an evaluator, applying the principles of petroleum occurrence from worldwide experience, may make use of subjective judgment to provide a basis for the assumption of whether or not potential hydrocarbons are present. As the data base grows due to increased knowledge accompanying exploration and as the results of geophysical surveys, drilling, and geochemical data become available, methods incorporating more objective data should be utilized. The methods used may evolve to the level of dealing with exploration plays, or they may focus on making estimates of undiscovered prospects. When abundant and detailed data are available, the choice of the method used may become more dependent on other factors, such as availability of the estimator's time, costs and efforts involved, the purpose and use of the resource estimate, and concerns regarding the credibility of the estimates. The credibility of the estimate, however, is dependent on the quality of the geologic data, the studies on which it is based, and the geologic experience and expertise of the estimators.

Areal and Volumetric Yield Methods in Combination with Geologic Analogy

Areal and volumetric yield techniques have been used in a wide variety of ways in making petroleum resource estimates. These techniques range from the use of worldwide average yields expressed in barrels of oil or cubic feet of gas per cubic mile of sedimentary rock, or per square mile of surface area (assuming constant rock thickness) applied uniformly over a sedimentary basin, to more sophisticated analyses in which the yields from a geologically analogous basin have been used to provide a basis of comparison. The pioneer works of Weeks (1950), Zapp (1962), and Hendricks (1965) are illustrative of early techniques (Meyer, 1978; Miller, 1979).

Some highly sophisticated refinements have been made on the basic volumetric techniques that are used in almost all present-day appraisal methods. In fact, close scrutiny reveals very few methods of petroleum resource appraisal that are not in some way, directly or indirectly, dependent on the basic concepts of volumetrics. The few exceptions may be the performance or behavioristic extrapolation methods and some of the purely subjective methods.

Geologic analogy enters into nearly every resource appraisal method. It is the key to the comparative input needed for the volumetric yield techniques. An example of its use was in the 1975 USGS national assessment of petroleum resources (Miller et al., 1975), in which the records of the oil and gas yields from well-explored areas within 75 North American basins were compiled to establish a scale of hydrocarbon yields for geologically analogous basins. The 1972 regional assessments for the provinces of Canada are another example of the use of volumetric yield methods based on a range of representative basin yields (McCrossan and Porter, 1973).

The accuracy of the volumetric yield and geologic analogy methods depends on the validity of the analog chosen before a hydrocarbon yield is selected to make a forecast for the potential of the unexplored basin or parts of a basin. The usual procedure is to select one analogous yield and derive a single value estimate. An approach I favor for this method, however, is the selection of a representative range of analogous basins and their respective yields with probabilities assigned to determine a minimum and maximum estimate for the potential resource. These estimates are made on the basis of the minimum and maximum favorable conditions for the occurrence of petroleum. The results obtained from the volumetric yield method can be useful on a broad regional basis or in a reconnaissance-type estimate of the resource potential, particularly in evaluation of frontier or unexplored geologic areas or as a cross check of resource estimates in areas where other methods have been used.

Delphi or Subjective Consensus Assessment Methods

In the Delphi approach, the estimation of the petroleum resources is the consensus of a team of experts. A group of experts usually reviews all the geologic information available in an area or basin, which sometimes includes detailed geologic basin analyses and the results of any previous

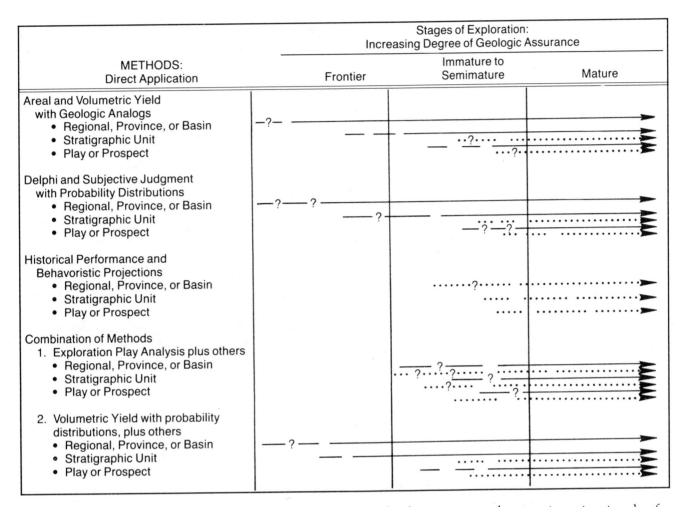

Figure 1. Resource appraisal methods applicable for the various stages of exploration in a petroleum province, given in order of an increasing degree of geologic assurance. Modified from figure 9 of Miller (1983). The solid lines indicate long-term applications and dotted lines short-term applications. Where lines are continuous it indicates that the variations in the resource methods are applicable at different stages in exploration; dashed lines indicate that the application of the methods may generate less reliable estimates at the respective stages in exploration; question marks indicate concern over the application of the method at that stage of exploration with questionable resource estimates.

assessments by other estimators and/or by other methods. Usually each member of the team constructs his or her own probability distribution of the estimated potential resources. The group reviews all the individual results and makes modifications where considered necessary. The final probability distributions are determined either by consensus of the group or by averaging the individual probability distributions (Miller et al., 1975; Dolton et al., 1981). A special National Petroleum Council (NPC) study on the petroleum assessment of the Arctic region of the United States was derived as the consensus of a group of geologists representing over 20 oil companies, a good example of the use of this methodology (National Petroleum Council, 1981).

The main advantages of the Delphi technique are that (1) it can be used at all levels of data availability, from frontier to maturely drilled basins; (2) it is basically a fairly rapid and simple procedure (but dependent on the amount of effort that goes into the compilation of data on the assessed area); and (3) the results can be expressed as probability distributions, which reflect the uncertainties in the

estimates. The biggest disadvantage in the method is the lack of documentation of the data input, assumptions, and basic logic that are used throughout the crucial steps of the Delphi process and for the resulting estimates that are reported as probability distributions. One of the major concerns of many estimators and users of these estimates is that "one must know how expert are the experts in order to assess the assessment" (White and Gehman, 1979, p. 2186), This issue is important regardless of the method of assessment being used, but it is rarely addressed in any resource assessment publications.

Historical Performance or Behavioristic Methods

Performance or behavioristic methods are based on the extrapolation of historical data, such as discovery rates, drilling rates, productivity rates, and known field size distributions. Historical data are fitted into logistic or growth curves by various mathematical derivations that extrapolate past performances into the future. These techniques are not directly applicable to unexplored or nonproducing areas or to any area that is not a geologic and economic analog of the historical model. They are generally most applicable to the later stages of exploration in a maturely explored area. Well-known examples of these models are Hubbert's (1962, 1974) growth curve projections; Arps and Roberts (1958) Denver–Julesburg basin study; Moore's (1962, 1966) rate of discovery curves; Kaufman's (1965) and Kaufman et al.'s (1975) field size studies; and the NPC's (1973) projected price and supply studies.

Most published studies to date have based projections of estimated resources primarily on statistical studies of historical data and have purposely included little or no geologic information. This emphasis on historical drilling data and more recently on field size distributions rather than geologic data is due in part to the very large sample areas that generally have been evaluated by this method, such as the entire conterminous United States or other large geographic regions. To improve on finding rate and field size distribution methods so that they can be applied more realistically to resource assessment work, we should use them only in specific cases: (1) where they can be directly related to the geologic settings that control the field size distributions and finding rates; (2) where the area of study is limited to well-defined geologic basins or petroleum provinces or; (3) better yet, where the area is limited to specific stratigraphic units or geologic sections in a basin or province. If the finding rate projections or projected field size distributions were related more directly to the geology of the basin or province, more reliable analog projections could be provided for application to lesser explored and frontier areas. This is particularly true where discoveries and production take place in formations of differing lithologies and trapping mechanisms in a basin.

The major shortcoming of the finding rate and field size distribution techniques for projecting remaining resource estimates is that they can be applied directly only to semimature and maturely drilled producing areas. They are considered to be conservative techniques for estimating resources, because they do not allow for any surprises in petroleum exploration in new areas or plays within a basin or from geologic zones where production had not yet been found previously. They do not allow for improvements in exploration technology or economics. Many classical examples can be cited where new discoveries within an area or basin have completely changed the entire resource assessment outlook for that area. The recent discoveries in the Michigan basin, the Wyoming overthrust belt, and the North Sea are just a few of many examples.

Encouraging developments in the use of performance or behavioristic methods in more recent resource assessment work include the following: (1) efforts to relate the finding rates or field size distributions to the geology of the basins (Ivanhoe, 1976a, b, c; Klemme, 1971, 1983); (2) efforts to evaluate the economic factors relative to finding rates and field size distributions (USGS, 1980); and (3) the incorporation of historical data from analogous geologic basins or producing areas as input into the more detailed resource estimating methods combining geologic and statistical models. The third development will be discussed more fully in the section entitled *Combined (Integrated) Methods*.

Geochemical Material Balance Methods

The geochemical material balance methods are a special type of volumetric resource appraisal procedure by which one can estimate the amount of hydrocarbons generated in the source rocks, the amount of the hydrocarbons involved in migration, the probable losses of hydrocarbons during the migration process, and the quantity of hydrocarbons that have accumulated in the deposits. This approach has been utilized by Russian geologists (Neruchev, 1962; Semenovich et al., 1977), and its use has been illustrated by McDowell (1975).

Although geochemists have made major contributions in our understanding of the processes of hydrocarbon generation (e.g., Hunt and Jamieson, 1956; Philippi, 1956, 1976; Vassoyevich and Neruchev, 1964; Tissot et al., 1971, 1974; Hood et al., 1975; Dow, 1977), much remains to be learned and understood regarding the theory of hydrocarbon generation and the mechanics of hydrocarbon migration in buried strata. This lack of understanding of the major fundamentals of petroleum generation, migration, and entrapment has resulted in limited use of this method in petroleum assessments in the United States. New applications of geochemical methods to quantified hydrocarbon predictions that seem promising are reported by Demaison (1984), Sluijk and Nederlof (1984), Welte and Yukler (1984), Bishop et al. (1984), Ungerer et al. (1984), and Kontorovich (1984). If and when we have a better understanding of these fundamentals, geochemical methods may gain wider acceptance and increased application for assessing large regional areas.

Combined (Integrated) Methods

Combined methods are based on an amalgamation or integration of some or all of the methods that are described above and that incorporate geologic and statistical models. These methods consist of more sophisticated techniques that usually require larger amounts of data as well as more complicated mathematical and computer methods for handling the information. Combining many of the previously discussed methods is becoming the more frequently used approach to resource estimation procedures. The various combinations of methods are too numerous to describe here. For the most part, they involve (1) geologic basin analysis, in which geologic models and basin classification techniques are used; (2) play analysis or prospect analysis techniques; (3) statistical, economic, and supply projection models; and (4) more comprehensive petroleum province analog systems.

Combined methods can range from the simple combination of a performance or behavioristic method (i.e., field size distributions) and a geologic basin classification system (Ivanhoe 1976a, b, c; Klemme, 1983), to the well-documented methodology of the Potential Gas Committee (PGC) (1984) that combines volumetric yields with the estimated probabilities for trap accumulations and risk factors. Combined methods also include the presently popular play analysis or prospect analysis techniques, which may incorporate some or all aspects of the resource appraisal methods discussed above.

I will address only the exploration play analysis methods, which have been designed to assess conventional petroleum resources in identified or conceptual exploration plays in a basin or province (Miller, 1981a). They are usually applied to smaller areas of appraisal than are the previously described methods, areas such as a geologic trend consisting of a reef play or a channel or bar sand. However, many variations of the play analysis definition and the basic assumptions applied to play concepts have been used by resource estimators when applying play analysis techniques. In some studies the play analysis procedure has been applied to an entire stratigraphic unit or geologic zone. Although the resource estimators may call their procedure a "play analysis," the basic concepts are no longer those of the original definition. Thus, there are some extreme variations in the assumptions and resulting estimates used in this method from one assessor to another. The basic techniques, however, require more detailed data than do the volumetric yield methods. The play analysis methods utilize all the data in the field size distribution approach, as well as the additional geologic data on the individual fields within a play, plus the basic information on the reservoir characteristics in these fields.

An estimate of conventional petroleum resources is usually expressed as an equation in a play analysis procedure that relates a series of geologic and reservoir variables to the amount of potential oil or gas within the reservoir. Probability values are assigned to the favorability of a play and to the probable exploration success of the prospects within the play. Most of the geologic and reservoir variables are described by subjectively derived probability functions that are based on the judgment of the estimators; others are described by use of selected analogs. The data formats are commonly designed for computer processing. The estimates of the resource are derived by means of the equation for each play using Monte Carlo methods. The total resource estimate for the area or basin is determined by aggregating the potential of all plays, or all prospects within the plays, also by using Monte Carlo techniques. The estimate is in the form of a probability distribution for the total resource assessment.

The play analysis approach simplifies, or appears to simplify, the task of the geologist in evaluating the resource of an area because it provides a fixed format for the variables he or she must evaluate, and because the actual resource assessment is determined directly by means of a mathematical computation using a computer model. Such sophisticated computerized procedures, however, do not necessarily mean that the accuracy in the resource assessments has been increased over those assessments made by means of other resource appraisal methods.

The following comments are based on events experienced by geologists in the USGS during the application of play analysis procedures to four separate petroleum assessments of the National Petroleum Reserve of Alaska (NPRA) during 1978–1980, and to an assessment in 1980 of the Arctic Wildlife Range of Alaska (Mast et al., 1980; Miller, 1981a, b). Geologists concerned over the results from their input to these programs became increasingly concerned over weaknesses in the assumptions and mathematical manipulations within the computer system. Frequently these systems are designed by technical personnel who are not familiar with the basic concepts regarding petroleum geology (Miller, 1981a). One such weakness in the play analysis model is the assumption that all the variables assessed in

each play, as used in the Monte Carlo simulation, are independent. Many of the geologic and reservoir variables are directly correlated. This often creates a conflict for the geologists who are asked to assign the values to each variable and to assign the degree of risk or success for the occurrence of a favorable play and a favorable prospect in that play.

Another area of concern in resource assessment studies is the application of play analysis techniques to frontier areas where limited data are available and where geologists must base their subjective evaluations for the attributes only on comparative analogs. The resulting resource assessment can be only as good as the geologic analogs selected by the geologists, which may or may not match the frontier basin. When the play analysis method is used in a frontier basin, all the potential plays are assumed to have been identified and adequately described. However, after further exploration, even in fairly well-explored basins, many unanticipated resources in unidentified plays have been found. Thus, the initial assessments in those basins were too conservative. In frontier areas where the subsurface geology is little known and highly speculative, the identification and adequate description of all probable plays are often difficult to make, if not impossible (Miller, 1981b).

The current literature shows that applications of the play analysis method range from the fairly simple to the highly sophisticated. Probably one of the most publicized play analysis methods has been that of the Geological Survey of Canada (Canadian Department of Energy, Mines and Resources, 1974, 1977; Porter and McCrossan, 1975; Roy, 1975; Lee and Wang, 1983; Proctor et al., 1984). Play analysis and prospect analysis techniques used by the petroleum industry have been reported by Mobil Oil Corporation and Exxon Production Research (Roadifer, 1975; White et al., 1975; White, 1980; Baker et al., 1984). The applications by the USGS of play analysis to petroleum resource assessments for the NPRA and the Arctic Wildlife Range of Alaska have been published by the U.S. Department of the Interior (1979), White (1979), and the USGS (Mast et al., 1980; Miller, 1981a, b). Additional play analysis concepts were first modified and used by the USGS (Conservation Division) in 1978 for the assessment of presale offshore continental shelf tract evaluations (G. L. Lore, personal communication, 1985).

A SYSTEMATIC APPROACH TO THE SELECTION OF RESOURCE ESTIMATING PROCEDURES

In estimating resources, as in any other technical endeavor, the estimator should make a conscientious effort during the early planning stages of the assessment to review information input, to define the limits of the study, and to consider the purpose of the resource assessment and the nature of the output. These steps should be done before the finally selection of the resource appraisal methods to be used in the study. The estimator should follow a systematic approach to the selection of the resource appraisal methods by reviewing a checklist of the basic requirements and criteria to be considered for the specified project. It is not feasible to develop a complete checklist here of all the criteria an estimator should consider relative to a particular assignment. Some of the key issues that should be a part of such an approach are as follows.

1. Establish the short- and long-term purpose(s) of the resource assessment. Short-term uses may include, for example, projected future supplies relative to price and quantity under short-term time constraints, such as 5–10 years or less; projected supplies under economic and technological constraints; exploration planning purposes; and presale offshore lease estimates. Long-term uses of estimates (> 10 years) may include, for example, long-term supply development; resource base information, with assessments free (as much as possible) of economic, technical, and political constraints; problems of resource exhaustion; and determination of long-range national energy policy.

2. Identify the area to be assessed by defining the area, with boundary limitations, and by determining all levels of knowledge relative to the area (e.g., stage of exploration and history, geologic and geophysical data, well data, field data, and production and reserve data).

3. Set consistent standards of terminology to be used for commodity classifications and reserve and resource classifications.

4. Define, clarify, and establish all basic assumptions to be used consistently throughout study.

5. Establish resources available to the study, including human resources, financial resources, time available to deadlines, computer facilities, and information data systems (field data and well data).

6. Determine the resource appraisal methods that best fit requirements based only on available data considerations.

7. Determine the resource appraisal methods that best meet the purposes of the resource assessment.

8. Determine the resource appraisal methods that are feasible with available human and financial resources under the time constraints imposed on the assessment.

9. Determine which of the resource appraisal methods best meet all the requirements of the assigned resource assessment project.

10. Determine what compromises have to be made in the final selection of the assessment methods.

11. Determine whether a credible resource assessment can be made that will meet the basic criteria for the project.

12. Prepare documentation on all of the above issues.

If the estimator has the freedom to meet the considerations in items 1 and 2 reasonably without having to compromise too much for the probable limitations of item 5, he or she could make a rational selection of resource appraisal methods that would accomplish the specific assignment of assessing the petroleum resources.

The problem for the geologist is how to maximize the input of all available geologic data, personal experience, and the developing knowledge of the fundamental principles that control the generation, migration, and accumulation of petroleum. The purposes to be served by the resource assessment, however, may modify the approach to the assessment procedures. The purposes and methods of a resource assessment developed for short-term uses in the exploration and economic planning of an oil company, for example, are considerably different from those of an organization such as the USGS or the Geological Survey of Canada, which are involved in national assessments of the long-term resource base.

I emphasize that it is just as important for the users interpreting the results of these resource assessments to be cognizant of the criteria, limitations, and purposes that influenced the selection of the assessment methods as it is for the estimator in making the assessments. There is a strong need for clear and complete documentation of the methodology in resource appraisal work. Thoroughly adequate documentation may eliminate a great deal of the misinterpretation and misuse of resource estimates.

Figure 1 is a simplistic guide to the selection of the basic resource assessment methods at various stages of exploration and different levels of knowledge for designated areas. The figure shows selections of the basic resource methods as they relate to stages of exploration for a basin or province, a stratigraphic unit, or a play or prospect and as they relate to either a long- or short-term application. On the basis of my experience in resource assessment work, some methods are better suited and provide more credible results for long-term applications and others are better for short-term applications. Also, some methods are better suited to assessments of frontier and immaturely explored areas, while still others are better for the semimature and maturely explored areas. Each resource estimator should assume the responsibility of a similar exercise for his or her assignment, as demonstrated in Figure 1, to aim for the optimum credibility essential in resource assessment work.

DIFFERENCES REFLECTED IN RESOURCE ESTIMATES DEPENDENT ON METHODOLOGY

In this section, I would like to examine the following question: Is the outcome of the resource assessment significantly dependent on the selection of the methodology?

Comparisons of resource assessments and appraisal methods from published resource estimates during the last four decades or more for any specific area are difficult to make for many reasons. The following are just a few of the basic problems: lack of documentation; disagreement on definitions and terminology; lack of specified boundaries or variations in boundaries for given areas selected by estimators; different areas included for offshore and in varying water depths; differences in commodities included in the estimates; differences in assumptions on geology, economics, and technology regarding recovery efficiency; and different resource appraisal methodology (Miller, 1979; Thomsen, 1979; Dolton et al., 1981).

At the 1979 AAPG Annual Convention, H. L. Thomsen presented the results of a preliminary study comparing 14 different resource assessments for oil in the United States completed in the period 1965–1978. He separated the assessments into three categories according to the basic resource appraisal methods used (Table 1). These three categories of methods were defined as (1) geologic analysis, which assumes that the appraisal is backed by a large amount of pertinent geologic information; (2) volumetric yield methods, which assume that the estimate was made primarily by relating ultimate recovery per unit area or per unit volume of tested areas to untested areas; and (3) projection analysis, which assumes that the primary method involves projection of past experience in discovery and drilling. Thomsen (1979) concluded that, "The most pessimistic estimates are those made by the projection method, the highest are the ones in the volumetric box [category], and the geologic estimates are in between. In my opinion these are probably the best estimates available [referring to the 14 assessments] and when categorized by method, they show a remarkable consistency" (Thomsen, 1979, p. 9).

To answer the question posed above, I have made a comprehensive survey of the literature on oil and gas resource estimates as a followup to Thomsen's work in an attempt to determine whether patterns of differences in resource assessments significantly depend on the assessment methodology used. My objective was to find a series of resource estimates that have been conducted over a period of time on a given area, either by the same group of estimators using different methods or by different estimators using different methods. Three areas were selected for case studies. They are reviewed here as examples that illustrate the complexities in making resource assessments, the difficulties in making comparative studies, and some explanations for the differences among the resource estimates that depend significantly on the methodology used in the assessment. The three studies are the petroleum resource estimates for Canada, the United States, and the Permian basin of western Texas and southeastern New Mexico.

Assessing Canada's Oil and Gas Resources from 1969 to 1983

Estimates of undiscovered oil and gas potential for Canada have been prepared by the Canadian Petroleum Association (CPA), the Canadian Society of Petroleum Geologists (CSPG), and the Geological Survey of Canada (GSC). The Department of Energy, Mines and Resources in the GSC began an inventory of Canada's undiscovered oil and gas in 1971. The first GSC estimates were published in "An Energy Policy in Canada—Phase 1" (Canadian Department of Energy, Mines and Resources, 1974). Three detailed earlier resource estimates of Canada's oil and gas potential are available for comparison with the GSC 1973 and later estimates (Canadian Department of Energy, Mines and Resources, 1974). These estimates were prepared by the CPA in 1969, the GSC in 1972, 1973, 1976, and 1983, and by the CSPG in 1973. Each of these approaches involved different methods and different data; they are each presented here in some detail.

Canadian Petroleum Association, 1969

Although there had been earlier attempts, the 1969 estimates by the CPA were one of the first published. The method of estimating the potential resources was purely volumetric. It involved multiplying figures representing the

Table 1. Ultimately recoverable crude oil resources (billion bbl) for the United States for selected estimates, 1965–1978.[a]

| Year | Source | Methods Used | | |
		Geologic Analysis	Volumetric Yield	Projection Analysis
1965	Weeks	230[b]	—	—
1965	Hendricks	—	320	—
1968	Inst. Gas Tech.	—	—	225
1970	NPC (AAPG)	259	—	—
1971	Cram (AAPG)	224	—	—
1971	Moore	—	—	188
1972	Theobald et al.	—	517[b]	—
1974	McKelvey	—	423[b] (mean)	—
1974	Mobil	230[b]	—	—
1974	Hubbert	—	—	213
1975	NAS	236[b]	—	—
1975	Miller et al	250	—	—
1975	Exxon	244[b]	—	—
1978	Shell	211	—	—
Averages		236	420	209

[a]The estimates are separated into three categories according to the basic resource appraisal methods used for the assessments. Courtesy of H. L. Thomsen (1979).
[b]Crude oil assumed to be 85% of total liquids.

volumes of sediment covering eight large regions (each of which included one or more sedimentary basins) by a yield factor in barrels of oil per cubic mile. The yields used were selected as representative of the average for such large heterogeneous regions. The resulting ultimate potential oil resources (expressed in barrels) were multiplied by a conversion factor to give an ultimate potential gas resource in cubic feet. The gas–oil factor was considered to represent the average proportion of gas to be discovered in each region; an average of 6,000 cu ft of gas to 1 bbl of oil was used. The study included the sedimentary volume to a depth of 25,000 ft and, offshore, to a water depth of no greater than 600 ft (Canadian Department of Energy, Mines and Resources, 1974). The results of the study are shown in Tables 2 and 3.

Geological Survey of Canada, 1972

The GSC embarked on a systematic program of estimating the petroleum potential of Canada that was completed by the end of February 1972. The study, which represented the GSC's first departures from a strictly volumetric approach, divided Canada into 32 basins for regional analyses. The volumetric method consisted of the determination of the volume of the sedimentary rocks found in each basin, multiplied by a yield factor, to arrive at an oil and gas potential estimate for each basin. The yield factor for each basin was determined on the basis of knowledge of the basin with respect to a worldwide basin classification scheme. The GSC used a preliminary basin analysis approach for the characterization of each basin. They then used the resulting information to qualify the sedimentary volume and yield factor and initiated the use of a probabilistic approach by determining minimum, maximum, and "best estimate" figures of potential for each basin.

The GSC had confidence in its method, and eventually the process led to the systematic basin analysis, the

Table 2. Comparison of liquid hydrocarbon potential estimates of Canada, crude oil and natural gas liquids (in billions of bbl).

	CPA 1969	GSC 1972	GSC 1973	CSPG 1973A	CSPG 1973B	GSC 1976[a] 90%	50%	10%	GSC 1983[b] High (95%)	Average (50%)	Speculative (5%)
Ultimate recoverable oil potential[c]	120.9	134.4	99.2	85.2	23–98	25	30	43	29.4	49.8	76.7
Estimated cum. production liquid hydrocarbons[d]	4.66	6.413	7.168	7.168	7.168		8.499			12.382	
Established reserves:[e]											
Crude oil	10.495	9.603	9.018	9.018	9.018		7.842			6.433	
Natural gas liquids	1.746	1.575	1.498	1.498	1.498		1.523			1.247	
Total liquids	12.241	11.178	10.516	10.516	10.516		9.365			7.680	
Total discovered liquid hydrocarbon resources	16.901	17.591	17.684	17.684	17.684		17.864			20.062	
Undiscovered liquid hydrocarbon resources[f]	104	117	82	68	5–80	7	12	25	9	30	57[g]

[a]GSC 1976: offshore estimates of continental slopes were not included. Accessible offshore areas included in assessments: Atlantic Shelf south, East Labrador and Newfoundland Shelf to 1,500 ft water depth, and MacKenzie Delta–Beaufort Sea to 600 ft water depth. All other offshore areas were considered inaccessible and not included in assessment.

[b]GSC 1983: maximum offshore water depths include from 660 ft west and Arctic coast, to 1,300 ft for the Labrador and Newfoundland shelves; the Scotian shelf and Georges Banks were included to 4,920 ft and Baffin Bay–Lancaster Sound with an estimated 6,500 ft of water depth.

[c]Ultimate recoverable oil and gas potential includes cumulative production, established reserves and undiscovered resources. All were taken as direct assessments from published reports except GSC 1983. See footnote g.

[d]Estimated cumulative production: all figures for crude oil and natural gas liquids were taken from Canadian Petroleum Association (1982) to report a consistent series of figures. Some figures were not the same as those published in the Geological Society of Canada reports.

[e]The expression "established reserves" is used by the GSC to describe those reserves that, on the basis of identified economic considerations and within a specified time frame, are recoverable with a high degree of certainty from known reservoirs. This does not include discovered reserves in frontier areas that are not fully delineated (Procter et al., 1984).

[f]Calculated undiscovered potential equals the ultimate recoverable oil or gas potential minus the discovered resources (cumulative production and established reserves). All the estimates cited, except the 1983 study, were assessed as the ultimate recoverable potential resource.

[g]Undiscovered potential resources were directly assessed in the 1983 study. Ultimate recoverable potential is undiscovered plus discovered liquid hydrocarbon resources or discovered gas resources.

identification of plays, and the probabilistic curve determination for potential resources used in its 1973 estimates. The GSC results in 1972 did not differ greatly from those of the CPA in 1969 for oil, although there were differences between specific regions. The GSC 1972 gas estimates, however, were considerably higher than the CPA 1969 figures for gas. Areas included for the estimates, however, were different: the CPA 1969 study included only areas in water depths of 600 ft or less, whereas the GSC 1972 (and 1973) estimates included the entire area of the Atlantic offshore continental slope (Canadian Department of Energy, Mines and Resources, 1974) (see Tables 2 and 3).

Canadian Society of Petroleum Geologists, 1973

In the meantime, in 1969, the Canadian Association of Petroleum Geologists, which is now the Canadian Society of Petroleum Geologists (CSPG), was meeting to plan a study of Canada's petroleum and natural gas resources. It was felt that this study would complement the project being undertaken by AAPG for the evaluation of the petroleum potential of the United States. The AAPG study was published as Memoir 15, entitled "The Future Petroleum Provinces of the United States—Their Geology and Potential" (Cram, 1971).

The CSPG study was described by 27 authors and covered approximately 38 sedimentary basins in at least 7 distinct geologic categories. Because of the variations in the approaches to the studies by the different geologists, a complete synthesis of the principal observations of the various contributors and their estimates of the petroleum potential in each of the basins was prepared by McCrossan and Porter (1973). The detailed results of the basin studies and the potential estimates were published by the CSPG in 1973 as "The Future Petroleum Provinces of Canada—Their Geology and Potential" (McCrossan, 1973).

The CSPG released its results of the potential estimates in a press conference in Calgary on March 19, 1973. They were based on a variety of methods, primarily volumetric yield techniques and basin classification systems. Each estimate was based on a sound knowledge of the geology (known at the time) for each basin. Individual estimates were compiled and modified to achieve overall consistency with a basin classification scheme modified from Klemme (1971). This first set of CSPG estimates is given as CSPG 1973A in Tables 2 and 3.

At the 1973 press conference in Calgary, it was reported that there was incomplete agreement within the CSPG

Table 3. Comparison of natural gas potential estimates of Canada (in trillion cu ft).[a]

	CPA 1969	GSC 1972	GSC 1973	CSPG 1973A	CSPG 1973B	GSC 1976 90%	GSC 1976 50%	GSC 1976 10%	GSC 1983 High (95%)	GSC 1983 Average (50%)	GSC 1983 Speculative (5%)
Ultimate recoverable gas potential	724.8	906.2	782.9	577.5	157–655	229	277	378	301.5	483.9	793.7
Estimated cum production natural gas	17.0	24.948	28.146	28.146	28.146		34.399			56.726	
Established reserves: natural gas	57.833	60.786	61.022	61.022	61.022		78.749			91.464	
Total discovered natural gas resources	74.833	85.734	89.168	89.168	89.168		113.148			148.190	
Undiscovered natural gas resources	650	820	694	488	68–566	116	164	265	153	336	645

[a]See footnotes to Table 2 for further details.

regarding the estimates. Subsequently, in a submission to the Science Council of Canada, dated March 27, 1973, the CSPG presented a second set of figures (reported in Tables 2 and 3 as CSPG 1973B) that were apparently based on an entirely different method. By this latter approach, CSPG attempted to estimate potential volumes of oil and gas in place using 10 and 90% confidence limits. No estimates of recoverable potential were provided. Using recovery factors of 33% for oil and 85% for gas (Canadian Department of Energy, Mines, and Resources, 1974) and applying them to the CSPG 1973B estimates, one has a rough basis for comparison of the potential estimates. This would result in a potential ultimate recoverable oil range of 23–98 billion bbl and a potential ultimate recoverable natural gas range of 157–655 trillion cu ft. These ranges were considered too broad at the time by the GSC to provide a basis for further analysis. However, under the assumption of a 50% probability of occurrence of potential on the order of 61 billion bbl of oil and 406 trillion cu ft of gas, these estimates were less than any of the previous ones. Because the methodology of estimating the resources was not discussed in the CSPG brief and a breakdown of potential areas was not presented, it is not possible to analyze the differences between the two methods or the estimates. More is said later on methods for estimating in-place potential resources.

Geological Survey of Canada, 1973

According to the GSC, their 1973 study was an improvement on their original estimate. It incorporated all geologically conceivable "exploration plays" or groups of "plays" in a given basin, it adopted a probabilistic approach, it compared all new estimates with estimates derived by "volumetric" analysis, and it was based on comprehensive basin analysis studies, including geochemistry and geophysics. Both the GSC 1972 and 1973 estimates included Atlantic offshore slope sediments, and thus they encompassed a greater volume of sediments than in either the CPA 1969 or CSPG 1973 studies.

For the first time, consideration was being given in the methods to the size of the oil and gas pools that would be anticipated with each play. A lognormal distribution was assumed for each play, and the largest pool size for each play was estimated; then an array of pool sizes for each basin was calculated (Canadian Department of Energy, Mines and Resources, 1974). The results of the GSC 1973 estimates are shown in Tables 2 and 3.

A comparison of the estimates for the undiscovered oil and gas potential, as determined by the GSC in 1972 and 1973, show that the 1973 estimates were smaller. The 1973 estimates for oil (but not for gas) were also smaller than the CPA 1969 estimates. Both the oil and gas estimates of the CSPG 1973A and 1973B studies were less than either the CPA 1969 study or the GSC 1972 and 1973 studies. The CSPG 1973B study was the smallest estimate reported through 1973.

Geological Survey of Canada, 1976

In 1975, the Canadian Department of Energy, Mines and Resources (Geological Survey of Canada), through a continuation of its successive basin analysis studies and resource estimating procedures, completed a new assessment of the oil and gas resources for nine regions of Canada. The offshore inaccessible areas were not included in the assessments—these were the continental slopes and rises off the Scotian shelf, Grand Banks, northeast Newfoundland and Labrador shelves, the Baffin Bay shelf and slope, and the Arctic Coastal Plain shelf. The west coast shelf and slope were assessed on the basis of the available data, but the assessments were not included in the calculation of the total Canadian oil and gas estimate as reported in the publication, "Oil and Natural Gas Resources of Canada 1976" (Canadian Department of Energy, Mines and Resources, 1977).

In the GSC 1976 study, the play analysis technique was more clearly defined and documented in the report. All exploration plays were defined in a given area for each basin. The estimate of hydrocarbon potential was expressed by one of two equations: (1) the "volumetric" type, which made use of analogs of the basin, area, or rock unit under consideration; and (2) the "exploration play" equation, which calculated the prospect potential and which was basically a reservoir engineering equation for determining the volume of oil or gas in a reservoir. The volumetric approach was used as a check on the exploration play method, and where few data were available, the volumetric approach was considered the best alternative. The geologic parameters in the exploration play equation were described, where necessary, by subjectively derived cumulative distribution functions based on the judgment of the estimators. These distributions were "risked" by applying marginal and conditional probabilities assigned by the experts for all the equation variables. Pool size distributions were assumed to be lognormal and were generated for each play. This enabled a direct tie-in with economic studies that were to be conducted later by the GSC. The estimates of individual plays within a sedimentary basin were summed by means of a statistical technique known as the Monte Carlo method. The estimates for nine separate geologic–geophysical regions of Canada were then aggregated for a single cumulative probability curve for total oil and total gas. The results of the study are reported in Tables 2 and 3.

Estimates of both oil and gas resources for Canada were substantially lower in the GSC 1976 study than those published in 1973. One of the major reasons stated for the reduction was that the 1976 estimates excluded offshore areas considered to be inaccessible using known industrial technology in 1976, whereas the 1973 estimates included such areas as the continental slopes and rises and the offshore Arctic Coastal Plain. This exclusion was thought by the GSC to be responsible for approximately one-third of the reduction from the 1973 estimate. They reported in the 1976 study that the "decrease also results from the new and predominantly disappointing flow of information that has been generated by exploration in the interim period. Certain of the changes in estimates have resulted from an increased capability to process information and improvements in methodology" (Canadian Department of Energy, Mines and Resources, 1977, p. 4).

Geological Survey of Canada, 1983

In 1984, the GSC published a summary of the "Oil and Natural Gas Resources of Canada 1983," which is drawn from their continuing petroleum resource evaluation program in which the detailed appraisals of individual basins or regions are periodically reviewed. The petroleum resource evaluation activities of the GSC were focused primarily on the undiscovered or potential components of conventional oil and gas resources. Compiled data from the literature and provincial government agency publications on Canada's nonconventional resources were also reported by the GSC (Procter et al., 1984).

The GSC began in 1972 with estimates from rather simplistic volumetric calculations, which have now evolved to what are described in the current study as "a probabilistic methodology conducted at the exploration play level, incorporating both objective data and informed geological

opinion. A few years ago, this would have been referred to as the Monte Carlo approach. However, methodology has now advanced well beyond the Monte Carlo stage with more rigorous and more powerful mathematical procedures being incorporated into the system" (Procter et al., 1984, p. 7). In addition to estimates of basin potential, the present operating methodology used by GSC can produce an array of hypothetical pools with attached reservoir characteristics consistent with the input geology and data. Additional information on the probabilistic methods was published by Lee and Wang, 1983.

The GSC publication of 1983 continued to report the estimates in probabilistic terms, although they now expressed a range of values as having a "high confidence" (95% probability), "average expectation" (50% probability), or "speculative estimate" (5% probability), rather than giving values at the 90, 50, and 10% probability ranges used in their 1976 study (Procter et al., 1984, Figure 2.1). The resource estimates included sediments out to water depths of 600 ft in the Beaufort Sea and MacKenzie Delta areas and off the west coast of British Columbia, and to water depths of 1,200 ft on the Newfoundland and Labrador shelf areas. These offshore areas included in this report are comparable to those of the 1976 estimates. The results of the 1983 GSC resource estimates are shown in Tables 2 and 3.

In the 1983 assessments, the "average expectation" for undiscovered liquid hydrocarbons of 30 billion bbl is higher than the average estimate of 12 billion bbl in the 1976 estimate. For undiscovered gas, the 1983 average of approximately 336 trillion cu ft is more than double the 1976 average estimate of 164 trillion cu ft. The major differences between the 1976 and 1983 estimates appear to reflect the enthusiasm generated by the oil and gas discoveries off the east coast and the Beaufort and MacKenzie Delta areas.

The 1983 estimates (just as the 1976 estimates), however, are considerably less than the 1973 and earlier estimates for Canada's undiscovered resources. In light of the continuing decline in Canada's established reserves of crude oil and natural gas liquids (from more than 8 billion bbl in 1976 to 7 billion bbl in 1980 and a slight increase to more than 7 billion bbl in 1982), the estimates for the remaining oil resources reflect that no major changes are expected in exploration concepts. Because of the new offshore discoveries, however, estimates of natural gas reserves continued to increase from 75 trillion cu ft in 1976 to more than 90 trillion cu ft in 1982, and the increase in the gas resource estimates between 1976 and 1983 are consistent with these changes (Figure 2). The maximum (or "speculative estimate") for the undiscovered natural gas resources of 645 trillion cu ft in the 1983 study was influenced by the new gas discoveries and is thus now back in the range of the single value estimate of the CPA 1969 study of 650 trillion cu ft, which was determined by the all-inclusive volumetric yield method.

Results

I have reviewed in some detail the background of the oil and gas resource estimates for Canada, assessed from 1969 to 1983, beause these seven assessments document the evolution of the resource appraisal methods used in these

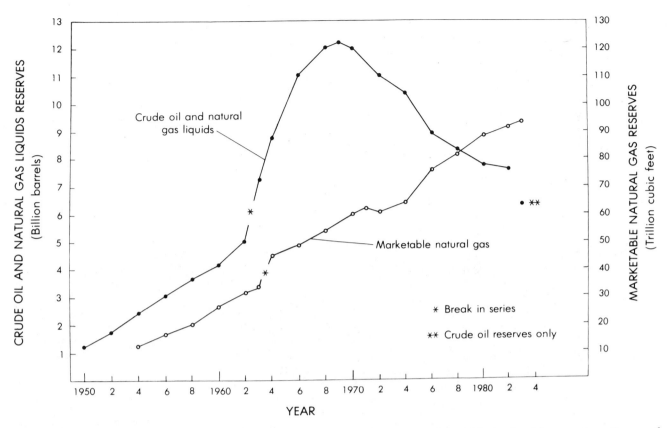

Figure 2. Established petroleum reserves in Canada for crude oil and natural gas liquids and for marketable natural gas. Source of 1950–1982 data was CPA (1982) and of 1983 data, Oil and Gas Journal (1984).

estimates and the actual exploration in Canada that took place during this period in the offshore and Arctic discoveries, particularly those of natural gas resources.

The categories of resource appraisal methods in these studies basically evolved from the purely volumetric yield methods, through modifications of these methods that incorporated probability distributions, basin classifications, and subjective probability (or Delphi) techniques, to a simplified play analysis technique. In the latest study, the methods have evolved to a more sophisticated play analysis technique incorporating field size distributions and economic considerations. No time limitations are documented for the earlier resource studies; however, the 1983 study refers to petroleum supplies through the 1990s.

To compare resource estimates is very difficult, and in some cases impossible and often misleading, but it is a necessary and unavoidable exercise. All the liquid hydrocarbon potential estimates and natural gas estimates were converted either to the ultimate recoverable potential or to the remaining undiscovered recoverable potential, and comparisons were made for those commodities (Tables 2 and 3).

The 1972 study of the GSC (Canadian Department of Energy, Mines, and Resources), which used a volumetric method, has the highest estimates for the petroleum resource potential. The lowest petroleum estimates were generated by the GSC in their 1976 study, in which a well-developed play analysis methodology was used. Reasons for the reduction in

the resource estimates from the GSC 1972 and 1973 studies are documented as follows. First, the 1976 study excluded offshore areas considered to be inaccessible using known technology, whereas the 1972 and 1973 estimates included such areas as the continental slopes and rises and the Arctic Coastal Plain offshore. The exclusion was reported to be responsible for approximately one-third of the reduction of the 1973 estimates. Second, the decrease was also reported to be from the new and disappointing flow of information generated by exploration in the interim period. Third, it was reported that certain changes in estimates resulted from an increased capability to process information and improvements in methodology (Canadian Department of Energy, Mines and Resources, 1977).

Improvements have been made in methodology, but I think that the changes in the methods of assessing the resources from the 1972 volumetric yield study to the 1976 play analysis study account for the greater part of the differences between the earliest and latest estimates.

The 1983 GSC estimates incorporated additional modifications to the play analysis method and showed some increase in the oil resource potential and a significant increase in the gas resource potential compared to the 1976 study. These increases are probably due to certain offshore areas being reincorporated into the total assessments and to newly acquired geologic data from exploration and discovery of new plays on the offshore east coast, in Arctic areas, and in western sedimentary basins (Figure 2).

From the comparisons of the Canadian resource estimates, the following conclusions seem reasonable. First, volumetric-yield methods usually generate more optimistic resource appraisals than other methods. Expertly derived geologic analogs, however, have much to offer for the regional appraisals of resource bases and for the all-encompassing, reconnaissance appraisals, as long as both the estimator and the users of the assessments understand the strengths and weaknesses of the methodology. Second, play analysis techniques usually generate more pessimistic resource appraisals. Incorporated within the method are truncated field size distributions for the smaller field sizes and various economic limitations. Play analysis methods emphasize detailed geologic or reservoir input and concepts on the probabilities of occurrence of the resource and the risks of finding it, but commonly do not allow much imagination in predicting the occurrence of new exploration plays in unexplored areas within partially explored or frontier basins. They are well suited to the local and short-term projections of resource availability and supply, and for application and input to economic studies.

Are these patterns or similar patterns of resource estimates versus resource methods evident on comparable studies in other areas? The following case study on some resource assessments for the United States is briefly discussed, and similar patterns are evident on comparison of the resource estimates.

A Review of Resource Estimates for the United States

Published estimates of undiscovered oil and gas resource potential for the United States date back nearly 80 years. One of the early estimates for the United States was published by D. T. Day (1909) of the USGS. Since then, many estimates have been published, and a number of authors have attempted to compare these estimates to explain disparities in the quantities reported for the undiscovered recoverable petroleum resources of the United States. Some of the more comprehensive comparisons of resource estimates have been made by McCulloh (1973), Miller et al. (1975), Potential Gas Committee (PGC) (1977), Thomsen (1979), and Dolton et al. (1981). However, the authors making the comparisons even differ in their interpretations of the various estimators' work.

Direct comparisons among resource estimates of the United States are extremely difficult, and the problems have been discused in detail before by many authors (Miller et al., 1975; Miller, 1979; PGC, 1977; Thomsen, 1979; Dolton et al., 1981).

Many of the comparative studies published on the resource estimates for the United States have concentrated on the estimates of crude oil resources. In 1977, however, the PGC published a comprehensive study comparing 18 estimates of the ultimately recoverable natural gas in the United States during the period from 1961 to 1974. The report includes a discussion of the methodology, definitions, limiting factors, and other important aspects of each estimate (PGC, 1977).

Continuing with the objective of reviewing resource estimates with respect to their methods of assessment, I used the results of the PGC (1977) report for this case study and updated them with more recent national assessments for the

natural gas resources of the United States, which extended the comparisons through 1980 (Tables 4 and 5).

Tables 4 and 5 record a comparative summary of the statistical information derived from 19 estimates of the natural gas resources of the United States. I took 13 of the estimates, (through 1974) from the PGC study (1977), and I compiled 6 from more recent publications.

The basic resource appraisal methods documented for Table 4 can be summarized into the following categories:

- Variations on the areal and volumetric yield methods in which some form of geologic analogs are used: report numbers 1, 2, 3, 5, 6, 8, and 9.
- A combination of volumetric yield and geologic analogs and probability distributions: report numbers 17 and 18.
- Combined volumetric yield and geologic analogs, with historical projections, Delphi techniques, and probability distributions: report numbers 13 and 15.
- Subjective judgment (Delphi), a group consensus: report number 10.
- Combined methods: play analysis, volumetric yields, historic projections, probability distributions, and economic truncations for field sizes: report numbers 11, 12, 14, and 19.
- Historical projections: report numbers 4, 7, and 16.

The mean ultimately recoverable resource estimates (known plus undiscovered resources) are recorded in Table 5 as reported from Table 4. These estimates are listed in the order of their basic resource appraisal methods as described above. A reasonable pattern or progression in the magnitude of the resource estimates is shown in Table 5 that is fairly consistent with the basic methods that were used for the assessments.

Results

A general pattern of correlation exists between the magnitude of the estimated ultimate resources and the choice of the resource appraisal methods used for the assessment of oil and gas in the United States. Although only the estimates of natural gas resources are reported here, the same patterns between methods and results are also evident in oil resource estimates. The more purely volumetric yield methods produce the larger resource assessments, whereas those methods based on historical projections (behavioristic or field size distributions) produce the most conservative estimates. The play analysis approach, as identified by Mobil and Exxon oil companies (Moody and Geiger, 1974, 1975; Exxon, 1976) is highly dependent on projected field or pool size (or prospect size) distributions, The restrictions imposed, assuming risk and economic conditions, limit the minimum size and number of the prospects included in the resource assessment, and they thus provide assessments closer to the more conservative estimates. The various combinations of resource appraisal methods (geologic and statistical) provide a range in the resource estimates from those of the purely volumetric yield techniques to those of the purely historical projection techniques.

Even where the scope of the comparisons is reduced to the resource assessments having the same offshore water depths, the pattern of the methods used and the magnitude of estimates still holds true. For example, estimates from offshore studies, where water depths were 200 m (656 ft), range from 2,388 trillion cu ft (volumetric yield method) to 1,184 trillion cu ft (historical projection method). At 2,500-m (8,200-ft) water depths, the estimates range from 2,781 trillion cu ft (Theobold et al., 1972, USGS Circular 650, volumetric yield methods) to 1,542 trillion cu ft (Dolton et al., 1981, USGS Circular 860, combined methods). The pattern is consistent for each of the five water depths represented in 16 of the 19 studies; the National Research Council (1975) and studies done by Shell (Shell, 1978; Rozendal, this volume) did not report maximum water depth used.

A Review of the Resource Assessments for the Permian Basin

I have demonstrated significant association between the particular resource appraisal methods chosen for an assessment and the relative magnitude of the resource estimates for two nationwide assessments that cover millions of square miles in the United States and Canada. Does the same correlation hold when a resource assessment is made for a smaller area consisting of several basins or for a single province? The following case study was selected for the Permian basin, or province, in western Texas and southeastern New Mexico where numerous resource assessments have been published during the last 12 years.

It is difficult to find in the literature any comprehensive sequence of resource assessments for smaller areas or basins where all the different resource appraisal methods were used. If a series of assessments can be found, such as for the Permian basin, the size and boundaries of the designated areas may vary considerably from study to study. Keeping these difficulties in mind, I compiled Tables 6 and 7 showing the results of oil and gas resource assessments for the Permian basin area over the last decade.

In Table 6, the results of seven estimates of undiscovered recoverable oil resources for the Permian basin are shown for comparison, along with the methods used to make the assessments. In Table 7, the results of eight resource estimates for natural gas are shown along with the basic methods of assessment. Three of the assessments (McDowell, 1975; Dolton et al., 1979; USGS, 1980) were originally assessed as in-place estimates. Recovery factors have been applied to these in-place estimates to make them comparable to the other assessments reported as recoverable estimates.

The size of the areas assessed for the Permian basin (or province) varied, sometimes resulting in variations in the cumulative production and proved and inferred reserves used. There were additional variations introduced in some studies by economic limitations and by estimates based on future enhanced recovery. Because of these variations, comparisons summarized in Table 8 are based only on the reported undiscovered recoverable resource estimates along with their methods of assessments.

Results

In spite of the variations in the studies discussed above, the same general pattern of correlation between the magnitude

of the estimated undiscovered resources and the choice of the resource appraisal method is again observed. The results of the volumetric yield and geologic analog techniques are, for the most part, higher than those generated by the other methods. The geochemical material balance method is a special volumetric yield approach, but it was used in assessing the Permian system only within that province (McDowell, 1975). However, the estimate of 3.4 billion bbl for the recoverable undiscovered oil is considerably higher than the USGS assessment for only the Permian system, which was 0.76 billion bbl of recoverable oil (Dolton et al., 1979). The USGS estimate relied on a combination of methods with heavy emphasis on historical performance procedures.

Again, the resource estimates are significantly more conservative when derived by relying heavily on historical projections, as was the USGS study (Dolton et al., 1979), or where they are primarily historical projections with economic limitations incorporated, as were the USGS (1980) study and Nehring's (1981) estimates. No known application of play analysis in the Permian basin was available in the literature for use in this comparison.

An interesting comparison between resource estimates and methods is illustrated by the USGS study (Dolton et al., 1979) for the Permian basin, where the USGS group relied heavily on historical performance techniques, and the USGS study (Dolton et al., 1981), where the same group assessed the Permian basin by using a combination of methods. Both the oil and the gas undiscovered resource estimates were more than doubled from the 1979 study to the 1981 study, when the methodology was changed.

Another important factor that I have observed frequently when conducting resource assessments can be illustrated by the same two studies. When a group of geologists is requested to make subjective judgments for directly estimating undiscovered oil and gas resources as in-place rather than as recoverable estimates, the in-place estimates (when converted to recoverable resources) are almost always more conservative than estimates made directly for recoverable amounts. The USGS 1979 study was done as an in-place estimate, and recovery factors were reported later in a related study (USGS, 1980) of 25% for oil and 75% for gas for the same area. These conversion factors were used in Table 8. The more recent USGS assessment for the Permian basin (Dolton et al., 1981) directly assessed the recoverable resource. These 1981 estimates are nearly double the previous estimates reported in the 1979 study.

These two studies provide an interesting illustration, because the same group of geologists used basically the same data but placed a different emphasis on the methodology, making in-place estimates in one study and recoverable resource estimates in the other. It is impossible to determine which contributed more to the conservative USGS estimate of 1979, the emphasis on the historical projection methodology or the phenomenon of a direct in-place assessment method. The argument in the past for providing in-place assessments for a resource estimate is that they are more compatible to economic analyses and can be directly modified by recovery factors and price changes. Some workers believe (mistakenly, I think) that in-place assessments are free from economic assumptions. Petroleum

Table 4. Comparison of statistical information derived from recoverable natural gas resource estimates for the United States (in trillion cu ft).[a]

Report Number	Reports[b]	Estimate as of Year End	Demonstrated Resources (API, 1979)		Future Potential Resources			Ultimately Recoverable
			Cumulative Production	Proved Reserves	Growth of Known (Probable/Inferred)	Undiscovered Resources (Possible)	(Speculative)	Demonstrated Plus Future Potential Resources
1	USGS Circular 522 (Hendricks, 1965)	1961	231	266	—	1,503		2,000
2	Potential Gas Committee (1969)	1968	346	287	260	335	632	1,860
3	NPC-AAPG Future Petroleum Provinces (1970) (Cram, 1971)	1968	345	287	260	651		1,543
4	C. L. Moore (NPC) in Cram (1971)[c]	1969	366	275	Ult. additions 848	Mean 906	Ult. Dis. 964	1,489 Mean 1,547 1,605
5	USGS Circular 650 (Theobald et al., 1972)	1970	391* (includes stored gas)	290	—	2,100		2,781
6	Potential Gas Committee (1971)	1970	388	291	257	387	534	1,857
7	Hubbert (1974) Senate Comm. Report	1971	409	279	135	361		1,184
8	USGS News Release (McKelvey, 1974)	1972	432	266	High 250 / Low 130	2,000		2,948
9	Potential Gas Committee (1973)	1972	433	266	266	384	496	1,828 / 1,845
10	National Academy of Sciences (NRC, 1975)	1973	454	250	125	530		1,359
11	Mobil (Moody and Geiger, 1974, 1975)	1973	454	250	52	90% 320 / Expected Value 485	10% 700	90% 1,076 / Expected Value 1,241 / 10% 1,456
12	Exxon (1976) (Langston, 1976)	1974	477	237	Almost 0% prob. 321	942		1,857
	Full Inventory[d]				Mean 111	582		1,407
					Almost 100% prob. 56	342		1,137
	Attainable				Almost 0% prob. 321	657		1,577

Table 4. (continued)

	Year			Almost 100% prob.	Low	Mean	High	Low	Mean	High
13 USGS Circular 725 (Miller et al., 1975)	1974	481* (includes stored gas)	237	202+ ; 56 ; Mean 111	95% 322 (287)	Mean 484 (127)	5% 655	95% 1,242 (917)	Mean 1,404 (1,112)	5% 1,575
14 Shell (Oil and Gas Journal, 1978)[e]	1977	552	210	—	Min. 150	Mean 315	Max 500	Min. 912	Mean 1,077	Max 1,262
15 USGS Circular 860 (Dolton et al., 1981)[f]	1979–1980	578	192	178	95% 475	Mean 594	5% 739	95% 1,423	Mean 1,542	5% 1,687
16 R. Nehring (1981)[g]	1979–80 (prior to 1975)	------751------		90% 29 / Mean 71 / 10% 131	90% 143	Mean 170	10% 209	90% 923	Mean 992	10% 1,091
17 Potential Gas Committee (1981)	1980	596[h]	199	193	358		362			
18 Potential Gas Committee (1983)	1982	615[i]	202	192	355		329			
19 Shell, 1983[j] (Rozendal, this volume)	1982	633	(202)	(70)		Mean 310			Mean 1,215	

[a] Modified from PGC (1977), Table 1, p. 6–8. For more detailed explanations see original publication.

[b] The offshore areas included in the resource assessments vary for these 19 assessments according to maximum water depths. These include six categories and are listed according to report number: 200 m (660 ft), reports no. 1, 7, 8, and 13; 457 m (1,500 ft), reports no. 2, 3, 4, 6, and 9; 1,000 m (3,300 ft), reports no. 16, 17, and 18; 1,830 m (6,000 ft), reports no. 11 and 12; 2,500 m (8,200 ft), reports no. 5 and 15; and maximum water depths unknown, reports no. 10, 14, and 19.

[c] C. L. Moore (in Cram, 1971) reports future potential in three categories: ultimate discoveries 964 trillion cu ft (tcf), mean discoveries 906 tcf, and ultimate additions 848 tcf. These estimates are not additive.

[d] Exxon (1976) emphasized the distinction between full resources inventory and attainable resources inventory. The attainable resources are those which can be exploited assuming normal technological growth and no significant change in economic incentives. On these terms all the probable and possible supply was also considered attainable, except for the offshore portion in waters deeper than 2,500–3,000 ft, or out to 6,000 ft.

[e] Shell (1978) estimates reported in Oil and Gas Journal, (1978). Methodology for resource assessment and areas included offshore not documented.

[f] Dolton et al., 1981: offshore province boundaries were extended to 2,400 m (7,870 ft) off Alaska and 2,500 m (8,200 ft) off the conterminous United States in this assessment. This is in contrast to the USGS study (Miller et al., 1975) that included water depths only to 200 m (660 ft).

[g] Nehring (1981): total cumulative production and reserves are reported as "all natural gas in fields discovered before 1976 produced or known to be recoverable as of 1979." Reserve growth is "anticipated additions to recoverable amounts in all fields discovered before 1976." The undiscovered future potential is "anticipated recovery from all fields discovered after 1975." Areas covered by assessment are out to 1,000 m water depth.

[h] PGC (1981): cumulative production and proved reserve figures for year ending Dec. 31, 1980, are estimated using data from Office of Oil and Gas, DOE/EIA. Cumulative production figures include gas in underground storage; gas in underground storage is not included in proved reserves volumes.

[i] PGC (1983): cumulative production and proved reserves estimates are based on year end 1981 data from the Office of Oil and Gas, DOE/EIA; 1982 data not available at time of PGC compilation.

[j] Shell, 1983: figures reported for cumulative production, proved and unproved reserves, and future discoveries are quoted from R. A. Rozendal's, "Conventional U.S. oil and gas remaining to be discovered: estimates and methodology used by Shell Oil Company," reported in this publication. I have assumed the year-end estimate to be 1982, and the resource estimates to be "mean" estimates.

Table 5. Comparison of estimated mean of the ultimately recoverable natural gas resources in the United States by resource appraisal methodology (in trillion cu ft).

Method	Report Number	Mean Ultimately Recoverable Resource	Maximum Water Depth Included (m)
Volumetric yield and geologic analogs	1	2,000	200
	2	1,860	475
	3	1,543	475
	5	2,781	2,500
	6	1,857	475
	8	2,388[a]	200
	9	1,845	475
	Average	2,039	
Volumetric yield, geologic analogs, and probability distributions	17	1,708	1,000
	18	1,693	1,000
	Average	1,701	
Combination volumetric yield and geologic analogs, and probability distributions, Delphi, and historical projections (USGS)	13	1,404	200
	15	1,542	2,500
	Average	1,473	
Subjective (Delphi) group consensus	10	1,359	Not reported
Combination volumetric yield and play analysis, historical projections, probability distributions, and economic limitations (major oil companies)	11	1,241	1,830
	12	1,112	1,830
	14	1,077	Not reported
	19	1,215	Not reported
	Average	1,161	
Historical (behavioristic) projections	4	1,238[b]	475
	7	1,184	200
	16	992	1,000
	Average	1,138	

[a]Estimated mean between a reported high of 2,948 tcf and a reported low of 1,828 tcf (McKelvey, 1974).
[b]An ultimately recoverable resource of 1,547 tcf was reported by Moore (1971) for 100% discovery and 100% ultimate recovery. An 80% ultimate recovery, more in keeping with the other estimates, was calculated at 1,238 tcf.

geologists, however, have access to information on recoverable data, such as production and reserve records, and mainly rely on recoverable conventional resources and not in-place resources when they assess the potential of a geologic province. In-place assessments may have their own inherent set of problems that question their credibility.

RECOMMENDATIONS FOR FURTHER STUDIES IN RESOURCE APPRAISAL METHODS

A review of the various published resource estimates, including those discussed here and many others too numerous to include, indicates that there is no totally systematic approach to the selection of the resource appraisal methods and procedures that takes into consideration the purposes of the assessments and all of the basic criteria and requirements necessary for specific resource assessment projects. This seems particularly true for those assessments conducted on a nationwide or worldwide level, which should be designed to consider all types of geologic basins and areas, at all levels of information for all stages of exploratory effort.

No doubt many improvements and various modifications of the basic procedures in resource assessment methods have been made and used during the last several decades, including the use of the computer for incorporating information from large data banks and applying more complex statistical techniques. Unfortunately, most estimators tend to lock onto one resource appraisal method and are reluctant to consider the merits of any other method. Even more important, little has been done in the last decade to promote research for the development of new resource appraisal methodology.

In 1975, M. F. Searl (1975) expressed his concern that few systematic efforts had been made at establishing resource appraisal methodology and that it was a much neglected topic of research in the United States. Today these concerns still exist, and the need is as great, if not greater, when national policies hinge on energy resource factors, particularly on our future energy supplies.

White and Gehman (1979, p. 2191) stated that the problem of estimating oil and gas resources "is complicated but not hopeless." The hope will lie in the development of more effective approaches to resource appraisal methodology. Not only must future methodology address all of the

Table 6. Comparison of statistical information derived from recoverable oil resource estimates for the Permian basin (in billion bbl).

Reports	Est. as of Year End	Demonstrated Resources — Cumulative Production	Proved Reserves	Future Potential Resources — Growth of Known Resources (Probable/Inferred)	Undiscovered Resources (Possible)	(Speculative)	Ultimately Recoverable — Demonstrated and Future Potential Resources	Areas of Study (sq mi)	Basic Resource Appraisal Methods Used in Estimates
AAPG future petroleum provinces (Galley, 1971)	1967	15.9	7.0	21.5	15.6		60	193,000	Volumetric yield and geologic analogs
USGS Circular 725 (Miller et al., 1975)	1974	21.4	9.1	1.6	95% 4; Mean 8	5% 14	95% 36; Mean 40; 5% 46	193,400	Combination methods
USGS Open-File Report 79-838 (Dolton et al., 1979)[a]	1974	23.8		NA	95% .86; Mean 1.65	5% 2.7	95% 24.7; Mean 25.5; 5% 26.5	82,000	Historical projections, pool sizes, discovery rates, subjective probability
USGS Circular 828[b] (USGS, 1980)	1974	24		1.1/1.3	$10/BOE 0.26; $40/BOE 1.28		$10/BOE 24.3; $40/BOE 25.3	80,000 / 86,000 / 100,000	Historical projections, field size, discovery rates, economic supply projections
McDowell (1975)[c] Permian system only	1974	NA	NA	NA	3.4		Mean 19.5	20,000	Geochemical material balance (Permian system only) for basin
USGS Circular 860[d] (Dolton et al., 1981)	1979–1980	Region 25.2; Basin 20.2	6.7; 5.4	4.0; 3.2	Region: 95% 2.7; Mean 5.4 — Basin: 1.0; 2.9	5% 9.4; 6.2	Region: 95% 39; Mean 41; 5% 45.3 — Basin: 30; 32; 35	172,000; 86,000	Combination methods
Nehring (1981)[e]	Pre-1976 as of 1979	26.9		3.2; 7.2; 13.0	90% 0.2; 50% 0.4	10% 0.7	90% 30; 50% 34; 10% 40	Not reported	Historical projections

[a] USGS Open-File Report 79-838: The resource estimates for this study were done as in-place quantities of undiscovered oil, dissolved/associated gas, and nonassociated gas; estimates were made of size and number of pools. Although volumetric-yields and other subjective probability procedures were used in this study, historical pool-size distributions and the historical finding rates were emphasized. No estimates were made of the inferred (growth) reserves of known fields. The amounts reported in the tables have been converted to recoverable quantities from the reported in-place estimates. Recovery factors used were 26% for oil and 75% for natural gas. This study was completed as a part of the Interagency Oil and Gas Supply Project (USGS, 1980). The area assessed in this USGS 1979 study was only 42% of the area covered in the USGS 1975 study, but this study used the same cumulative production and reserve data for the discovered resources.

[b] USGS Circular 828: A report of the Interagency Oil and Gas Supply Project (DOI/DOE). The study includes supply projections of oil and gas from presently known fields through additional drilling and enhanced-recovery techniques and from fields not yet discovered. The procedures are based primarily on historical projections of field size distributions and find rates using economic assumptions for rate of return and prices per BOE. Estimates reported in the table are at a 15% assumed rate of return and a price range of $10/BOE to $40/BOE. The actual area used in this study is uncertain because three different areas are quoted in the report: 80,000, 86,000, and 100,000 sq mi.

[c] McDowell (1975): The geochemical material balance technique was used to generate the in-place oil estimates for the Permian system only in the Permian basin. New discoveries of 13 billion bbl in-place and 75 billion bbl in-place for ultimate goal were converted by using a 26% recovery factor. The USGS (Dolton et al., 1979) assessed the mean undiscovered recoverable oil for the Permian system only as 0.76 billion bbl.

[d] USGS Circular 860: This study used 172,000 sq mi for the total region compared with 193,400 sq mi used in USGS Circular 725. The Permian Basin was reported separately in Circular 860 using an area of 86,000 sq mi compared with the USGS Open-File Report 79-838 of 82,000 sq mi.

[e] Nehring (1981): The demonstrated resources are reported as known recovery and include all oil or gas in fields discovered before 1976 and produced or known to be recoverable as of 1979. The reserve growth constitutes anticipated additions to recoverable amounts in all fields discovered before 1976. The undiscovered is anticipated recovery from all fields discovered after 1975.

Table 7. Comparison of statistical information derived from recoverable natural gas resource estimates for the Permian basin (in trillion cu ft).[a]

Reports	Est. as of Year End	Demonstrated Resources — Cumulative Production	Demonstrated Resources — Proved Reserves	Future Potential Resources — Growth of Known Resources (Probable/Inferred)	Future Potential Resources — Undiscovered Resources (Possible)	Future Potential Resources — (Speculative)	Ultimately Recoverable — Demonstrated and Future Potential Resources	Areas of Study (sq mi)	Basic Resource Appraisal Methods Used in Estimates
AAPG future petroleum provinces (Galley, 1971)	1967	41.2	27.6	25.7	79.12		173.6	193,000	Volumetric yield and geologic analogs
USGS Circular 725 (Miller et al., 1975)	1974	58.7	24.6	23.3	95% 35 / Mean 70 / 5% 101		95% 142 / Mean 177 / 5% 208	193,400	Combined methods: volumetric yield and geologic analogs, Delphi, probability
USGS Open-File Report 79-838 (Dolton et al., 1979)	1974	58.7	24.6	22.9	95% 9.7 / Mean 16.4 / 5% 25.4		116 / 123 / 132	82,000	Historical projections: pool size distribution, finding rates, volumetric, probability
USGS Circular 828 (USGS, 1980)	1974	74(89)[b]		7.1 (8.9–13.7 EGR)		$10/BOE 4.98 / $40/BOE 15.81	Without EGR 101 / 112; With EGR 110 / 126	80,000 86,000 100,000	Historical projections: field size distribution, discovery rates, economic supply projections
USGS, Circular 860 (Dolton et al., 1981)	1979–1980	Region 70.7 / Basin 59.3	15.9 / 14.6	18.1 / 15.4	22.4/13.9 / 42.8/33.3 / 75.2/65.3		127.1/103.5 / 147.5/122.9 / 179.9/154.8	172,000 / 86,000	Combined methods: volumetric yield and geologic analogs, Delphi, historical projections, probability
Potential Gas Committee (1981)	1980	63	14	16	34	1	128	Est. 85,000	Volumetric yield and geologic analogs by reservoirs and probabilities
Potential Gas Committee (1983)	1982	67	16	15	33	1	132	Est. 85,000	Volumetric yield and geologic analogs by reservoir and probabilities
Nehring (1981)	Pre-1976 as of 1979	73.4	1.8	4.4 / 9.6	2.4 / 5.2	9.8	78 / 83 / 93	Not reported (Est. 86,000)	Historical projections: field size distributions

[a] See footnotes to Table 6 for further details.
[b] In 1974, Texas had only 74 tcf; with southeastern New Mexico added value is 89 tcf (API, 1980).

Table 8. Comparison of estimated mean of the undiscovered recoverable oil and natural gas resources for the Permian basin by resource appraisal methodology.

Basic Methods by Report	Estimated Mean Undiscovered Recoverable Oil (billion bbl)	Estimated Mean Undiscovered Recoverable Gas (tcf)	Area Included in Study (sq mi)
Volumetric yield			
AAPG (Galley, 1971)	15.6	79	193,000
PGC (1981)	NA	35	Est. 85,000
PGC (1983)	NA	34	Est. 85,000
Geochemical material balance			
McDowell (1975)	3.4[a]	NA	20,000
Permian system *only*			
Combination methods			
USGS (Miller et al., 1975)	8	70	193,400
Region			
USGS (Dolton et al., 1981)	5.4	43	172,000
Region			
USGS (Dolton et al., 1981)	3	33	86,000
Permian basin only			
Historical performance			
USGS (Dolton et al., 1979)	1.65	16.4	82,000
USGS (1980)	~0.77	~10.39	80,000–100,000
Nehring (1981)	0.4	5.2	Est. 86,000

[a]For comparison: USGS (Dolton et al., 1979) assessed the mean undiscovered oil for the Permian system *only* as 0.76 billion bbl.

problems that have been mentioned in this paper, but it must also address the issues of (1) the fundamentals of the physical or geologic occurrence or nonoccurrence of the resources that contribute to the estimates of the resource base; (2) new concepts in geology and technology, new exploration tools, and new recovery techniques; (3) economic analysis to provide information useful in the development of supply curves and short-term energy needs; and (4) application of new statistical and computer techniques.

The computer revolution of the 1980s, with the new developments in computer hardware, software, and microcomputer applications, opens up the promising application of artificial intelligence techniques to potential resource appraisal research. The concepts of artificial intelligence are now used, for example, as rule-based programs for medical diagnosis that are adaptable to consultation systems to assist geologists in mineral exploration and assessment (PROSPECTOR is one example) (Gaschnig et al., 1981). Other uses include a diagnostic program to examine the causes for failures in producing oil wells (Tompkins, 1982), and a program that can monitor and advise a log analyst in the correct interpretation of a suite of wire-line logs (Weiss et al., 1982; Ennis, 1983). Fertl (1984) discussed similar applications of knowledge engineering to the field of automated well logging and well interpretation. Such knowledge engineering, based on computer programs referred to as expert systems, allows applied utilization of artificial intelligence.

Acquisition systems and interpretive concepts used by geologists in their analysis of basins for petroleum deposits are promising areas for research in artificial intelligence applications to resource appraisal methods and resource assessments. Knowledge engineering systems can be focused on resource assessment problems by incorporating and interpreting fundamental knowledge, principles, experience, and judgment of such experts as geologists, log analysts, and geophysicists knowledgeable in basin assessment procedures. These expert systems could capitalize on the experience and expertise of an entire cadre of geologists and other specialists in industry, government, and academic institutions. With the new wave of computer applications and better knowledge acquisition, I believe that the application of expert systems to resource appraisal methodology holds great promise for more reliable resource assessments in the future.

CONCLUSIONS

In three case studies I have reviewed the nationwide resource assessments for Canada and the United States, and the regional resource assessments for the Permian basin of western Texas and southeastern New Mexico. The review enables us to compare the results of selected resource assessments from several geographic areas that were obtained by different methods so that we can answer certain fundamental questions. What should an estimator consider in making the selection of a method or combination of methods for an assessment of the petroleum resources of a specified area? What limitations of the estimates used to generate the assessments should the user be aware of? Is the outcome of the assessment significantly affected by the particular methodology that was chosen? Has any progress been made in the last decade toward developing a systematic approach to a total resource appraisal methodology that addresses all the basic issues of the petroleum potential of a particular area?

The difficulties in comparing assessments of oil and gas resources are numerous and involve most of the following situations: (1) lack of documentation; (2) differences in geographic boundaries and areas included or excluded for assessments; (3) differences in data bases, terminology, commodities assessed, assumptions, qualifications, and limitations; (4) differences in economic assumptions and constraints and in recovery factors; (5) differences in the purposes of the assessments or motivations and biases of the estimators; and (6) differences in the methods of assessing the remaining undiscovered resources. This paper focuses on the last issue, that of addressing the different methods of assessing the remaining oil and gas resources, and how the method affects the outcome of the resource estimate.

Recognizing the difficulties of comparing resource estimates, I believe that (1) resource estimates depend significantly on the resource method chosen for the assessment, (2) consistent patterns do occur in the magnitude of the estimates, and (3) the general magnitude of the estimates are predictable depending on the specific method of assessment used. The volumetric yield and geologic analog techniques invariably generate the most optimistic resource estimates, whereas the historical performance projections generate just as consistently the most pessimistic estimates of remaining undiscovered resources. A combination of several different appraisal methods usually generates an estimate of the undiscovered resources that falls somewhere between these two extremes. The magnitude of the estimate and its position in this range of extremes is dependent on the choice of the methods used in the combination method approach and on the varying assumptions of the geologic and economic constraints that the estimators use in the assessment procedures. These, in turn, contribute to a variety of limitations that affect the resource estimates.

Many (but not all) of the more recently published resource assessments attempt to incorporate a combination of methods that tend to maximize the geologic input while utilizing the historical data. Because the uncertainties of assessing the unknown is recognized, the estimator applies risk analysis and probability distributions in reporting the resource estimates. This combination of methods for a systematic approach to resource assessment appears to use more of the available data and seems to be the most reasonable approach. These combination methods also provide the most professional effort for assessing the remaining undiscovered resources. The choice of the methods of assessment, however, should be flexible and appropriate to the status of the area being assessed. No single method is applicable to all provinces, basins, stratigraphic sequences, or stages of exploration effort (Haun, 1975). One additional recommendation should be considered: more than one method of assessment, providing more than a single estimate, should always be applied to a specific area so that reliability can be enhanced and reasonableness of estimates can be tested and cross-checked (Semenovich et al., 1977; USDOE/USGS, 1979; Miller, 1979). This also applies to the use of a single combination of methods that provides only one estimate of the resource potential. Several assessment procedures should be used to provide more than one estimate of the resources as a means of cross-checking on the assumptions used and on the methods of assessment applied in an area. It is rarely reported in the literature, however, that more than one method with one set of assumptions is employed in a resource appraisal effort that would provide

more than one estimate of the resources for a specific area as a cross-check.

Obviously, improvements and various modifications of the basic procedures in resource assessment methods have been used during the last several decades, including the use of the computer to incorporate large information data banks and apply complex statistical techniques. But little has been done in the past decade to promote significant research programs in the development of new concepts in resource appraisal methodology.

Perhaps the encouragement of research will produce more effective approaches to assessing our resource potential. Future methodology must address the various problems that have been mentioned in this paper and the issues of (1) understanding the fundamentals of the geologic occurrences of the resources that contribute to the estimates of the resource base; (2) researching new concepts in geology and technology, new exploration tools, and new recovery techniques; (3) providing economic analysis useful for the development of supply curves and short-term energy needs; and (4) applying new statistical and computer techniques, particularly by geologists in their resource assessments, directly utilizing microcomputers.

ACKNOWLEDGMENTS

I thank my reviewers whose comments were most helpful: W. Dewitt, Jr., C. D. Masters, R. F. Meyer, T. W. Offield, H. L. Thomsen, D. White, and the U.S. Geological Survey Technical Editing staff.

REFERENCES

American Petroleum Institute, American Gas Association, and Canadian Petroleum Association, 1980, Reserves of crude oil, natural gas liquids, and natural gas in the United States and Canada as of December 31, 1979: Washington, D.C., American Petroleum Institute, v. 34, June, 253 p.

Arps, J. J., and T. G., Roberts, 1958, Economics of Drilling for Cretaceous Oil on East Flank of Denver–Julesburg Basin: AAPG Bulletin, v. 42, n. 11, p. 2549–2566.

Baker, R. A., H. M. Gehman, W. R. James, and D. A. White, 1984, Geologic field number and size assessments of oil and gas plays: AAPG Bulletin, v. 68, n. 4, p. 426–437.

Bishop, R. S., H. M., Gehman, Jr., and A. Young, 1984, Concepts for estimating hydrocarbon accumulation and dispersion, in G. Demaison and R. J. Murris, eds., Petroleum geochemistry and basin evaluation: AAPG Memoir 35, p. 41–52.

Canadian Department of Energy, Mines and Resource, 1974, An energy policy for Canada—Phase 1: Information Canada, Ottawa, v. II, appendices p. 31–56.

——— , 1977, Oil and natural gas resources of Canada, 1976: Report EP 77-1, 73 p.

Canadian Petroleum Association, 1982, Statistical Handbook 1979, with 1981 and 1982 updates: Canadian Petroleum Association, Calgary, Alberta.

Cram, I. H., ed., 1971, Future petroleum provinces of the United States—their geology and potential: AAPG Memoir 15, v. 1 and 2, 1496 p.

Day, D. T., 1909, The petroleum resources of the U.S., in Papers on the conservation of mineral resources: U.S. Geological Survey Bulletin 394, p. 30–61.

Demaison, G., 1984, The generative basin concept, *in* G. Demaison and R. J. Murris, eds., Petroleum geochemistry and basin evaluation: AAPG Memoir 35, p. 1–14.

Dolton, G. L., A. B., Coury, S. E. Frezon, K. Robinson, K. L. Varnes, J. M. Wunder, and R. W. Allen, 1979, Estimates of undiscovered oil and gas, Permian basin, west Texas and southeast New Mexico: U.S. Geological Survey Open-File Report 79-838, 72 p.

———, K. H. Carlson, R. R. Charpentier, A. B. Coury, R. A. Crovelli, S. E. Frezon, A. S. Khan, J. H. Lister, R. H. McMullin, R. S. Pike, R. B. Powers, E. W. Scott, and K. L. Varnes, 1981, Estimates of undiscovered recoverable conventional resources of oil and gas in the United States: U.S. Geological Survey Circular 860, p. 87.

Dow, W. G., 1977, Kerogen studies and geological interpretations: Journal of Geochemical Exploration, v. 7, p. 79–99.

Ennis, S. P., 1983, Expert Systems—an emerging computer technology: Oil and Gas Journal, v. 81, n. 30, p. 184–188.

Exxon Company, U.S.A., 1976, U.S. Oil and gas potential: Report, Exploration Department.

Fertl, W. H., 1984, Advances in well logging, well interpretation: Oil and Gas Journal, v, 82, n. 16, p. 85–90.

Galley, J. E., 1971, Summary of petroleum resources in Paleozoic rocks of Region 5—north, central, and west Texas and eastern New Mexico, *in* I. H. Cram, ed., Future petroleum provinces of the United States—their geology and potential: AAPG Memoir 15, v. 1, p. 726–737.

Gaschnig, J., J. Reiter, and R. Reboh, 1981, Development and application of a knowledge-based expert system for uranium resource evaluation: Final Report, SRI Project 2225, Stanford Research Institute International, 163 p.

Haun, J. D., 1975, Methods of estimating the volume of undiscovered oil and gas resources: AAPG Studies in Geology, n. 1, 206 p.

Hendricks, T. A., 1965, Resources of oil, gas, and natural gas liquids in the United States and the World: U.S. Geological Survey Circular 522, 20 p.

Hood, A., C. C. M. Gutjahr, and R. L. Heacock, 1975, Organic metamorphism and the generation of petroleum: AAPG Bulletin, v. 59, p. 986–996.

Hubbert, M. K., 1962, Energy Resources—a report to the Committee on Natural Resources of the National Academy of Sciences and National Research Council: National Academy of Sciences and National Research Council Publication, n. 1000-D, 141 p.

———, 1974, U.S. energy resources, a review of 1972, Part 1, *in* A national fuels and energy policy study: U.S. 93rd Congress, 2nd Session, Senate Committee Interior and Insular Affairs, Serial No. 93-40, (92-75), 267 p.

Hunt, J. M., and G. W. Jamieson, 1956, Oil and organic matter in source rocks of petroleum: AAPG Bulletin, v. 40, n. 3, p. 477–488.

Ivanhoe, L. F., 1976a, Evaluating prospective basins: oil/gas potential in basins estimated: Oil and Gas Journal, v. 74, n. 49, p. 154–155.

———, 1976b, Evaluating prospective basins: foreign prospective basins evaluated: Oil and Gas Journal, v. 74, n. 50, p. 108–110.

———, 1976c, Evaluating prospective basins: economic feasibility appraisal for petroleum search in remote regions: Oil and Gas Journal, v. 74, n. 51, p. 82–84.

Kaufman, G. M., 1965, Statistical analysis of the size distribution of oil and gas fields, *in* Symposium on petroleum economics and evaluation: AIME, p. 109–124

———, Y. Balcer, and D. Kruyt, 1975, A probabilistic model of oil and gas discovery, *in* J. D. Haun, ed., Methods of estimating the volume of undiscovered oil and gas resources: AAPG Studies in Geology, n. 1, p. 113–142.

Klemme, H. D., 1971, The giants and supergiants: Oil and Gas Journal, pt. 1, pt. 1, v. 69, n. 9, p. 85–90; pt. 2, pt. 2, v. 69, n. 10, p. 103–110; pt. 3, pt. 3, v. 69, n. 11, p. 96–100.

———, 1983, Field size distribution related to basin characteristics: Oil and Gas Journal, v. 81, n. 52, p. 168–176.

Kontorovich, A. E., 1984, Geochemical methods for the quantitative evaluation of the petroleum potential of sedimentary basins, *in* G. Demaison and R. J. Murris, eds., Petroleum geochemistry and basin evaluation: AAPG Memoir 35, p. 79–109.

Langston, J. D., 1976, A new look at the U.S. oil and gas potential: Proceedings of the 14th Institute on Exploration and Economics of the petroleum industry, Southwestern Legal Foundation, Dallas, Texas, p. 33–50.

Lee, P. J., and C. C. Wang, 1983, Probabilistic formulation of a method for the evaluation of petroleum resources: Journal of the International Association for Mathematical Geology, v. 15, n. 1, p. 163–181.

Mallory, W. W., 1975, Accelerated national oil and gas resource appraisal (ANOGRE), *in* J. D. Haun, ed., Methods of estimating the volume of undiscovered oil and gas resources: AAPG Studies in Geology, n. 1, p. 23–30.

Mast, R. F., R. H. McMullin, K. J. Bird, and W. P. Brosge, 1980, Resource appraisal of undiscovered oil and gas resources in the William O. Douglas Arctic Wildlife Range, U.S. Geological Survey Open-File Report 80-916, 62 p.

McCrossan, R. G., ed., 1973, The future petroleum provinces of Canada—their geology and potential: Canadian Society of Petroleum Geologists, Memoir, n. 1, 720 p.

———, and J. W. Porter, 1973, The geology and petroleum potential of the Canadian sedimentary basins—A synthesis, *in* R.G. McCrossan, ed., The Future Petroleum Provinces of Canada—their geology and potential: Canadian Society of Petroleum Geologists, Memoir 1, 720 p.

McCulloh, T. H., 1973, Mineral resource estimates and public policy, *in* D. A. Brobst and W. P. Pratt, eds., United States mineral resources: U.S. Geological Survey Professional Paper 820, p. 9–19.

McDowell, A. N., 1975, What are the problems in estimating the oil potential of a basin?: Oil and Gas Journal, v. 73, n. 23, p. 85–90.

McKelvey, V. E., 1974, USGS releases revised U.S. oil and gas estimates: U.S. Department of the Interior News Release, March 26, 4 p., 1 table.

Meyer, R. F., 1978, The volumetric method for petroleum resource estimation, Journal of the International Association for Mathematical Geology Bulletin, v. 10, n. 5, p. 501–518

Miller, B. M., 1979, The evolution in the development of the petroleum resource appraisal procedures in the U.S. Geological Survey and a summary of current assessments for the United States: SPE–AIME Eighth Hydrocarbon Economics and Evaluation Symposium, Dallas, Texas, SPE 7720, 1979, p. 79–90.

———, 1981a, Application of exploration play-analysis techniques to the assessment of conventional petroleum resources by the U.S. Geological Survey: 1981 Economics and Evaluation Symposium of the Society of Petroleum Engineers of AIME, Dallas, Texas, SPE 9561, p. 101–109.

———, 1981b, Methods of estimating potential hydrocarbon resources by the U.S. Geological Survey—case studies in resource assessment in the National Petroleum Reserve in Alaska and the William O. Douglas Arctic Wildlife Range, *in* New ideas, new methods, new developments: Proceedings of the Southwestern Legal Foundation, Exploration and Economics of the Petroleum Industry, v. 19, p. 57–95.

———, 1982, Application of exploration play analysis techniques to the assessment of conventional petroleum resources by the U.S. Geological Survey: Journal of Petroleum Technology, v. 34, n. 1, p. 55–64.

———, 1983, Methodology of the U.S. Geological Survey Resource Appraisal Group for gas resource estimates in the United States, in C. Delahaye and M. Grenon, eds., Conventional and unconventional world natural gas resources: Proceedings of the Fifth IIASA Conference on Energy Resources, Laxenburg, Austria, p. 237-266,

Miller, B. M., H. L. Thomsen, G. L. Dolton, A. B. Coury, T. A. Hendricks, F. E. Lennartz, R. B. Powers, E. G. Sable, and K. L. Varnes, 1975, Geological estimates of undiscovered recoverable oil and gas resources in the United States: U.S. Geological Survey Circular 725, 78 p.

Miller, B. M., B. M., R. S. Pike, R. B. Powers, E. W. Scott, A. S. Khan, K. H. Carlson, B. T. Vietti, and G. D. Lambert, 1978, Pilot Area: Gulf of Mexico—Oil and Gas Resource Appraisals, Probability Distributions for size and estimated numbers of undiscovered fields: Administrative Report prepared for USGS/DOE Interagency Oil and Gas Supply Project, U.S. Geological Survey, n. 2, 115 p.

Moody, J. D., and R. E. Geiger, 1974, Potential resources of natural gas reported, in R. Gillette, ed., Oil and gas resources: Did USGS gush too high?: Science, v. 185, n. 12, p. 127–130.

———, 1975, Petroleum resources: How much oil and where?: Technology Review, v. xx, n. xx, p. 39–45.

Moore, C. L., 1962, Method for evaluating U.S. crude oil resources and projecting domestic crude oil availability: Washington, D.C., Department of the Interior, Office of Oil and Gas, 112 p.

———, 1966, Projections of U.S. petroleum supply to 1980: U.S. Department of the Interior, Office of Oil and Gas, 13 p, with appendix.

———, 1971, Analysis and projection of historic patterns of U.S. crude oil and natural gas, in I.H. Cram, ed., Future petroleum provinces of the United States—their geology and potential: AAPG Memoir 15, v. 1, appendix F, p. 50–54.

National Petroleum Council, 1970, Future Petroleum Provinces of the United States—A summary: Committee on Possible Future Petroleum Provinces of the United States, Washington, D.C., National Petroleum Council, 138 p.

———, 1973, U.S. energy outlook, oil and gas availability: Washington, D.C., National Petroleum Council Committee on U.S. Energy Outlook, 786 p.

———, 1981, U.S. Arctic oil and gas: Washington, D.C., National Petroleum Council, 284 p,

National Research Council, 1975, Mineral resources and the environment: Commission on Natural Resources, Committee on Mineral Resources and the Environment (COMRATE), National Academy of Sciences, 348 p.

Nehring, R., 1981, The discovery of significant oil and gas fields in the United States: Report R-2654/1-USGS/DOE, Santa Monica, California, Rand Corporation, v. 1, 236 p.

Neruchev, S. G., 1962, Oil generating suites and the migration of oil: Leningrad, Gostoptekhizdat, p. 217–221.

Oil and Gas Journal, 1984, Canadian exploration outlay up; reserves rise: v. 82, n. 27, p. 25–26

Philippi, G. T., 1956, Identification of oil source beds by chemical means: Proceedings XX, International Geology Congress, Mexico City, v. 111, p. 25–38.

———, 1976, On the depth, time, and mechanism of origin of the heavy to medium-gravity naphthenic crude oils: Geochimica et Cosmochimica Acta, v. 41, p. 33–52.

Porter, J. W., and R. G. McCrossan, 1975, Basin consanquinity in petroleum resource estimation, in J. D. Haun, ed., Methods of estimating the volume of undiscovered oil and gas resources: AAPG Studies in Geology, n. 1, p. 50–75.

Potential Gas Committee, 1969, Potential supply of natural gas in the United States (as of December 31, 1968): Potential Gas Agency, Colorado School of Mines, 39 p.

———, 1971, Potential supply of natural gas in the United States (as of December 31, 1970): Potential Gas Agency, Colorado School of Mines, 41 p.

———, 1973, Potential supply of natural gas in the United States (as of December 31, 1972): Potential Gas Agency, Colorado School of Mines, 52 p.

———, 1977, A comparison of estimates of ultimately recoverable quantities of natural gas in the United States: Gas Resource Studies, Potential Gas Agency, Colorado School of Mines, n. 1, 27 p.

———, 1981, Potential supply of natural gas in the United States (as of December 31, 1980): Potential Gas Agency, Colorado School of Mines, 119 p.

———, 1983, Potential supply of natural gas in the United States (as of December 31, 1982): Potential Gas Agency, Colorado School of Mines, 74 p.

———, 1984, Definitions and procedures for estimation of potential gas resources (2nd ed.): Potential Gas Agency, Golden, Colorado, 16 p.

Procter, R. M., G. C. Taylor, and J. A. Wade,, 1984, Oil and natural gas resources of Canada, 1983: Geological Survey of Canada, Department of Energy, Mines and Resources, Paper 83-31, 59 p.

Roadifer, R. E., 1975, A probability approach to estimate volumes of undiscovered oil and gas: in M. Grenon, ed., First IIASA Conference on Energy Resources, Laxenburg, Austria, n. CP-76-4, p. 333–343.

Roy, K. J., 1975, Hydrocarbon assessment using subjective probability and Monte Carlo methods, in M. Grenon, ed., First IIASA Conference on Energy Resources, Laxenburg, Austria, n. CP-76-4, p. 345–359.

Searl, M. F., 1975, Resource assessment and supply curve development: toward better methodologies, in M. Grenon, ed., First IIASA Conference on Energy Resources, Laxenburg, Austria, n. CP-76-4, p. 71-83.

Semenovich, V. V., N. I. Bayalov, V. N. Kramarenko, A. E. Kontorovich, Yu. Ya. Kuzentsov, S. P. Maksimov, M. Sh. Modelevsky, and I. I. Nesterov, 1977, Methods used in the U.S.S.R. for estimating potential petroleum resources, in R. F. Meyer, ed., The future supply of nature-made petroleum and gas technical reports: New York, Pergamon Press, p. 139–153,

Shell Oil Company, 1978, Alaska holds 58% of future U.S. oil finds: Oil and Gas Journal, v. 76, n. 47, p. 214.

Sluijk, D., and M. H. Nederlof, 1984, Worldwide geological experience as a systematic basis for prospect appraisal, in G. Demaison and R. J. Murris, eds., Petroleum geochemistry and basin evaluation: AAPG Memoir 35, p. 15–26.

Theobald, P. K., S. P. Schweinfurth, and D. C. Duncan, 1972, Energy resources of the United States: U.S. Geological Survey Circular 650, 27 p.

Thomsen, H. L., 1979, Oil and gas resource appraisal—state of the art (abs.): AAPG Bulletin, v. 63, n. 3, p. 176–177.

Tissot, B., Y. Califet-Debyser, G. Dervo, and J. L. Oudin, 1971, Origin and evolution of hydrocarbons in early Toacian shales, Paris basin, France: AAPG Bulletin, v. 55, n. 12, p. 2177–2193.

———, B. Durand, J. Espitalie, and A. Combaz, 1974, Influence of nature and diagenesis of organic matter in formation of petroleum: AAPG Bulletin, v. 58, n. 3, p. 499–506.

Tompkins, J. W., 1982, PROBWELL: an expert advisor for determining the cause of problems with producing oil wells: IBM Engineering and Scientific Study Conference Proceedings, Poughkeepsie, New York.

Ungerer, F. B., P. Y. Chenet, B. Durand, E. Nogaret, A. Chiarelli, J. L. Oudin, and J. F. Perrin, 1984, Geological and geochemical models in oil exploration; principles and examples, in G.

Demaison, and R. J. Murris, eds., Petroleum geochemistry and basin evaluation: AAPG Memoir 35, p. 53–77.

U.S. Department of the Interior, 1979, Final report of the 105(b) economic and policy analysis, alternative overall procedures for the exploration, development, production, transportation, and distribution of the petroleum resources of the National Petroleum Reserve in Alaska (NPRA): Office of Minerals Policy and Research Analysis Report, 145 p.

U.S. Department of Energy and U.S. Geological Survey, 1979, Report on the petroleum resources of the Federal Republic of Nigeria: DOE/IA-0008, 63 p.

U.S. Geological Survey, 1974, USGS releases revised U.S. oil and gas resource estimates: Department of the Interior News Release, March 26, 1974, 5 p.

———, 1980, Future supply of oil and gas from the Permian basin of west Texas and southeastern New Mexico, a report of the Interagency Oil and Gas Supply Project: U.S. Department of the Interior and U.S. Department of Energy, U.S. Geological Survey Circular 828, 57 p.

Vassoyevich, N. B., and S. G. Neruchev, 1964, Origin, evolution, and primary migration of micro-oil: Proceedings of the 22nd International Geologic Congress, New Delhi, India, part I, p. 71–85.

Weeks, L. G., 1950, Concerning estimates of potential petroleum reserves: AAPG Bulletin, v. 34, p. 1947–1953.

Weiss S. C., C. Kulitzer, C. Apte, M. Uschold, J. Patchett, R. Brigham, and B. Spitzer, 1982, Building expert systems for controlling complex programs: Proceedings of the National Conference on Artificial Intelligence, Los Altos, California, p. 322–326.

Welte, D. H. and M. A. Yukler, 1984, Petroleum origin and accumulation in basin evolution—a quantitative model, in G. Demaison and R. J. Murris, eds., Petroleum geochemistry and basin evaluation: AAPG Memoir 35, p. 27–39.

White, D. A., 1980, Assessing oil and gas plays in facies-cycle wedges: AAPG Bulletin, v, 64, n. 8, p. 1158–1178.

———, R. W. Garrett, Jr., G. R. Marsh, R. A. Baker, and H. M. Gehman, 1975, Assessing regional oil and gas potential, in J. D. Haun, ed., Methods of estimating the volume of undiscovered oil and gas resources: AAPG Studies in Geology, n. 1, p. 143–159.

———, and H. M. Gehman, 1979, Methods of estimating oil and gas resources, AAPG Bulletin, v. 63, n. 12, p. 2183–2192.

White, L. P., 1979, A play approach to hydrocarbon resource assessment and evaluation, in Conference on the Economics of Exploration for Energy Resources: New York University, New York.

Zapp, A. D., 1962, Future petroleum producing capacity in the United States: U.S. Geological Survey Bulletin 1142-H, 36 p.

Geologic Field Number and Size Assessments of Oil and Gas Plays

R. A. Baker, H. M. Gehman, W. R. James, and D. A. White
Exxon Production Research Company
Houston, Texas

Assessments of undiscovered oil and gas potentials for a group of geologically related, untested prospects can be effectively made from an estimate of the possible ranges in number and size of potential fields, assuming that the play exists, coupled with an evaluation of geologic risks that it might not exist. Field size distributions are constructed from known field reserves in geologically similar plays, from assessments of representative prospects in the play, or from simulations of distributions of the play's prospect areas, reservoir parameters, and potential hydrocarbon relationships. The field size distributions are truncated at both ends: at a practical minimum and at the largest size reasonably expected in the play. The possible range of the number of potential fields is estimated from the counted and postulated number of untested prospects in conjunction with a success ratio, or it is estimated from look-alike field densities. The chance that the play exists is the chance that there is at least one field of at least the minimum size assessed. The final assessment curves, developed in a Monte Carlo simulation, give the exceedance probability versus the range of possibly recoverable hydrocarbon potentials.

INTRODUCTION

A straightforward way to assess regional undiscovered oil and gas resources is to estimate geologically the number and size distributions of potential fields in exploration plays. A play is a group of field prospects with geologically similar source, reservoir, and trap controls on oil and gas occurrence. In its simplest form (Figure 1), the method, as pioneered by Atwater (1956), is to multiply a prospect count by an assumed success ratio to estimate the number of potential fields. This number times the potential average field size in ultimately recoverable barrels gives a single-valued oil assessment. Belov (1960) and Semenovich et al. (1977) reported similar approaches. Ivanhoe (1976), Nehring (1978), and Momper (1979) have illustrated how effectively counts and sizes of fields can be used in assessments.

The Geological Survey of Canada (GSC) (Roy et al., 1975; Energy, Mines, and Resources Canada, 1977; Procter et al., 1982) made significant advances in methodology by using ranges of values for both prospect numbers and field sizes and then combining these ranges in a Monte Carlo simulation to produce the assessment. Ranges are important for showing the uncertainties inherent in any such assessments and for producing the final probability curves (D. A. White et al., 1975). Building on the GSC model, L. P. White (1979) incorporated marginal as well as conditional probabilities along with the analyses of prospect number and field size

distributions. Our own similar approach, which we have been developing since 1972, is outlined in this paper.

The field size play assessment method is appealing because it deals directly with the natural units of petroleum exploration—prospects and fields—in a versatile way useful for both geologic and economic analyses. The ideal method of assessing a play is to aggregate all the individual prospect assessments (Gehman et al., 1975, 1980). Lack of time or data, however, often dictates use of the shortcut play approach, which essentially is a form of prospect summation.

The requirements for this play assessment method, as discussed in the next section, are geologic estimates of the following: (1) the likely field size distribution, in potentially recoverable barrels or cubic meters, with specified minimum and maximum cutoffs; (2) the numbers of potential fields, usually based on counts of undrilled prospects considered along with postulated field success ratios; and (3) the chance of the play's existence, that is, the chance of occurrence of at least one field of minimum size. The basic decisions are geologic, and this method is distinct from the approaches that depend on statistical extrapolations of field numbers and sizes. These latter methods are very useful where abundant data exist, and they provide instructive examples of field size distributions. Although not further discussed here, a few examples are the extrapolations of Arps and Roberts (1958), Cozzolino (1972), Kaufman et al. (1975), Menard and Sharman (1975), and Drew et al. (1982).

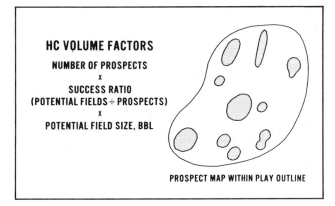

Figure 1. Play assessment from field number and size distribution. Prospect outlines are shown within mapped boundaries of the play. From White and Gehman (1979).

In practice, the geologic play assessment procedure involves two steps. First, the assessor assumes that the play exists and models what it would be like in terms of numbers and sizes of fields. Second, the assessor judges the chance that this model is basically right—that the play really does exist.

PLAY DELINEATION

It is critical first to outline the play geologically. Typically, the delineated prospects form a group areally, have the same basic trap type, have the same type of reservoir facies, and presumably have the same hydrocarbon source. Thus they share some common elements of risk relating to the possible occurrence of oil and gas. Lumping distinctly different prospect types can cause serious problems in the later sizing and risking steps. Wherever a significant areal change in geologic controls is anticipated, it is often best to define a new play. Such judgments should be based, where possible, on exploration maps of trap anomalies, reservoir facies distributions and characteristics, seal thicknesses, source rock qualities and maturations, and hydrocarbon shows.

For practical purposes, a large geologic play area can be broken for separate assessments along arbitrary lines such as concession blocks, international boundaries, or bathymetric contours. The resulting subplay assessments will have interdependent risks that should be considered if the results are aggregated (Gehman et al., 1975, 1980). Different trap types that are areally interspersed can likewise be assessed as separate plays and then aggregated.

Estimates of field numbers and sizes can be used for assessments at every knowledge level. If assessments are required of areas where data are minimal, the estimates (although little more than guesses guided by experience) serve to document current thinking in a way that can be scaled and compared with other assessments and known field populations. The initial postulates can readily be revised as new data arrive. The method is most applicable where potential structural traps are seismically identifiable. Stratigraphic traps are much more difficult to assess than structural ones, but this is true for any approach.

FIELD SIZE DISTRIBUTIONS

For compelling practical reasons, we use field size distributions truncated at both ends and plotted on log probability graphs (Figures 2 through 5). The importance of selecting practical minimum and maximum values has been emphasized by Ivanhoe (1976 and this volune). Klemme (1971, 1975 and this volume) for a long time has pointed out the overwhelming significance of the larger fields.

The distribution of ~ 14,000 U.S. oil fields (Figure 2), which is a partial sample of those in the conterminous United States, illustrates the importance of the larger fields. (The sample includes almost all larger fields as known in 1970 and excludes many tiny fields and all of the most recent discoveries.) The lower dotted line in Figure 2 was constructed from 13,985 points representing the fields ordered according to increasing size of reserves. Only 440 fields, or ~ 3%, are major ones larger than 50 million bbl. The upper curve represents the percentage of the total oil volume occurring in fields greater than each size. From this curve we can see that the major fields, constituting only 3% of the total number, contain 80% of the total amount of oil. In this type of distribution one can thus account for the bulk of the oil by assessing only the largest field possibilities.

Selecting an effective minimum field size cutoff is very important, since it affects every major factor in the assessment: the prospects to be counted, the success and risk levels, and the average field size. Normally, the minimum size is taken at or just below the assumed economic minimum for the area. This approach ensures that all prospects of real interest are included. It also avoids time that would otherwise be wasted on hundreds or thousands of fields that are inconsequential to early exploration stages. Furthermore, the comparative data base for assessing subeconomic fields is very weak, since the true sizes of these fields have never been scaled. If desired, one can assess the small fields by statistical extrapolation or by estimating a lump sum proportion from a volume curve like that of Figure 2.

Economic limits always truncate the lower ends of observed field size distributions (Arps and Roberts, 1958; Kaufman et al., 1975; Grender et al., 1978; Vinkovetsky and Rokhlin, 1982; Drew et al., 1982). In nature's distribution, numbers of deposits probably increase progressively in successively smaller sizes down to droplets and molecules; such a distribution is not lognormal. But we deal exclusively with artificially truncated distributions whose plots almost invariably curve upward near the low-side truncation point (upper curve, Figure 3). Our U.S. distribution (lower curve, Figure 2) does not include any data for fields smaller than 1,000 barrels, and many of the data points for fields smaller than 10,000 barrels, (where the graph ends) are questionable. If the plot were continued to the left, it would ultimately curve upward at the point of economic truncation beyond which there are no data.

We use the computational convenience of the lognormal distribution, appropriately truncated, but we would not argue that this scheme is better or worse than other computational ones for strongly right-skewed distributions that have many more little fields than big fields. Some investigators (e.g., Ivanhoe, 1976; Folinsbee, 1977; Coustau, 1980) plot field size bilogarithmically against rank order. For our assessment

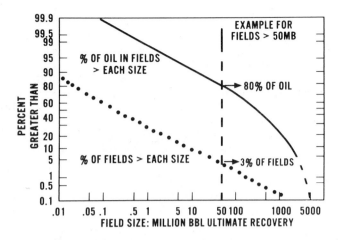

Figure 2. Field size distribution for 13,985 oil fields of the conterminous United States. Lower curve shows the percentage of the number of fields greater than each size, and the upper curve shows the percentage of the total oil volume in fields greater than each size. As an example, major fields of 50 million bbl or more of oil, which represent only ~3% of the fields, contain 80% of the total oil.

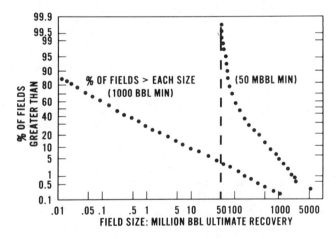

Figure 3. Field size distribution for 13,985 oil fields larger than 1000 bbl in the conterminous United States (lower curve), with a truncated distribution of 440 major oil fields larger than 50 million bbl or more of oil each (upper curve).

approach, we must normalize field numbers at this stage by plotting "percentage greater than" against log size. Depending on purpose and data, we can express field size as recoverable volumes of oil or gas, or of oil plus gas on an energy equivalent basis.

Plays that differ geologically commonly have different field size distributions (Coustau, 1980). Of the three different Alberta basin reef plays (Figure 4), the Keg River fields have the steepest distribution line, reflecting smaller fields and relatively uniform sizes within the truncated range. The Beaverhill Lake fields have the flattest distribution, reflecting larger fields and more diverse sizes. Coustau (1980) classes these as dispersed and concentrated habitats, respectively. We call them "splitters" and "lumpers," the

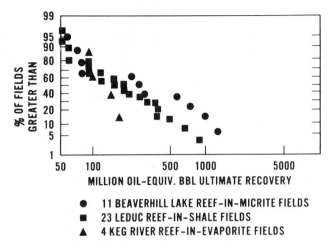

Figure 4. Field size distributions of three Devonian reef plays, Alberta basin.

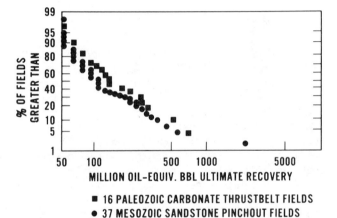

Figure 5. Field size distribution of Paleozoic carbonate thrustbelt fields compared with that of Mesozoic sandstone pinchout fields, Alberta basin.

latter lumping greater proportions of oil and gas in the largest fields.

Plays differing geologically may also have virtually identical field size distributions. Figure 5 compares Albertan Paleozoic carbonate thrustbelt and Mesozoic sandstone pinchout fields. Their size distributions are similar, with the exception of the giant one-of-a-kind Pembina Sandstone field, which has ultimate reserves of ~2 billion bbl. Such one-of-a-kind giants that do not fit the distribution of the other fields are best assessed as individual prospects.

Field size distributions for play assessments can be constructed in at least three ways. First, a look-alike known play can be selected and its fields plotted, as in Figures 4 and 5. Second, representative prospects in the play can be assessed, and the mean assessments plotted as before. Third, the distributions of prospect areas, reservoir parameters, hydrocarbon fill fractions, and recovery factors can be combined in a Monte Carlo simulation to produce a distribution of possible field sizes (Roy et al., 1975; Energy, Mines, and Resources Canada, 1977; Procter et al., 1982).

This last approach is more time consuming, but it has the advantage of providing many of the detailed data on prospects required for thorough economic analyses.

The ultimate key to selecting any distribution is that it should be tied to the largest field anticipated in the play, as emphasized by Ivanhoe (1976). The best way to do this is to identify and assess the largest prospect in the play. It will, for example, make a big difference in the assessment whether the largest field is 600 versus 2,000 million bbl in a sandstone pinchout play (Figure 5), or 200 versus 1,100 million bbl in different reef play models (Figure 4). The distribution should not be cut short of the largest reasonably foreseeable size. It should not, however, be extended far beyond this size, or serious overestimates may result. Nature truncates all distributions ultimately by limiting effective closure space or source rock capabilities. The assessor should judge these factors and truncate accordingly. Where data are limited, this truncation may have to be based only on look-alike experience and judgment.

FIELD NUMBER DISTRIBUTIONS

Where data permit, potential field numbers can be estimated from counted and/or postulated untested prospects in conjunction with success ratios (Atwater, 1956; Energy, Mines, and Resources Canada, 1977; White, 1979). The only prospects to be counted or postulated are untested ones large enough to hold the minimum size selected for the assessment's field size distribution. Prospect densities (numbers of prospects per unit area) from known look-alike plays can be helpful in postulating.

The success ratio equals the expected number of fields of at least minimum size divided by the number of prospects capable of holding that size. Success ratios, which can be drawn from known look-alike plays, reflect the independent geologic risks among prospects. For example, some prospects in the group may have locally poor reservoirs, and others may have broken seals or may have been flushed. These are prospect-specific attributes conditional on the play's existence (White, 1979). They reflect the almost universal observation that, even in richly productive plays, not all prospects of adequate size contain adequate fields. These success ratio attributes must be treated separately from the play chances or marginal probabilities, which reflect play-specific risks that could wipe out productive chances for the group as a whole (White, 1979). It is not always easy to sort the independent and group risks geologically without hitting the same risk twice, but it must be done to preserve realistic hydrocarbon volume versus probability relationships in individual as well as aggregated assessments.

Success ratio levels are relative. If a high graded play includes only the best prospects, the success ratio is apt to be high. If an area is not high graded and includes many poor prospects, the success ratio is apt to be low.

Where it is not realistically possible to estimate the number of prospects and a success ratio, the assessor can estimate numbers of fields directly by using look-alike field densities—number of fields per unit area (Grossling, 1977). This method is useful not only in virgin frontiers but also in highly mature areas where the presently very small prospect objectives cannot be identified and mapped on a play-wide scale.

Computer programs can easily calculate the binomial distribution of the number of prospects and the success ratio (Roy et al., 1975; White, 1979). Alternatively, field number distributions can be input in a variety of ways.

PLAY CHANCES

The chance that a play exists is the chance that there is at least one field of at least the minimum size in the field size distribution. This is the second-step judgment about whether the first-step assumptions of field numbers and sizes are right or not. The assessor decides on a value between 0 and 1 on the basis of an analysis of the group or marginal geologic risks that could deny the existence of any fields whatsoever. For example, the regional hydrocarbon source, migration path, timing, or reservoir facies may be inadequate or lacking throughout the play area.

The total play chance should recognize risks related both to the regional geology and to the number of prospects, if that number is limited. The regional chance by itself would be the same as the play chance given an unlimited number of opportunities. Where prospects are few, however, there is an additional risk that, even if the possible regional problems do not materialize as the play develops, all available opportunities could prove unsuccessful as a result of an unlucky combination of prospect-specific factors. Lee and Wang (1983) discuss the use of the binomial distribution to calculate risk related to limited opportunities.

The play chance is tied specifically to achieving at least the minimum single field potential. It can often be taken as 1.0 in active productive plays where one more discovery is essentially assured. The principle of "risking the minimum" is pointed out by Roy (1975) and Gehman et al. (1975, 1980). White and Gehman (1979) further discuss risking mechanics, and White (1980) summarizes play chance studies for more than a 1,000 major field plays in 80 basins. Play chances are relative, varying not only with the geology but also with the size of the chosen minimum field.

As noted by L.P. White (1979), the average prospect chance for the play equals the success ratio (independent or conditional probability) times the regional chance (group or marginal probability). If one keeps this key relationship straight, it is possible to make the results of play assessment compatible with those achieved by summing individual prospect assessments.

PLAY ASSESSMENT EXAMPLE

In this simplified example, we are assessing potential field numbers, sizes, and chances in a small offshore area on a delta whose stratigraphy and structure remind us of south Texas. Minimum assessed field size here given as 50 million bbl. (Any practical minimum size can be chosen.) Look-alike field densities suggest the most likely possibility of three fields. A reasonable range for uncertainty gives a 1–3–5 triangular distribution, which can be converted to a histogram and an exceedance probability curve (Figure 6).

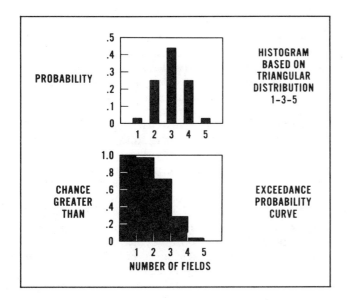

Figure 6. Triangular field number distribution shown as histogram and exceedance probability curve.

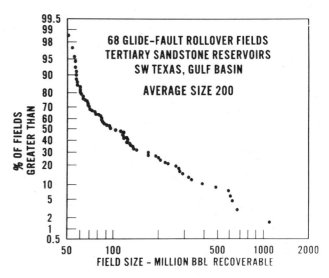

Figure 7. Field size distribution of rollover traps at down-to-basin faults, south Texas, Gulf Basin. Units are in barrels of oil equivalent.

The minimum at one field shows that at this point we are assuming the play exists. In this example, the direct estimate of field numbers bypasses explicit determination of a prospect count and success ratio.

For the field size distribution (Figure 7), we plot the estimated ultimately recoverable reserves of 68 south Texas fields. These fields all are on anticlinal rollovers associated with down-to-basin faults, the same trap type anticipated in our new play area. We also expect that the largest field in the new area could approach but will probably not exceed 1,000 million bbl, about the same size as the largest field in south Texas. The average reserves size of all the south Texas fields is 200 million bbl.

The field number and field size distributions (truncated at 50 and 1,000 million bbl, respectively) are entered in the computer for a Monte Carlo simulation. For each of 5,000 trials, the computer at random selects a potential number of fields from our specified distribution from one to five (Figure 6). If it selects three, the most likely value, the computer then randomly samples three different field sizes from the truncated field size distribution (Figure 7). The computer adds the three values and stores the results as one possible assessment, and then goes on to repeat the process in the next trial.

The "unrisked" assessment curve (Figure 8) is simply the result of all 5,000 trials plotted proportionally from smallest to largest. It shows, for example, that about 70% of the trials (3,500) assessed more than 400 million bbl. Typically, the computer picked the most likely three fields, which commonly averaged about 200 million bbl each, the average of the whole field size distribution. Thus the average assessment is three times 200, or about 600 million bbl. Occasionally the computer picked the minimum of one field, and at least once it assigned the smallest possible size to that one field, giving the minimum assessment of 50 million bbl. Occasionally the computer picked the maximum of five fields, and one group of five sampled mostly from the large end of the field size distribution, and gave the maximum

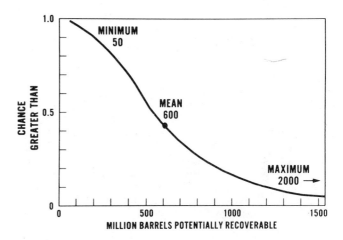

Figure 8. "Unrisked" assessment probability curve derived by Monte Carlo simulation from the field number and size distributions of Figures 6 and 7.

assessment of 2,000 million bbl. The range of minimum through mean to maximum shown in Figure 8 thus gives a picture of what the play might contain, if it exists and if our field number and size assumptions are correct.

Next we judge the play's chance of having at least one field with potential reserves of at least 50 million bbl (Figure 9). We assume the chances are 0.9 that the faulted structures as a group will have adequate seals, 0.8 that they will have adequate reservoirs, and 0.7 that they will have adequate access to a mature source. ("Adequate" as stipulated in this assessment means the capability of source, reservoir, and seal to generate, store, and hold, respectively, at least 50 million bbl of oil.) The overall play chance for at least 50 million bbl is the product of these individual chances, or 0.5, which is essentially the regional chance. Plays like this would typically have a dozen prospects, thus the element of risk related to limited opportunities would be negligible.

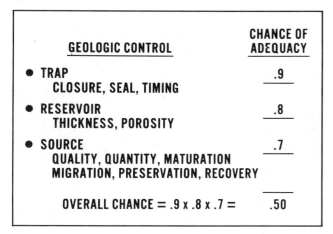

Figure 9. Estimation of play chance for the assessment of Figures 6, 7, and 8.

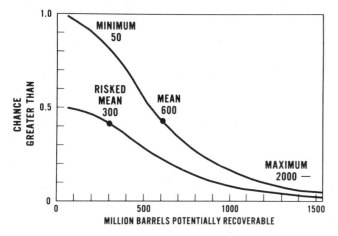

Figure 10. "Risked" and "unrisked" assessment probability curves for the assessment of Figures 6, 7, 8, and 9.

The assigned play chance says that, if there were 100 plays like this one, only 50 would be productive and the other 50 would be effectively dry. Since half the final risked results are zero, the risked curve (Figure 10) shows that the probability of exceeding the given amount of any barrel is cut in half relative to the unrisked curve. The risked mean includes all the zeros and is therefore half of the unrisked mean.

The unrisked curve is used for economic modeling of the play's rewards if it pans out. The risked curve is used for aggregating play assessments into basin assessments, care being taken to handle any dependencies correctly (Gehman et al., 1975, 1980).

SUMMARY

Geologic estimates of possible field sizes and numbers provide a natural basis for play assessment, and the results can closely approximate summations of individual prospect assessments. Key points in the procedure are as follows:

1. First, we model the play as if it exists, and then estimate the chance that this model is right.

2. The play is then delineated as a geologically coherent group of prospects.

3. Next, we establish a practical minimum field size (in terms of recoverable oil and gas) to be included in the assessment, thereby deemphasizing multitudes of insignificant fields and emphasizing the fewer large ones that contain most of the hydrocarbons; the excluded small fields can always be assessed separately as a lump sum.

4. Field size distributions are then constructed from populations of known look-alike field reserves, from representative prospect assessments in the play, or from simulations of distributions of the play's prospect areas, reservoir parameters, and hydrocarbon proportions.

5. We plot the field size distribution on a log probability graph truncated not only at the selected minimum size but also at the largest size reasonably expected in the play.

6. We estimate the possible range of numbers of potential fields from counted and postulated numbers of untested prospects on the basis of a success ratio, or from look-alike field densities.

7. We then assign the play chance that there is at least one field of at least the minimum size in the assessed field size distribution, keeping in mind that the average prospect chance equals the success ratio (conditional probability) times the play chance (marginal probability).

8. Finally, we compute in a Monte Carlo simulation the final assessment curves that portray probability versus the range of possibly recoverable hydrocarbon potential.

The whole approach focuses on the geology of the play itself, and look-alike data are carefully selected and appropriately modified to fit. The postulated numbers and sizes of fields can readily be compared with those of productive plays elsewhere, providing a judgment check on the results. The requirements are fundamentally simple and direct, and the method can be used at any knowledge level. The risk-related probabilities can be guided by experience but will always have an unavoidably subjective cast. As a result, it is still possible to get the wrong answer with the right method, just as the right answer occasionally falls out of the wrong method. The worst danger for any assessment is that a new play possibility will be overlooked entirely. On balance, field size play assessment seems capable of giving a better tie with reality over the long term than other commonly used approaches to regional assessment, and it contains the basin components needed for economic screening.

ACKNOWLEDGMENTS

We are indebted to D. S. McPherron, Esso Exploration, Inc., who originally suggested the utility of this assessment approach to us in 1972. We thank Exxon Production Research Co. for permission to publish, and D. O. Smith for reviewing the manuscript.

REFERENCES CITED

Arps, J. J., and T. G. Roberts, 1958, Economics of drilling for Cretaceous oil on east flank of Denver-Julesburg Basin: AAPG Bulletin, v. 42, p. 2549-2566.

Atwater, G. I., 1956, Future of Louisiana offshore oil province: AAPG Bulletin, v. 40, p. 2624-2634.

Belov, K. A., 1960, Geological prospects of discovery of new oil and gas fields in the Stavropol district and Kalmyk ASSR: Petroleum Geology, v. 4, p. 185-192.

Coustau, H., 1980, Habitat of hydrocarbons and field size distribution, a first step towards ultimate reserve assessment, *in* Assessment of undiscovered oil and gas: Bangkok, United Nations ESCAP, CCOP Technical Publication 10, p. 180-194.

Cozzolino, J. M., 1972, Sequential search for an unknown number of objects of nonuniform size: Operations Research, v. 20, p. 293-308.

Drew, L. J., J. H. Schuenemeyer, and W. J. Bawiec, 1982, Estimation of the future rates of oil and gas discoveries in the western Gulf of Mexico: U.S. Geological Survey Professional Paper 1252, 26 p.

Energy, Mines, and Resources Canada, 1977, Oil and natural gas resources of Canada, 1976: Ottawa, Report EP 77-1, 76 p.

Folinsbee, R. E., 1977, World's view—from Alph to Zipf: Geological Society of America Bulletin, v. 88, p. 897-907.

Gehman, H. M., R. A. Baker, and D. A. White, 1975, Prospect risk analysis *in*, J. C. Davis et al., conveners, Probability Methods in Oil Exploration: AAPG Research Symposium Notes, Stanford University, p. 16-20.

——— , 1980, Assessment methodology, an industry viewpoint, *in* Assessment of undiscovered oil and gas: Bangkok, United Nations ESCAP, CCOP Technical Publication 10, p. 113-121.

Grender, G. C., L. A. Rapoport, and Y. Vinkovetsky, 1978, Analysis of oil-field distribution for sedimentary basins of United States: AAPG Bulletin, v. 62, p. 518 (abstract).

Grossling, B. F., 1977, A critical survey of world petroleum opportunities, *in* Project Interdependence: U.S. 95th Congress, House of Representatives Committee on Energy and Natural Resources, Print 95-33, p. 645-658.

Ivanhoe, L. F., 1976, Evaluating prospective basins: Oil and Gas Journal, December 6, p. 154-156; December 13, p. 108-110; December 20, p. 82-84.

Kaufman, G. M., Y. Balcer, and D. Kruyt, 1975, A probabilistic model of oil and gas discovery, *in* John D. Haun, ed., Methods of Estimating the Volume of Undiscovered Oil and Gas Resources: AAPG Studies in Geology Series 1, p. 113-142.

Klemme, H. D., 1971, What giants and their basins have in common: Oil and Gas Journal, March 1, p. 85-90; March 8, p. 103-110; March 35, p. 96-100.

——— , 1975, Giant oil fields related to their geologic setting a possible guide to exploration: Bulletin of Canadian Petroleum Geology, v. 23, p. 30-66.

Menard, H. W., and G. Sharman, 1975, Scientific uses of random drilling models: Science, v. 190, p. 337-343.

Momper, J. A., 1979, Domestic oil reserves forecasting method, regional potential assessment: Oil and Gas Journal, August 13, p. 144-149.

Nehring, R., 1978, Giant oil fields and world oil resources: Report R-2284-CIA, Rand Corporation, Santa Monica, California, 162 p.

Procter, R. M., P. J. Lee, and G. C. Taylor, 1982, Methodology of petroleum resource evaluation: Geological Survey of Canada manual, Petroleum Resource Assessment Workshop and Symposium, Third Circum-Pacific Energy and Mineral Resources Conference, Honolulu, 59 p.

Roy, K. J., 1975, Hydrocarbon assessment using subjective probability and Monte Carlo methods, *in* First IIASA Conference on Methods and Models for Assessing Energy Resources: New York, Pergamon Press, p. 279-290.

Roy, K. J., R. M. Procter, and R. G. McCrossan, 1975, Hydrocarbon assessment using subjective probability, *in* J. C. Davis, J. H. Doveton, and J. W. Harbaugh, conveners, Probability methods in oil exploration: AAPG Research Symposium Notes, Stanford University, p. 56-60.

Semenovich, V. V., et al., 1977, Methods used in the USSR for estimating potential petroleum resources, *in* R. F. Meyer, ed., The future supply of nature-made petroleum and gas: New York, Pergamon Press, p. 139-153.

Vinkovetsky, Y., and V. Rokhlin, 1982, Quantitative evaluation of the contribution of geologic knowledge in exploration for petroleum, *in* Predictive Geology: New York, Pergamon Press, p. 171-190.

White, D. A., 1980, Assessing oil and gas plays in facies-cycle wedges: AAPG Bulletin, v. 64, p. 1158-1178.

White, D. A., et al., 1975, Assessing regional oil and gas potential, *in* John D. Haun, ed., Methods of Estimating the Volume of Undiscovered Oil and Gas Resources: AAPG Studies in Geology Series 1, p. 143-159.

White, D. A., and H. M. Gehman, 1979, Methods of estimating oil and gas resources: AAPG Bulletin, v. 63, p. 2183-2191.

White, L. P., 1979, A play approach to hydrocarbon resource assessment and evaluation: U.S. Department of Interior, Office of Minerals Policy and Research Analysis, Memorandum, May 14, 30 p.

Evaluation of Petroleum Resources from Pool Size Distributions

P. J. Lee
Institute of Sedimentary and Petroleum Geology, Geological Survey of Canada
Calgary, Alberta, Canada

P. C. C. Wang
Department of Statistics, The University of Calgary
Calgary, Alberta, Canada

The PRIMES (Petroleum Resources Information Management and Evaluation System) has been developed for assessments of oil and gas plays. The basic elements for PRIMES are a pool size distribution and a number of pools distribution. From these elements, individual pool sizes can be estimated. The remaining play potential can then be obtained from the undiscovered individual pool sizes.

The superpopulation play model approach is employed by PRIMES for making resource estimations. One of its main capabilities is enabling assessors to match reserves against estimated pool sizes; this provides a basis from which to estimate remaining potentials and to validate the input data. The Rimbey–Meadowbrook reef play from Alberta is examined in this paper and used as a guide to the procedure employed by PRIMES.

INTRODUCTION

Results of petroleum resources evaluations are normally presented as total play potentials. This aggregated potential does not provide enough information for economic studies and the development of exploration strategies because both of these processes require a knowledge of the number and sizes of pools.

Over the past several years, various methods have been proposed for estimating oil and gas resources. Kaufman et al. (1975) and Barouch and Kaufman (1977) formalized a probabilistic model of oil and gas discovery. Their model provides a method of estimating the number and sizes of pools yet to be discovered. This model is based on two assumptions: (1) that the pool size distribution is lognormal, and (2) that discovery is proportional to pool size and sampling without replacement. The second assumption is, for the purpose of modeling the observed phenomenon, that larger pools tend to be discovered earlier. Ivanhoe (1976), suggested that the logarithm of the size of the fields be plotted against the logarithm of the size rank of the fields. Such plots were prepared to provide quick estimates of the possible sizes and number of oil and gas fields in new basins. Roy and Ross (1980) used an empirical percent-rank equation to estimate the field sizes of a basin. An exposition on other methods of estimating oil and gas resources can be found in White and Gehman (1979) and Grenon (1979).

To achieve the requirement of blending available data from exploration with subjective opinions, the Hydrocarbon Assessment System Processor (HASP) was developed by the Geological Survey of Canada (Roy et al. 1975; Roy, 1979). A probabilistic formulation for HASP was given and extensions were made (Lee and Wang, 1983a, b, 1985). This formulation led to the development of the Petroleum Resources Information Management and Evaluation System (PRIMES) (Lee and Wang, 1984).

The approach taken in PRIME is based on the superpopulation play model. This model requires a play definition and treats it as if it were a random sample of size N from an infinite population. Geologic attributes, such as pool size, area of pool, and average net pay, average porosity, are regarded as random variables. Their underlying distributions can be considered as prior distributions reflecting the opinions of experts or as models describing the geologic process associated with the play. In particular, this play model assumes that the actual pool sizes in the play $y_1, ..., y_N$ are realized outcomes of the random variables $Y_1, ..., Y_N$, which are independent and identically distributed according to a pool size distribution $F(y)$.

If both N and $F(y)$ are known, the distributions of the order statistics of $Y_1, ..., Y_N$ can be obtained. A range, such

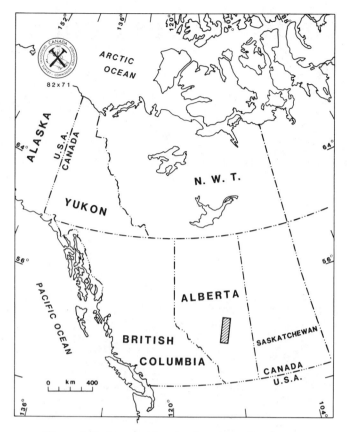

Figure 1. Index map showing locations of study area.

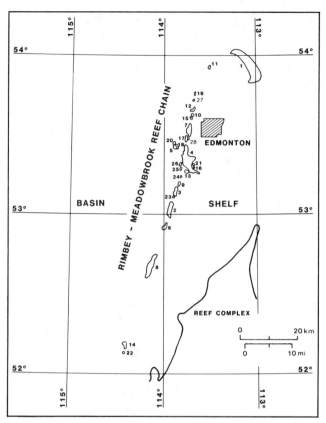

Figure 2. Map showing the areal extent of the Rimbey–Meadowbrook reef chain. The Number attached to each pool is the observed pool rank as shown in Table 1.

as the interquartile range, for each individual pool size can be plotted. These interquartile ranges can then be matched against the discovered pool sizes while taking into account the geologic information not expressed in the data. If a suitable match cannot be found, evidence is then provided against the given N and F(y). Otherwise, a plausible value of N and a pool size distribution F(y) are obtained for this play.

In this paper, some of the procedures in PRIMES are explained with the example of the Rimbey–Meadowbrook reef play (Figure 1). The following topics are discussed:

1. The role of a play definition as it relates to evaluations is described.
2. A method of obtaining initial estimates for N and the parameters of the lognormal pool size distribution is presented.
3. The interpretation of the order statistics of pool sizes and a discussion of the steps involved in the matching process are described.
4. Estimation of remaining potentials based on matching process is also described.
5. The performance of the matching technique is examined through a retrospective study.

GEOLOGIC MODEL AND DATA

To make an assessment of a play, a play model is required. By integrating geologic, geophysical, and geochemical data,

a model for a geologic setting can be formalized. In other words, a play model is a natural geologic population. From the play model, the definition can then be established. In cases where a play has not yet been explored, a model must be conceptually formalized. A play definition will lead us to collect relevant data or to establish subjective probability distributions for the play.

The formation of the Upper Devonian Rimbey–Meadowbrook reef chain (Figure 2) has been investigated by many workers (Belyea, 1964; Klovan, 1964; McGillivray and Mountjoy, 1975; McGillivray, 1977; Stoakes, 1980). A brief description of its geologic setting follows.

At the beginning of reef development, the Cooking Lake Formation (Figure 3) formed a broad carbonate bank. Subsequently, the Leduc reefs (Frasnian) grew on the margins of the banks as a reef chain. West and east of the reef chain, basin fill sediments of the Duvernay limestones and the Ireton shale were conformably deposited in interreef and offreef areas. The Leduc Formation consists of a basal unit and the main reef complex. The basal unit is composed of biostromal accumulations, poorly fossiliferous lime mudstones, and locally, brown or black mudstones. The main reef complex consists of forereef, reef, reef flat, and interior reef facies. The shale of the Ireton Formation now serves as the cap rock for the pools. The chain once extended to the north but has since been truncated by erosion. Some of the oil pools in the northern part of the chain may have been

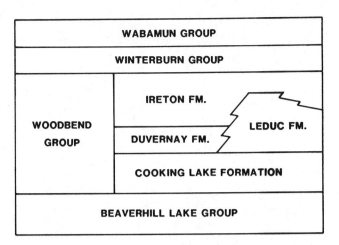

Figure 3. Upper Devonian Woodbend Group from central Alberta, Canada.

Table 1. Rank, pool name, discovery data[a] and pool data for the Leduc pools of the Rimbey–Meadowbrook reef play.

Rank	Pool Name	Discovery Date	Pool Size (million bbl)
4	Leduc-Woodbend A	02/12/47	351.5
1	Redwater	07/23/48	1295.4
5	Golden Spike A	09/06/49	290.8
24	Leduc-Woodbend G[b]	05/28/50	1.1
18	Golden Spike B[b]	07/11/50	4.0
7	Acheson	07/11/50	142.9
20	Godlen Spike C	02/23/51	2.6
3	Wizard Lake A	03/14/51	366.7
9	Glen Park	07/27/51	27.6
13	Leduc-Woodbend B	08/18/51	15.0
2	Bonnie-Glen	09/16/51	765.9
21	Leduc-Woodbend E[b]	03/10/52	2.0
26	Leduc-Woodbend D[b]	04/12/52	0.7
6	Westerose	05/07/52	169.6
25	Leduc-Woodbend C[b]	06/11/52	0.9
15	St. Albert B	10/31/52	8.6
11	Fairydell-Bon Accord	12/27/52	14.7
8	Homeglen-Rimbey	01/06/53	111.0
16	Leduc-Woodbend F	02/24/53	6.6
17	Yekau Lake A	06/07/55	4.7
27	Morinville A	08/25/55	0.6
10	St. Albert A	07/16/56	20.1
12	Morinville B	06/05/60	14.7
14	Sylvan Lake	10/14/61	14.2
19	Morinville C	10/07/63	3.4
23	Wizard Lake B	04/03/64	1.0
22	Lanaway	06/05/64	2.2
28	Yekau Lake B	07/17/67	0.3

[a]Discovery date is assigned as the commenced date of the discovery well.
[b]Not included in this study; see text for further details.

biodegraded into heavy oils. The regional dip of this chain is toward a southwest. Therefore, the Leduc conventional oil pools in this chain constitute the population or play used for the assessment in this paper.

The first discovery in the Rimbey–Meadowbrook reef play was made in 1947, and the last significant discovery was made in 1967. During this period, a total of 28 oil pools were discovered containing a total in-place reserve of over 3.6 billion bbl of oil. Geologic data from these pools, such as recovery factors, water saturation, oil shrinkage factors, area of pool, average net pay, average porosity, and pool reserves, are available from the Energy Resources Conservation Board of Alberta (1982). The discovery data for the play are listed in Table 1. In this paper, the Leduc–Woodbend pools C, D, E and G are considered to be a geologic part of the main reef, and the Golden Spike pool B to be a part of the main Golden Spike reef. These pools are footnoted Table 1 and are excluded from the study; hence the sample size is reduced to 23.

INITIAL ESTIMATION OF N AND F(y)

When discovery data are available, pool size distribution can be estimated by specifying a family of pool size distributions and then characterizing the process by which discoveries are made. A convenient and plausible model for the pool size distribution is given by the lognormal having the following probability density function:

$$f_\theta(y) = \frac{1}{y\sigma\sqrt{2\pi}} \, e^{-(\ln y - \mu)^2/2\sigma^2} \qquad \text{for } y > 0 \qquad (1)$$

where $\theta = (\mu, \sigma^2)$ is the population parameter to be estimated.

If the discovery data is a random sample from the play or if all the discoveries have been made, μ and σ^2 can be estimated by the sample mean and variance of the natural logarithms of the discovered pool sizes.

In reality, however, discovery is influenced by many factors, such as exploration techniques, drilling technology, and land availability. Furthermore, explorationists tend to test what is perceived as the best or largest prospect, which may not actually turn out to be the largest pool of the play. The tendency (drilling for what is perceived as the best prospect) leads to the statistical characterization of the discovery process as a sampling procedure—one that reflects the proportionality of the pool size and is without replacement, in other words, the same pool cannot be discovered twice (Barouch and Kaufman, 1977). A model of this procedure is expected to include the major component of a discovery process, and it can be applied to specific situations. Other factors are not involved in the model, but they could be indirectly incorporated by introducing a coefficient of exploration efficiency β into it (Bloomfield et al., 1979).

Specifically, this discovery model assumes that the probability of discovering the first n pools with sizes $Y_1, ..., Y_N$ in that order is

$$\prod_{\ell=1}^{n} \frac{Y_\ell^\beta}{Y_\ell^\beta + \cdots + Y_N^\beta} \qquad (2)$$

The β values, which range from 0 to 1 or greater, will be a measure of how the factor of proportionality to size plays a role. If $\beta = 0$, then the discovery process is characterized by

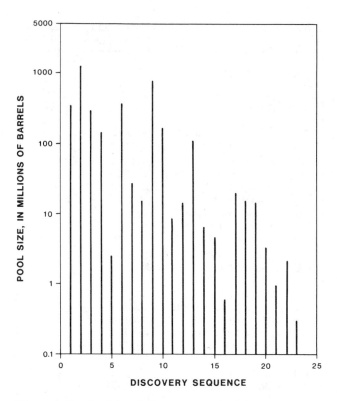

Figure 4. Graph of the pool sizes versus their discovery sequence for the Rimbey–Meadowbrook reef play.

Table 2. The maximum likelihood estimates $\hat{\mu}$ and $\hat{\sigma}^2$ of μ and σ^2 and the maximized log-likelihoods for different N values and β values.

N	β	$\hat{\mu}$	$\hat{\sigma}^2$	Maximized Log-likelihoods
30	0.0	3.0357	5.5552	−152.14
30	0.1	2.7968	5.6945	−150.55
30	0.2	2.5513	6.1903	−148.82
30	0.3	2.3329	6.9813	−146.98
30	0.4	2.1786	7.7310	−145.33
30	0.5	2.0873	8.2337	−144.15
30	0.6	2.0351	8.5310	−143.53
30	0.7	2.0035	8.7099	−143.44
30	0.8	1.9824	8.8277	−143.81
30	0.9	1.9669	8.9147	−144.58
30	1.0	1.9543	8.9147	−144.58
40	0.7	1.0298	10.9862	−144.11
50	0.7	0.2961	12.6105	−144.53
60	0.8	−0.3717	14.1655	−144.77
70	0.8	−0.8892	15.2966	−144.92
80	0.8	−1.3419	16.2951	−145.04

simple random sampling from the play. If $\beta = 1$, then the discovery is directly proportional to the pool size. If $\beta > 1$, then the pool sizes have even more impact on the order of discovery than when $\beta = 1$, because most of the large pools are discovered at a very early stage of exploration. Figure 4 shows a graph of the pool sizes versus their discovery sequence for the Rimbey–Meadowbrook play. This figure shows that pool sizes may influence the order of discovery.

By using the discovery model, a range of initial estimates of μ and σ^2 can be obtained, as well as estimates for the number of pools in the play. The steps involved in such estimation as applied to this play areas follows:

1. Let the number of pools, N, range from 30 to 80 with an increment of 10 (first column of Table 2).
2. Let β vary from 0.0 to 1.0 with an increment of 0.1 (second column of Table 2).
3. For each N and β, the discovery model is used to estimate μ and σ^2 such that the likelihood function of the discovery model is maximized (last column of Table 2). The estimates so obtained are denoted by $\hat{\mu}$ and $\hat{\sigma}^2$ for μ and σ^2, respectively (Table 2).

The results and discussions are summarized as follows. First, the last column of Table 2 contains the values of the maximized log-likelihoods. Each value can be regarded as an index to the likelihood of the particular combination of N and β, given the discovery sequence. The larger the value, the more plausible the combination for the given sequence of discovery pools.

Second, for purposes of illustration, all the β values for $N = 30$ are listed in Table 2. For $\beta = 0.0$, the estimates of μ and σ^2 are the same as in the random sampling except for the multiplicative factor $n/(n - 1)$ in $\hat{\sigma}^2$. We can see that for $\beta = 0.7$, the values of $\hat{\mu}^2 = 2.0035$ and $\hat{\sigma}^2 = 8.7099$ yield a maximum likelihood for $N = 30$. For other values of N, only those that yield the maximum likelihood are listed. The β values range from 0.7 to 0.8. These values suggest that pool sizes have exerted influence on the order of discovery and that the discovery process is related to size.

Third, based on our simulation study, the number of pools, N, in the play would normally be underestimated by the method of maximum likelihood. That is, the estimated N tends to be less than the true N. A plausible range, however, can be obtained and is found to be consistent with the exploration concept of the play. In this case, the number of pools should range from 30 to 80.

Finally, for each N, maximum likelihood estimates of the population parameters are given in Table 2. Because of the nature of the discovery model, the estimates might be conservative; therefore, estimates of μ and σ^2 from the discovery model would serve as lower and upper limits, respectively, of the population parameters. On the other hand, the estimation of μ and σ^2 can be achieved by the assumption of random sampling procedure, but this procedure tends to overestimate μ and underestimate σ^2; hence, $\hat{\mu}$ would serve as the upper limit and $\hat{\sigma}^2$ as the lower limit of the population parameters. Thus, for $N = 30$, interval estimates for μ and σ^2 are

$$2.004 \leq \mu \leq 3.036$$
$$\text{and} \quad 5.555 \leq \sigma^2 \leq 8.710 \quad (3)$$

The true population parameters could lie somewhere within these intervals.

These initial interval estimates for N, μ and σ^2 are used in the next section to construct hypothetical pool sizes that are matched against the discoveries.

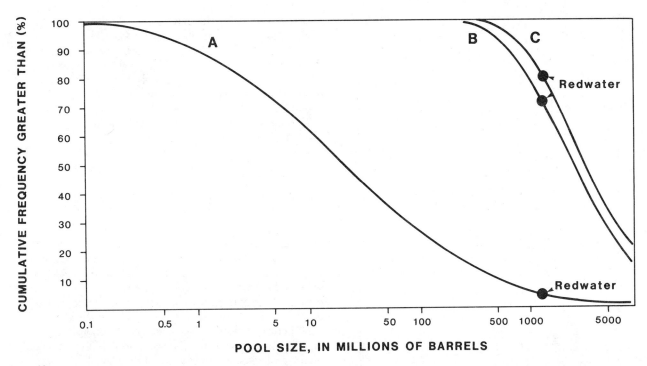

Figure 5. Pool size distribution (A) and distributions of the two largest pool sizes when $N = 30$ (B) and $N = 40$ (C), for the Rimbey–Meadowbrook reef play.

INDIVIDUAL POOL SIZE BY RANK

In resource evaluations, one of the most useful results is the collection of estimated pool sizes in order of rank, in other words, the largest pool first, the next largest pool second, and so on. In PRIMES, the procedure for estimating pool size by rank is called pool generation. This procedure is an assimilation of the combined geologic interpretations and statistical analyses. It can become a complicated and time consuming process; nevertheless, it can be quite useful.

The minimum information required to implement the pool generation procedure consists of the pool size distribution and the number of pools (N) in the play. The number of pools or their distribution is usually derived from the distribution of a number of prospects and the play risk, as described by Roy (1979). In the following discussions, we consider (1) the meaning of pool size by rank and (2) the matching process. The interrelationship between the parameters μ, σ^2, and N and their effects on pool size will also be discussed.

Interpretations of Pool Size by Rank

If $N = 1$, which means it is a single pool play, then the distribution of the largest and smallest pool is given precisely by the pool size distribution. More generally, $L_n(y)$, the greater-than distribution of the largest pool (assuming there are N pools), is

$$L_N(y) = 1 - F_\theta(y)^N \quad \text{for } y > 0 \qquad (4)$$

where $F_\theta(y)$ is the cumulative distribution function of the lognormal distribution with parameter $\theta = (\mu, \sigma^2)$.

There are two geologic interpretations of Equation (4). First, if all pools in a play were deposited by the same geologic model, then the more pools deposited, the more likely one of them is a relatively large one. Second, the size of the largest pool tends to change for different geologic settings, because the parameters μ and σ^2 vary for different models. Therefore, as empirically observed by Klemme (1975), the first few largest pool sizes are different for various geologic settings.

Let us examine the Rimbey–Meadowbrook reef play. As is indicated in Figure 5, the Redwater pool size is located in the fourth upper percentile. A 4% probability indicates the frequency of a Redwater sized pool or larger being in the superpopulation. The probability is much greater than 4% that the largest pool in the Rimbey–Meadowbrook play is as large as the Redwater pool, unless there is only one pool. In the case of more than one pool, the probability can be obtained from the distribution of the largest pool among N pools. The largest pool size distributions for $N = 30$ and $N = 40$ are displayed in Figure 5 along with the pool size distribution. The probabilities of having the largest pool size as large as Redwater are 70 and 80% for $N = 30$ and $N = 40$, respectively. Therefore, the probability distribution of the largest pool size would change, depending on the value of N, even for the same pool size distribution.

The difference in sizes between two adjacent pools is a function of σ^2 if N and μ are unchanged. In Figure 6, the medians of individual pool size distributions where $\sigma^2 = 6.0$, $\mu = 3.0$ and $N = 40$ are displayed by solid dots, whereas the medians of individual pool size distributions where $\sigma^2 = 0.5$ the same μ and N are shown by circles. Figure 6 clearly indicates that larger pool sizes decrease more rapidly when σ^2

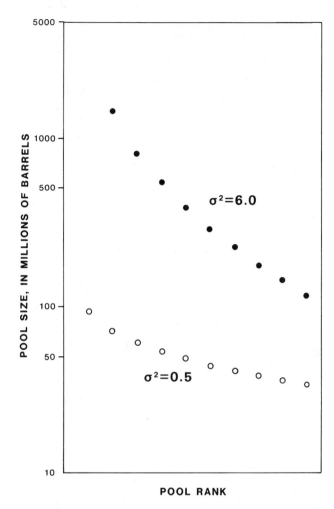

Figure 6. Individual pool sizes by rank with $\sigma^2 = 0.5$ and $\sigma^2 = 6.0$ where $\mu = 3.0$ and $N = 40$.

geologic models generally have different values for μ, σ^2 and N and thus correspondingly different pool sizes.

Finally, if a play has a pool size distribution with a large σ^2 value, then the major portion of the play's resources will be made up by the first few largest pools. If σ^2 is relatively small, however, then the pool sizes of the play would be approximately the same magnitude.

Thus far we have not concluded whether 40 pools in this play is more likely than 30 pools, or any other value; nor have we determined whether or not the Redwater pool is the largest pool or whether there is a larger pool yet to be discovered. The latter of these questions is discussed by Lee and Wang (1985). We now describe the matching technique and examine the first question.

Matching Process

The matching technique is a combination of geologic interpretations and trial and error procedures. The idea is to set N at different values according to geologic evidence and to obtain a set of hypothetical pool size distributions. The reserves of the play are then matched against the hypothetical pool sizes to see which N and pool size distribution yield good fits. The goodness-of-fit will depend on the geologists' interpretations of the play.

Distributions of individual pool sizes can be conveniently characterized by a few selected upper percentiles without much loss of information. From these upper percentiles, prediction intervals can be obtained. In the following, we use the 75–25% prediction interval as a statistical measure of goodness-of-fit, and we validate the statistical matchings by examination of their geologic implications. After each statistical matching, we observe whether or not the implications are in accord with the geologic model. Examples of the type of questions that one could ask after each matching include

1. Have we discovered the largest pool?
2. What are the sizes of the remaining largest pools?
3. What is the potential of the remaining undiscovered pools?
4. How do recent discoveries, which are not included in the analysis, fit into the prediction picture?

For the Rimby–Meadowbrook reef play, the matching process covered the interval $N = 30$ through 80. In this case, let μ and σ^2 range within their intervals as obtained from Equation (3). The result of the matching is summarized as follows. If $N \leq 30$ with all possible values of μ and σ^2, then the matching process suggests that the 23 reserves do not fall into the 75–25% prediction interval. Figure 7 illustrates one of the mis-match cases. The boxes indicate the 75–25% interval, whereas the dots indicate the reserves. Geologic evidence along with the lack of fit, here suggest that the reef chain should contain greater than 30 pools.

If the number of pools, however, is increased to 40 and $\mu = 2.4$ and $\sigma^2 = 5.75$, then all reserves can fit into the 50% prediction interval as shown in Figure 8. In this case, the first eight largest pools have been discovered, amounting to about 90% of the total resources of the play. These findings do agree with the current exploration concept of the play. Figure

is relatively large than when it is relatively small. For the lognormal pool size distribution, given unchanged values of μ and N, the larger the value of σ^2, the larger a big pool tends to be. Hence, the magnitude of the first few large pools among the N pools tends to be greater. The theoretical aspect of this tendency to have outliers has been studied by Neyman and Scott (1971) and Green (1974).

The above discussions can be summarized as follows. First, the distribution of pool size by rank should be computed from the distribution of the order statistics of a sample of N pools from the pool size distribution.

Second, the size of the largest pool increases as the number of pools increases. The amount of increase will depend on the magnitude of μ and σ^2. The larger the σ^2, the greater the increase as N increases.

Third, in resource evaluations, the parameters μ and N can be thought of as indicators of the richness of the play, whereas σ^2 together with N can be considered as indicators of the proneness toward outliers of the geologic model.

Fourth, for each hydrocarbon-bearing play, there is a unique set of μ, σ^2 (when lognormality is assumed), and N associated with the model that produced the play. Different

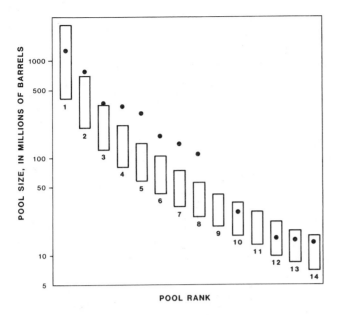

Figure 7. Example of an unacceptable match.

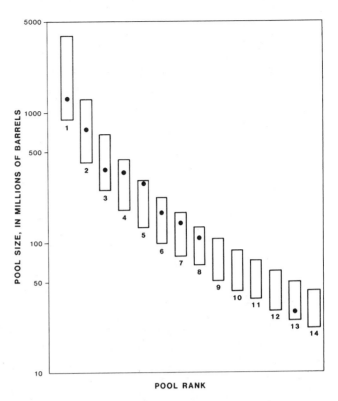

Figure 8. The predicted pool sizes by rank for the Rimbey–Meadowbrook reef play when $N = 40$. Dots indicate reserves of the pools, and boxes indicate values at the 25th and 75th upper percentiles.

8 also shows that the largest pool size is associated with the largest uncertainty. The prediction intervals between two consecutive pool sizes do overlap to some extent. The amount of overlap increases as pool rank decreases. The lognormal pool size distribution estimated by the matching process and the largest three pool size distributions are displayed in Figure 9.

By using the method given by Lee and Wang (1985), the ratio between two pool sizes of different ranks was also computed by using the same information: $N = 40$, $\mu = 2.4$, and $\sigma^2 = 5.75$. The first four ratio distributions of the play and the observed ratios are listed in Table 3. The information contained in this Table 3 lends more confidence to the estimates obtained from the matching process.

Pool Sizes Constrained to a Discovery Record

Once the "best matching" has been obtained, the remaining hydrocarbon potential of the play can be estimated by conditioning on the match. For the Rimbey–Meadowbrook reef play, the remaining pool sizes were estimated by constraining the pool sizes of the 23 discoveries and their ranks as suggested in Figure 8. The estimation was based on $N = 40$, $\mu = 2.4$, and $\sigma^2 = 5.75$. There are several results that can be seen from Figure 10.

1. The four largest remaining pool sizes range from 30 to 98 million bbl of oil in-place.
2. The width of the prediction intervals are smaller in Figure 10 than those of Figure 8. For example, the remaining largest pool sizes range from 70 to 98 million bbl (Figure 10) compared with 50 to 104 million bbl in Figure 8.
3. The overlapping length of two consecutive pool sizes is also much smaller than in Figure 8. For example, the overlapping length between the 9th and 10th pools is about 5 million bbl in Figure 10 compared with 35 million bbl in Figure 8.

Table 3. The first four ratio distributions of the Rimbey–Meadowbrook reef play.

Ratio	Observed Ratio	Upper Percentiles		
		75%	50%	25%
First/second	1.8	1.4	2.1	4.0
First/third	3.4	2.2	3.8	8.0
First/fourth	3.8	3.3	6.0	12.9
First/fifth	4.1	4.5	8.5	18.6

4. The degree of uncertainty in the prediction intervals is controlled by two factors: (a) the uncertainty inherited from the superpopulation and (b) the difference in reserves between the two nearest pools, as illustrated by the 9th to 12th pools of Figure 8.
5. The distributions of the pool sizes computed with a specified discovery record tend to be less skewed and more concentrated around the medians than those without any specified conditions.

The estimation of pool sizes constrained to a discovery record serves not only to estimate remaining resources but also to reduce uncertainty for the estimates. Furthermore, it can be used as a feedback mechanism for challenging geologic concepts. For example, geologists could ask what the largest pool size would be given that the Redwater pool is

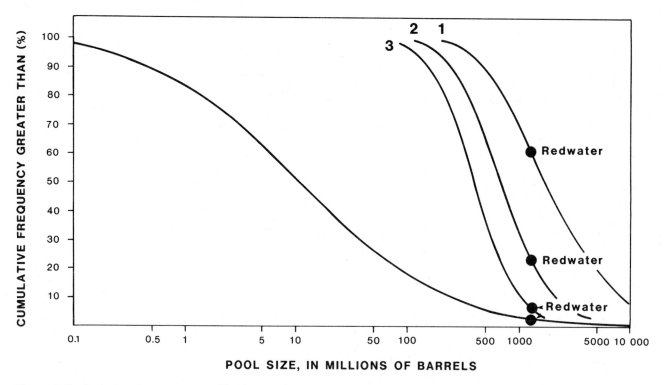

Figure 9. Pool size distribution estimated by the matching process for the Rimbey–Meadowbrook reef play. Number 1 indicates the largest pool size distribution, number 2 the second largest pool size, and so on.

the second largest. This type of feedback mechanism is useful in evaluations of frontier areas.

A RETROSPECTIVE STUDY

In this section we use pool data discovered prior to February 1951 from the Rimbey–Meadowbrook reef play to predict the present situation and to examine the performance of the matching technique in retrospect. The pool data consist of the first five discoveries listed in Table 1, excluding the annotated pools. The population parameters of the pool size distributions were estimated on the basis of the method discussed in the previous section. The initial interval estimates for μ, σ^2, and N are

$$-0.456 \leq \mu \leq 4.925$$
$$4.437 \leq \sigma^2 \leq 34.483$$
$$\text{and} \quad 10 \leq N \leq 80 \qquad (5)$$

In this case, the estimates contain much larger uncertainty because of the small sample size; therefore, geologic interpretation should play an important role in the matching process. The matching process suggests that there are several combinations of N, μ, and σ^2 that can reproduce the five discovered pool sizes. Their intervals are refined as:

$$2.0 \leq \mu \leq 3.0$$
$$4.5 \leq \sigma^2 \leq 6.0$$
$$\text{and} \quad 30 \leq N \leq 40 \qquad (6)$$

Table 4. The predicted first eight largest pool sizes from pool data gathered prior to February 1951 ($N = 40$, $\mu = 2.5$, $\sigma^2 = 6.0$).

Pool Rank	Predicted Pool Size Distribution (million bbl)			Present Reserve
	75%	50%	25%	
1	1062	2171	4907	1295[a]
2	489	848	1543	766
3	299	485	810	367
4	205	320	509	352[a]
5	151	228	352	291[a]
6	115	170	257	170
7	90	132	195	143[a]
8	72	105	152	111

[a]Pools discovered prior to February 1951.

Among the predictions from Equation (6), some of them can predict the present picture. Table 4 shows an example. Some of them, however, can only predict the picture in 1951. For example, one could interpret from the predictions presented in Table 5 that the second and third largest pools had been discovered. However, this interpretation can be challenged by exploration concepts that were current in 1951.

CONCLUSIONS

In summary, we have shown that estimations of a pool size distribution can be made by combining the discovery process

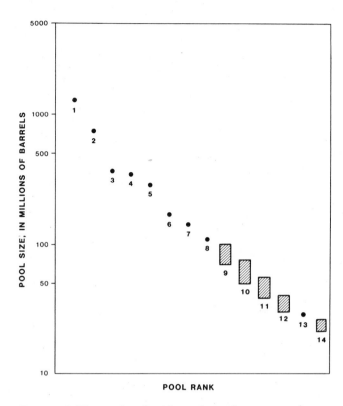

Figure 10. The predicted pool sizes by rank constrained to a discovery record for the Rimbey–Meadowbrook reef play when $N = 40$. Dots indicate reserves of the pools, and the shaded boxes indicate undiscovered pools at the 25th and 75th upper percentiles.

Table 5. The predicted first eight largest pool sizes ($N = 30$, $\mu = 2.0$, $\sigma^2 = 5.5$).

Pool Rank	Predicted Pool Size Distribution (million bbl)			Present Reserve
	75%	50%	25%	
1	392	802	1802	1295[a]
2	178	311	568	766
3	108	176	296	367
4	73	115	184	352[a]
5	53	81	126	291[a]
6	40	60	91	170
7	31	45	68	143[a]
8	24	35	52	111
[a]Pools discovered prior to February 1951.				

model and the matching process. The remaining play potential can then be expressed in terms of individual pool size by rank. The amount of uncertainty contained in the estimates can be reduced as more information, such as pool rank, is entered into the analysis. The framework thus produced can provide a feedback mechanism to challenge the input geologic data, the underlying concepts, or both. Finally, we have shown that the ratio between two consecutive pool sizes is not constant throughout the population.

ACKNOWLEDGMENTS

The authors would like to thank John Wade, Don Sherwin, Graham Campbell, Keith Williams, Red White, Bob Crovelli, and especially John Houghton for their comments and discussions.

REFERENCES CITED

Barouch, E., and G. Kaufman, 1977, Estimation of undiscovered oil and gas: Proceedings of Symposia in Applied Mathematics, v. 21, p. 77–91.

Belyea, H. R., 1964, Upper Devonian, Part II, Woodbend, Winterburn and Wabamun Groups; in R. G. McCrossan and R. P. Glaister, eds., Geological History of Western Canada: Alberta, Canada, Alberta Society of Petroleum Geologists, p. 66–68.

Bloomfield, P., K. S. Deffeyes, G. S. Watson, Y. Benjamini, and R. A. Stine, 1979, Volume and area of oil fields and their impact on order of discovery; in Resource Estimation and Validation Project: Princeton, New Jersey, Princeton University, Departments of Geology and Statistics, 53 p.

Energy Resources Conservation Board of Alberta, 1982, Alberta's reserves of crude oil, gas natural gas liquids, and sulphur: Alberta, Canada, Energy Resources Conservation Board, Tables 2–5.

Green, R. F., 1974, A note on outlier-prone families of distributions: The Annals of Statistics, v. 2, n. 6, p. 1293–1295.

Grenon, M., 1979, Proceedings of the First IIASA conference on methods and models for assessing energy resources: New York, Pergamon Press, 605 p.

Ivanhoe, L. F., 1976, Evaluating prospective basins: Oil and Gas Journal, v. 74, n. 49, p. 154–155, n. 50. p. 108–110, n. 51, p. 82–84.

Kaufman, G. M., Y. Balcer, and D. Kruyt, 1975, A probabilistic model of oil and gas discovery; in J. D. Haun, ed., Methods of Estimating the Volume of Undiscovered Oil and Gas Resources: AAPG Studies in Geology, n. 1, p. 113–142.

Klemme, H. D., 1975, Giant oil fields related to their geologic setting: a possible guide to exploration: Bulletin of Canadian Petroleum Geology, v. 23, p. 30–66.

Klovan, E. J., 1964, Facies analysis of the Redwater reef complex, Alberta, Canada: Bulletin of Canadian Petroleum Geology, v. 12, n. 1, p. 1–100.

Lee, P. J., and P. C. C. Wang, 1983a, Probabilistic formulation of a method for the evaluation of petroleum resources: Mathematical Geology, v. 15, n. 1, p. 163–181.

——— , 1983b, Conditional analysis for petroleum resource evaluations: Mathematical Geology, v. 15, n. 2, p. 353–365.

——— , 1984, PRIMES: a petroleum resources information management and evaluation system: Oil and Gas Journal, v. 82, n. 40, p. 204–206.

——— , 1985, Prediction of oil or gas pool sizes when discovery record is available: Mathematical Geology, v. 17, n. 2, p. 95–113.

McGillivray, J.G., 1977, Golden Spike D3-A pool, in I. A. McIlreath and R. D. Harrison, eds., The geology of selected carbonate oil, gas and lead–zinc reservoirs in Western Canada: Canadian Society of Petroleum Geologists, p. 67–88.

McGillivray, J. G., and E. N. Mountjoy, 1975, Facies and related reservoir characteristics, Golden Spike reef complex, Alberta:

Bulletin of Canadian Petroleum Geology, v. 23, n. 4, p. 753–809.

Neyman, J., and E. L. Scott, 1971, Outlier proneness of phenomena and of related distributions: optimizing Methods in Statistics: New York and London, Academic Press, p. 413–430.

Roy, K. J., 1979, Hydrocarbon assessment using subjective probability and Monte Carlo methods; *in* M. Grenon, ed., Procedings of the First IIASA Conference on methods and models for assessing energy resources: Pergamon Press, New York, p. 279–290.

Roy, K. J., and W. A. Ross, 1980, Apportioning estimates of basin potential to fields, *in* A. D. Miall, ed., Facts and principles of world petroleum occurrence: Canadian Society of Petroleum Geologists Memoir 6, p. 319–328.

Roy, K. J., R. M. Procter, and R. G. McCrossan, 1975, Hydrocarbon assessment using subjective probability: Probability methods in oil exploration; AAPG Research Conference, August 19–22, 1975, p. 56–60.

Stoakes, F., 1980, Nature and control of shale basin fill and its effect on reef growth and termination: Upper Devonian Duvernay and Ireton Formations of Alberta, Canada: Bulletin Canadian Petroleum Geology, v. 28, n. 3, p. 345–410.

White, D. A., and H. M. Gehman, 1979, Methods of estimating oil and gas resources: AAPG Bulletin, v. 63, n. 12, p. 2183–1292.

Finite Population Sampling Methods for Oil and Gas Resource Estimation[1]

G. M. Kaufman
Sloan School of Management, Massachusetts Institute of Technology
Cambridge, Massachusetts

Some recent developments in finite population sampling methods are applicable to oil and gas resource assessment. The objective of these methods is to provide an estimate of the empirical frequency function of magnitudes of undiscovered deposits in a petroleum play when the only observable data are deposit magnitudes in the order of discovery (observation). Our theme is the conceptual similarity of the well-known Arps and Roberts (1958) model for magnitudes of discoveries as a function of exploratory wells drilled to a finite population sampling scheme called "successive sampling" by statisticans. There are two types of approximately unbiased estimators for parameters of successively sampled finite populations: Gordon's (1983) moment matching estimator and Andreatta and Kaufman's (1983) anchored estimator. Both are similar in functional form to Arps and Roberts original estimator for the number of deposits of a given magnitude ultimately discoverable in a play, but arise from different assumptions about the nature of the sampling scheme. Using Western Gulf Miocene–Pliocene data and North Sea data, some estimates of empirical field size distributions generated by the three above-mentioned estimators are presented. When applied to Western Gulf Miocene–Pliocene data, the moment-matching estimator appears sensitive to how the sample is split. The anchored estimator replicates Smith and Ward's (1981) maximum likelihood estimates reasonably well.

INTRODUCTION

Our aim here is to discuss some developments in sampling methodology that bear on the problem of estimating the empirical frequency function of magnitudes of undiscovered deposits in a petroleum play when the only observable data are deposit magnitudes in order of discovery (observation). While it is clearly desirable to employ detailed geophysical, geologic and geochemical data in an attack on this problem, in many settings available data are limited to deposit magnitudes and deposit discovery dates. When such detailed physical data are available, the methods discussed here can also provide an alternative window of insight—a check on projections generated by more geologically specific approaches.

The pioneering analysis by Arps and Roberts (1958) on the discovery history of Denver–Julesberg basin field magnitudes in order of observation is based on the premise that petroleum deposit discovery can be modeled as a sampling process with two key features. We can characterize the order of discovery of deposits in a petroleum play as a sampling from a finite population of deposits (1) without replacement and (2) proportional to magnitude. (In the study of Arps and Roberts, the play is in the basin's one major producing interval, which is the D and J members of the Cretaceous Dakota Sandstone, and the magnitude is defined as the productive area of a deposit.) The primitive ideas guiding formulation of the model of Arps and Roberts (1958) have been adopted in several subsequent studies of petroleum exploration interpreted as a sampling process: e.g., Bloomfield et al. (1979), Gordon (1981, 1983); Drew et al. (1980), Kaufman et al. (1975), Kaufman and Wang (1980), Smith and Ward (1981), and Wang and Lee (1983a, b). Several of these studies model the order of discovery of petroleum deposits as sampling without replacement and proportional to magnitude from a finite population. This latter model is called a *successive sampling* model by statisticians.

In this paper we do the following: (1) discuss similarities of and differences between Arp–Roberts (A-R) and successive sampling (SS) models; (2) present some extensions of the simplest form of successive sampling that incorporate characteristics of oil and gas deposit data; (3) review

[1]This work was supported in part by the Energy Information Administration, U.S. Department of Energy. The opinions and conclusions appearing here are those of the author and do not represent policies or opinions of any U.S. Government agency.

43

properties of moment estimators for attributes of successively sampled finite populations; and (4) illustrate how these estimators can be used to estimate the empirical frequency function of undiscovered deposits using Western Gulf Miocene–Pliocene trend deposit data and North Sea deposit data.

ARPS–ROBERTS AND SUCCESSIVE SAMPLING MODELS

Connections of and distinctions between SS and A-R type models of petroleum discovery appear at two levels. On comparison, primitive assumptions underlying the structure of each model type show some similarities and some differences. Less obvious connections emerge only after analysis of sampling properties of each model. We begin with statements of assumptions. Model properties needed for inference about model parameters and for projections are presented unencumbered by proofs, which can be found in accompanying references. The emphasis here is heuristic, but elementary probabilistic arguments are used when informative.

Successive Sampling

We start with a description of the simplest form of successive sampling (SS) and then discuss possible elaborations designed to incorporate some features of petroleum exploration it does not capture.

Consider a *finite population* of N deposits in a petroleum play. Assign labels $1, 2, \ldots, N$ to these deposits and associate a *magnitude* $A_j > 0$ to deposit labeled $j, j = 1, 1, \ldots, N$. Define $U = \{1, 2, \ldots, N\}$ and $A = \{A_1, \ldots, A_N\}$. A successive sampling model is a probability law that assigns probabilities to the $N!$ possible orderings in which elements of U can be observed.

These probabilities depend on elements of A in the following fashion. Let $(i_1, i_2, \ldots, i_N)$ be any ordering of all elements of U.

SS: The probability of observing elements of U in the order $(i_1, i_2, \ldots, i_N)$ is

$$P\{(i_1, i_2, \ldots, i_n)|A\} = \prod_{j=1}^{N} A_{i_j}/(A_{i_j} + \ldots + A_{i_N}) \quad (1)$$

There is no loss in generality in what follows if we relabel and work with $P\{(1, 2, \ldots, N)|A\}$, which is what we shall do. Stated in words, SS gives that the sampling scheme generating an observed order $(1, 2, \ldots, N)$ is sampling proportional to magnitude and without replacement. At any stage, say at observation or discovery number j, successive sampling replicates itself. In particular, given that deposits labeled $1, 2, \ldots, j - 1$ have been discovered, the largest among undiscovered deposits $j, \ldots, N$ has the largest probability of being the next discovery.

A pictoral motivation for SS is provided by Figure 1. Here the magnitude of a deposit is defined to be the planar area of each labeled body in the figure. Imagine the figure being covered, and suppose that darts are thrown at it such that a dart is equally likely to hit any "small" rectangle within the

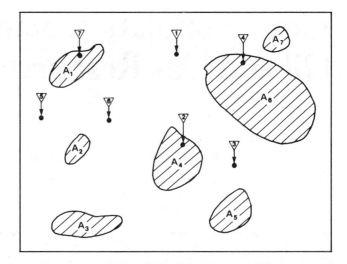

Figure 1. Hypothetical drilling area showing location of wells (triangles over dots) and areas (A) containing discoverable petroleum at depth (slashed areas). Following the example in the text, a series of randomly drilled holes would have the following history (where 0 indicates a dry hole): $0, A_4, 0, A_6, 0, 0, A_1$.

borders of the figure regardless of the location of the rectangle. Each dart throw is interpreted as an exploratory well searching for an undiscovered deposit. If the dart lands within a body labeled A_j deposit j is "discovered" and is erased from the figure. Dart hits not falling within a labeled area or body are ignored and are interpreted as dry holes. Discovery magnitudes are recorded in order of observation. This record is interpretable as a realization of the sampling scheme represented by SS. The choice of "area" as a definition of magnitude is adopted here as the simplest way of physically motivating a successive sampling scheme. Alternative definitions are possible.

Exploratory wells are not generally drilled in a spatially random fashion. Exploratory wells are located in light of geologic and geophysical information directed toward identification of prospects. An alternative motivation for SS that incorporates this characteristic of exploration is possible. Imagine Figure 1 modified to include an additional set of bodies with label set $U' = \{N+1, \ldots, N+M\}$ and magnitude set $B = \{B_{N+1}, \ldots, B_{N+M}\}$. The area of each of these latter bodies represents the magnitude of a prospect that is dry. Now if we cover the figure and throw darts as before, an exploratory well is considered to be drilled when a labeled body (labeled either A_j or B_k) is struck. We are now modeling exploration as unfolded by drilling prospects. This interpretation is particularly appropriate when the search is directed toward discovery of structural anomalies. The outcome of exploratory drilling is recorded by writing down each well outcome as an ordered pair (magnitude and type) where "type" is either a discovery or a dry anomaly. This record is interpretable as a realization of successive sampling via SS from a finite population composed of elements $U \cup U'$ with associated magnitudes $A \cup B$.

In practice, magnitudes (defined, e.g., as projective area) or dry prospects seldom, if ever, appear in deposit data records. However, under the conditions set down in this heuristic description of a model of exploratory drilling and discovery, the order in which deposits 1, 2, ..., N are observed still adheres to the probability law SS. This is not immediately obvious, but it can be demonstrated by interpretation of an alternative characterization of successive sampling presented by Gordon (1983). Gordon observed that the probability law SS is representable in terms of order statistics generated by mutually independent exponential random variables with varying means. Let $\bar{\xi}_1, ..., \bar{\xi}_N$ be N independent exponential random variables with identical means equal to one. Then

$$P\{(1, 2, ..., N)|A\} = P\left\{\frac{\bar{\xi}_1}{A_1} < \frac{\bar{\xi}_2}{A_2} < ... < \frac{\bar{\xi}_N}{A_N}\right\} \quad (2)$$

This reexpression of the more directly intuited probability law represented by Eq. (1) provides significant analytical leverage. An immediate by-product is that if a subset of $\xi = \{\bar{\xi}_1/A_1, ..., \bar{\xi}_N/A_N\}$ is deleted according to a rule that leaves the remaining elements of ξ independent, the observation of elements of ξ that remain adhere to successive sampling, that is, they obey SS.

Another example of this is the following: suppose that interest is focused on deposits in a prescribed depth interval in a sedimentary unit. If successive sampling is adopted as a model of discovery for deposits at all depths in this unit and deposit magnitudes can be assumed to be independent of depth, then either Eq. (1) or (2) effectively models the order of discovery within the prescribed depth interval alone. This, however, cannot be applied too broadly. For example, a natural gas play in a sedimentary unit that spans a large depth range may exhibit a positive correlation between depth and deposit magnitude measured in thousands of cubic feet in place. If drilling begins at shallow depths and moves deeper, then the order of discovery is correlated with depth.

The successive sampling scheme represented by Eqs. (1) and (2) can be enriched. Bloomfield et al. (1979) argued that the assumption that discovery is strictly proportional to magnitude, however magnitude is defined, is not always true. The "discoverability" of a deposit can possibly be more accurately represented as some function of magnitude. A simple generalization of Eq. (1) is to introduce a "discoverability parameter" δ of fixed value and recast this equation in the form

$$P\{(1, 2, ..., N)|A, \delta\} = \prod_{j=1}^{N} A_j^\delta / A_j^\delta + ... + A_N^\delta \quad (3)$$

When $\delta = 1.0$, Eq. (1) is recovered; when $\delta = 0$, Eq. (3) corresponds to hypergeometric sampling, and magnitude plays no role in determining the order of deposit discovery. The Eq. (3) model serves as a starting point for studying whether or not magnitude is a determinant of discovery order (see Bloomfield et al., 1979 Adelman et al., 1983).

More elaborate variations based on the reasoning of how deposits are actually discovered are possible. For example, $\delta(j, A_j)$ might be a prescribed function of discovery number and magnitude that incorporates "learning"; let $\delta(j, A_j)$ increase with increasing j and vary with magnitude.

The models represented by Eqs. (1), (2), and (3) are based on the assumption that from the outset, drilling of *all* prospects in a play is possible. External factors, however, can restrict access to prospects. For example, the U.S. government controls access to offshore U.S. acreage and only periodically releases increments of acreage for exploration. Onshore lease blocking also exists; an example is in the western Gulf of Mexico (Drew et al., 1982). These types of external forces influence the time pattern of observed magnitudes of discoveries, and if their effects are not explicitly incorporated into the model, the model is misspecified.

The effect of these factors on magnitudes of discovery in order of observation is structurally similar to that caused by mixing distinguishable populations of deposits. A straightforward modification of Eq. (1) or (2) can account for either of these external factors (see Adelman et al., 1983, for a description).

Variations other than those discussed thus far are possible. The model SS is both parsimonious and easily modified to account for a number of characteristics of deposit discovery. However, a theory of inference for this class of models is far from complete. As a prelude to a discussion of unbiased estimators of certain functions of elements of A that bear close resemblance to Arps and Roberts type estimators, we next present a probabilistic treatment of their model.

Arps–Roberts Model

The discovery model proposed in 1958 by Arps and Roberts is based on a scale of exploratory wells drilled. Their principal assumptions are (1) the probability that an exploratory well will discover a particular deposit with a projective area A is equal to the ratio of A times a constant c that measures the efficiency of exploration to the total search area T, and (2) discovery is without replacement, which means that discoveries are successively deleted from the initial population of deposits.

An intuitive interpretation of these assumptions is afforded by the dart throwing example if we modify the area of each body in Figure 1 by a multiplicative factor c and record dry holes and discovery magnitudes in order of observation.

The approach of Arps and Roberts (1958) in analyzing their model is to divide the range of possible deposit areas into a finite number of intervals and assign a representative area to each interval. In contrast to the successive sampling schemes discussed above (each of which mirrors joint effects of all discoveries), Arps and Roberts treat discoveries within each area interval or size class as independent of all other size classes, a tactic employed to simplify analysis. The postulates underlying their model do not demand this restriction, so our initial statement of them is unencumbered by it.

An orderly description of their model is aided by defining *well history* precisely. Labeling exploratory wells in the order drilled $w = 1, 2, ...$, we set $x_w = 1$ if the wth well is a discovery well and $x_w = 0$ if it is a dry hole. We let A_j denote the area of the jth discovery. Then we set $Z_w = 0$ if $x_w = 0$, and $Z_w = A_j$ if $x_w = 1$ and $\sum_{k=1}^{w} x_k = j$. We define the history of the first w exploratory wells to be

$$H_w = (Z_1, ..., Z_w) \quad (4)$$

Thus each zero element of H_w denotes a dry hole and each nonzero element denotes the area of a discovery. If the first w

wells discover n deposits with areas $A_1, \ldots, A_n$, then the set A_w of areas of deposits remaining undiscovered subsequent to a history H_w is created by deleting $A_1, \ldots, A_n$ from A. Letting $A_0 = A$ and H_0 be the (null) well history prior to the first well, Arps and Roberts' principal assumption can now be stated in the following form.

A-R: Given a well history H_w, the probability that the $(w+1)^{st}$ well discovers a deposit of area a is

$$P\{\bar{Z}_{w+1} = a | A, H_w\} = \begin{cases} ca/T & \text{if } a \in A_w \\ \\ 0 & \text{otherwise} \end{cases} \qquad (5)$$

The probability that the $(w+1)^{st}$ well is a dry hole given H_w is

$$P\{Z_{w+1} = 0 | A, H_w\} = 1 - \sum_{a \in A_w} ca/T \qquad (6)$$

This model ignores reductions in the effective search area generated by deletion of areas of discovered deposits and of areas condemned by dry holes ("area of influence of drill holes," see Drew et al., 1980). The heuristic argument for doing this is that if the area searched by w wells is small relative to the initial search area, then the total area condemned by drilling will also be small and the right-hand sides of Eqs. (5) and (6) are then reasonable approximations.

If the range of possible deposit areas is divided into a finite number of intervals or size classes and if each size class is studied independently of realizations of discoveries in all other size classes, the model represented by A-R takes on a very simple form. We partition $(0, \beta)$ into M intervals and assign a representative area to each interval. We then focus on one such interval or size class with representative area a and N deposits. Each well either discovers a deposit of area a or does not, so a well history $H_w = (Z_1, \ldots, Z_w)$ is composed of elements Z_j that take on value 0 if well j is a dry hole ($x_j = 0$) and value a if a discovery is made ($x_j = 1$).

A-R*: Given a well history H_w for which $\sum_{j=1}^{w} x_j = n$, the probability that the $(w+1)^{st}$ well discovers a deposit of area a is

$$P\{\bar{Z}_{w+1} = a | a, H_w\} = (N-n)p \qquad (7)$$

where $p = ca/T$ given H_w. The probability that the $(w+1)^{st}$ well is a dry hole is $1 - (N-n)p$. This restricted form of Arps and Roberts' principal assumption allows us to employ elementary probabilistic reasoning to obtain their "classic" equation, eschewing their deterministic approach to its derivation. A bonus is that measures of uncertainty are calculable. Our interest is in two uncertain quantities: the number $\bar{n}(w)$ of discoveries made by the first w wells and the number of wells $\bar{w}(n)$ that must be drilled to find n deposits.

We consider first the number $\bar{n}(w)$ of deposits of area a found by the first w exploratory wells drilled. Given A-R*, the probability that any particular deposit among N deposits of area a will not be discovered by one of the first w wells is $(1 - \{ca/T\})^w$. Consequently, the expected number $\bar{n}(w)$ of

discoveries made by the first w wells is

$$\bar{n}(w) = N\left[1 - \left(1 - \frac{ca}{T}\right)^w\right] \qquad (8)$$

If ca/T is small, $[1 - (ca/T)]^w \cong 1 - \exp\{-caw/T\}$ and

$$\bar{n}(w) \cong N(1 - \exp\{-caw/T\}), \qquad (9)$$

which is Arps and Roberts' classical equation.

Root and Scheunemeyer suggest a different definition of efficiency that is (asymptotically) equivalent to inflation of the target area of a deposit by a multiplicative factor c: the probability that a generic well will discover a particular deposit of area a is $1 - (1 - [a/T])^c$. When $c = 1$, drilling is "random"; $c > 1$ implies that drilling is probabilistically "more efficient" than random. If c is on the order of one (it is usually less than 3 or 4) and a/T is small (0.167×110^{-3} is a typical value for the Denver–Julesberg basin), then $1 - (1 - \{a/T\})^c \cong 1 - \exp\{-ca/T\}$, which leads back to Eq. (9).

Collecting our bonus, we find that the variance of $\bar{n}(w)$ is (see Appendix for derivation)

$$\begin{aligned} \text{Var}[\bar{n}(w)] = {} & N^2 \pi(w)[1 - \pi(w)] \\ & - N(N-1)[(1-p)^w - (1-2p)^w] \end{aligned} \qquad (10a)$$

and when w is large and p is small,

$$\text{Var}[\bar{n}(w)] \cong N\pi(w)[1 - \pi(w)] \qquad (10b)$$

with $\pi(w) = 1 - (1-p)^w$. Here is an example of use of Eq. (10). Using Arps and Roberts' Denver–Julesberg values, $T = 5.7 \times 10^6$ acres, $c = 2.0$, and, for 3,705 wildcats drilled by the year 1958, $caw/T = 0.0013a$. Representative productive acreages range from 2.5 to >8,000 acres. Their "class 8" assigns $a = 97.3$ acres, and thus for this class $p = ca/T = 0.3414 \times 10^{-4}$, $\pi(3,705) = 0.1188$; to four decimal places, $\pi(3,705)[1 - \pi(3,705)] = 0.1047$ and $(1-p)^{3705} - (1-2p)^{3705} = 0.1047$. Consequently, the model variance for this class is $0.1047N$. If we interpret the Arps–Roberts estimate of $\hat{N} = 379$ as if it were the true value of N, the model standard deviation for the number of discoveries in class 8 is about 6.3.

We are also interested in the number $\bar{w}(n)$ of wells that must be drilled to find $n \leq N$ deposits in the size class with representative area a. The waiting time $\bar{t}_1$ in wells drilled to the first discovery in a given class has geometric probability distribution $(Np)(1 - Np)^{t-1}$, where $t = 1, 2, \ldots$, so that the expectation $\bar{t}_1$ of $\bar{t}_1$ is $1/Np$. The waiting time $\bar{t}_r = \bar{w}(r) - \bar{w}(r-1)$ in wells drilled between the $(r-1)^{st}$ discovery up to and including the rth discovery has geometric probability distribution with a mean of $1/(N-r+1)p$, so the expectation of $\bar{t}_r$ is $\bar{t}_r = 1/(N-r+1)p$. Since $w(n) = t_1 + t_2 + \ldots + t_n$, the expectation of $\bar{w}(n)$ is $\bar{w}(n) = \bar{t}_1 + \ldots + \bar{t}_n$ or

$$\bar{w}(n) = \frac{1}{p}\left(\frac{1}{N} + \frac{1}{N+1} + \ldots + \frac{1}{N-n+1}\right) \qquad (11)$$

If N large and $n/N = h$ fixed and less than one,

$$\bar{w}(n) \cong \frac{1}{p} \log \frac{1}{N-n+1} \cong -\frac{1}{p} \log(1-h) \qquad (12)$$

or

$$\exp\{-p\bar{w}(n)\} \cong 1 - h \qquad (13)$$

Equation (13) is a waiting time analog of Arps and Roberts Eq. (9).

The waiting times $\bar{t}_1, \ldots, \bar{t}_n$ are independent, so the variance $\text{Var}[\bar{w}(n)]$ of $\bar{w}(n)$ is the sum of variances of $\bar{t}_1, \ldots, \bar{t}_n$. Since for $r = 1, 2, \ldots, n$,

$$\text{Var}(\bar{t}_r) = [1 - (N-r+1)p]/(N-r+1)^2 p^2,$$

$$\text{Var}(\bar{t}_r) < [1 - (N-n+1)p]/(N-r+1)^2 p^2 \qquad (14)$$

and

$$\text{Var}[\bar{w}(n)] < h|1 - N(1-h)p]/N(1-h)^2 p^2 \qquad (15)$$

Physical restrictions imply that $Np < 1$, so if n/N is fixed and N is large, the variance of $\bar{w}(n)$ will be of the order of $1/p$ with p being small. As an example, take Denver–Julesberg class 8 values presented earlier with $p = ca/T = 0.3414 \times 10^{-4}$, $n = 45$, and $N = 379$. An upper bound for the standard deviation of $\bar{w}(n)$ is about 583.

The discussion thus far has focused on a generic size class to keep the argument unencumbered by subscripts. Now we need them. For $i = 1, 2, \ldots, M$ we let N_i, a_i, and n_i denote the number of deposits in class i, the representative area for class i, and the number of deposits in class i discovered by w wells, respectively. We define

$$F\{da\} = \begin{cases} N_i \big/ \sum_{j=1}^{M} N_j & \text{if } a_i \in da \\[2mm] 0 & \text{otherwise} \end{cases} \qquad (16)$$

and $f = \sum_{i=1}^{M} n_i \big/ \sum_{j=1}^{M} N_j$. Here f is the fraction of *all* deposits found by w wells in a finite population of $N_0 \equiv \sum_{j=1}^{M} N_j$ deposits with an associated set of magnitudes (areas) $A = \{N_1 \text{ of areas } a_1, N_2 \text{ of area } a_2, \ldots, N_M \text{ of area } a_M\}$ and $F\{da\}$ is the relative frequency function generated by elements of A. Summing Eq. (9) over M size classes we obtain

$$\bar{f} = \sum_{i=1}^{M} \frac{N_i}{N_0}(1 - \exp\{-a_i w/T\})$$

$$= \int_0^\infty (1 - \exp\{-caw/T\})F\{da\} \qquad (17)$$

Connections Between SS and A-R

Arps and Roberts' equation Eq. (17) has a successive sampling analog, and as Gordon (1981) remarks, Eq. (9), interpreted as an estimator of the number N_i of deposits in

class i given that w wells have been drilled and n_i deposits in class i discovered, is identical to an (approximately) unbiased successive sampling estimator proposed by Holst (1973).

Explanation of the source of this similarity rests on properties of successive sampling *inclusion probabilities*. If a finite population U with N elements and an associated magnitude set A is successively sampled, what is the probability that element $k \in U$ will appear in an unordered sample of size $n \leq N$? In the vernacular of petroleum exploration, if N deposits in a play are labeled 1, 2, ..., N, what is the probability that the deposit labeled k will appear among the first n discoveries? Letting $s_n = \{i_i, \ldots, i_n\}$ denote an unordered sample composed of n distinct elements of U, we can write this probability as $P\{k \in \bar{s}_n\} = \pi_k(n)$. Then

$$\pi_1(n) + \pi_2(n) + \ldots + \pi_N(n) = n \qquad (18)$$

A more in-depth result was first proved by Rosén (1972) and improved by Gordon (1983). For each $k \in U$,

$$\pi_k(n) \cong 1 - \exp\{-\lambda A_k\} \qquad (19)$$

where λ is the unique solution to the equation

$$n = \sum_{k=1}^{N}(1 - \exp\{-\lambda A_k\}) \qquad (20)$$

With mild regularity conditions on $A_1, \ldots, A_N$, Gordon (1983) shows that the error in approximating $\pi_k(n)$ by $1 - \exp\{-\lambda A_k\}$ is at most on the order of $1/N$. Two numerical studies of Rosén's approximation show it to be a surprisingly good approximation (Hájek, 1981; Andreatta and Kaufman, 1983).

With no two fields having the same magnitude, we define $F\{dx\} = 1/N$ if $a \in dx$ and $F\{dx\} = 0$ otherwise and we let $f = n/N$. Then Eq. (20) can be recast in a form identical to Arps and Roberts' (17):

$$f = \int_0^\infty (1 - \exp\{-\lambda a\})F\{da\} \qquad (21)$$

While Eq. (21) is formally identical to Eq. (17), the parameter λ is not interpretable as an index of exploratory drilling effort. It is a function that shows how the distribution $\exp\{-\lambda A\}F\{da\}$ of magnitudes of undiscovered fields changes with changes in the fraction f of all N fields that have been discovered.

ESTIMATION

Exploratory drilling of w wildcats in a play results in n discoveries. Magnitudes of these discoveries in order of observation are $A_1, \ldots, A_n$. How can this data be used to estimate both the number of deposits in the play still undiscovered and the relative frequencies of undiscovered deposit magnitudes? Suppose that a geologic–volumetric analysis, coupled with knowledge of the results of drilling w wildcats, suggests a point estimate R_e of the sum of magnitudes of all deposits in the play. What relative frequency distribution of all deposits in this play is implied by R_e and observed data $A_1, ..., A_n$? If deposit magnitude data $A_1, ..., A_n$ are the only available data, what inferences can be made about the number of deposits and about the relative frequencies of magnitudes?

Given data consisting of a sample s_n of n deposit magnitudes (areas) discovered by w wells, Arps and Roberts' method for estimating the frequency N_i of deposits in the ith of M size classes is the following: if n_i deposits have been found by w wells, assume that the magnitude T of the search area and the efficiency parameter c are known with certainty and estimate N_i as

$$\hat{N}_i = n_i/(1 - \exp\{-cwa_i/T\}) \tag{22}$$

This approach requires knowledge of both w and the number $n_i, i = 1, 2, ..., M$, of discoveries made by these w wells.

When available data consist only of magnitudes of deposits in order of discovery, a different method of estimation must be employed. The relevant sampling model is then SS in place of A-R.

There is a close connection between two types of unbiased estimators for parameters of successively sampled finite population and the Arps and Roberts' estimator. Gordon (1981) was the first to observe that an estimator proposed in 1952 by Horvitz and Thompson (H-T) is identical in form to Eq. (22), which in turn is identical in form to an estimator for successively sampled finite population parameters proposed by Holst (1973).

The H-T estimator can be explained as follows. Suppose for the moment that all N elements of A are known with certainty. This (in principle) allows exact calculation of $\pi_k(n), k = 1, 2, ..., N$. Let $g(A)$ be a function of A with domain $(0, \infty)$ and range $(-\infty, \infty)$; for example, $\log A, A^d$, etc. If a successive sample $s_n = \{1, 2, ..., n\}$ from U is observed

$$\hat{g}_{H-T} = \sum_{j=1}^{n} g(A_j)/\pi(n) \tag{23}$$

is an unbiased estimator of $\sum_{k=1}^{N} g(A_k)$, the total sum of N population g values. The argument is a simple one: we let $I_k = 1$ if element k is in the sample, and $I_k = 0$ otherwise. Multiply $g(A_k)/\pi_k(n)$ by I_k. The expectation of

$$\sum_{k=1}^{N} \tilde{I}_k g(A_k)/\pi_k(n) \text{ is } \sum_{k=1}^{N} g(A_k) \text{ because the expectation of } \tilde{I}_k$$

is $\pi_k(n)$.

Upon replacing $\pi_k(n)$ with Rosén's approximation $\pi_k(n) \cong 1 - \exp\{-\lambda A_k\}$, the estimator $\hat{g}_{H-T}$ takes the form

$$\hat{g}_{H-T} \cong \sum_{j=1}^{n} g(A_j)/(1 - \exp\{-\lambda A_j\}) \tag{24}$$

and the analogy with Eq. (22) is evident.

A variety of useful functions of elements of λ can be estimated unbiasedly by proper choice of g: with

$$g(A) = A \qquad \sum_{k=1}^{N} A_k$$

$$g(A) = A^\delta \qquad \sum_{k=1}^{N} A_k^\delta$$

$$\qquad\qquad g_{H-T} \text{ estimates}$$

$$g(A) = \begin{cases} 1 & A \leq y \\ & \text{if} \\ 0 & A > y \end{cases} \qquad N(y)$$

$$g(A) = 1 \qquad N$$

Unbiased estimators of the variance of $\hat{g}_{H-T}$ are available (see, e.g., Cochran, 1953).

The key difficulty in applying an estimator $\hat{g}_{H-T}$ to petroleum data in the form of a sample s_n of deposit magnitudes is that $\hat{g}_{H-T}$ is a function of inclusion probabilities $\pi_k(n), k = 1, 2, ..., N$ each of which in turn is a function of *all* deposit magnitudes in A. Rosén's approximation is also a function of all elements of A. If all deposit magnitudes are known with certainty a priori, there is no inference problem!

Work on development of H-T type estimators that depend only on information in the sample has been done by Gordon (1983). Andreatta and Kaufman (1983) studied an estimator closely related to $\hat{g}_{H-T}$ (Murthy, 1957) that depends on sample information and knowledge of any one function of elements of A: the sum $\sum_{k=1}^{N} A_k$, the number N of elements in A, or a fractile of the frequency function of magnitudes generated by elements of A, for example. In the following two sections we discuss these estimators and give two examples of their use.

Anchored Estimation

If a geologic–volumetric analysis of a play yields a point estimate R_e of total barrels of oil equivalent (BOE) in place in a partially explored play or if detailed reconnaissance surveying results in a point estimate N_e of the number of deposits in the play, then either R_e or N_e can be used along with an incomplete sample s_n of individual deposit magnitudes to arrive at an approximately unbiased estimate of the empirical frequency function of undiscovered deposit magnitudes, if discovery is modeled as SS. For ease of exposition we shall assume that individual deposit magnitudes are measured in BOE, although this is not necessary. For example, deposit magnitude might be measured as productive area as long as deposit BOE can be assumed to be a known function of the deposit area.

Given N and a sample $s_n = \{A_1, ..., A_n\}$, we let λ be a solution to

$$N = \sum_{j=1}^{n} 1/(1 - \exp\{-\lambda_1 A_j\}) \tag{25}$$

Then

$$\hat{R}(N, s_n) = \sum_{j=1}^{n} A_j/(1 - \exp\{-\lambda_1 A_j\}) \tag{26}$$

is an approximately unbiased estimator of R. Given R, s_n, and λ_2, a solution to

$$R = \sum_{j=1}^{n} A_j / (1 - \exp\{-\lambda_2 A_j\}) \qquad (27)$$

then

$$\hat{N}(R, s_n) = \sum_{j=1}^{n} 1 / (1 - \exp\{-\lambda_2 A_j\}) \qquad (28)$$

is an approximately unbiased estimator of N.

Summing the right-hand side of Eq. (28) over indices of A_js in s_n that are less than or equal to y produces an estimate $\hat{N}(y|R, s_n)$ of the number of deposits in the population of magnitude less than or equal to y. Since $\hat{N}(y|R, s_n)$ is a function of R, the behavior of $\hat{N}(y|R, s_n)$ with alternative choices of R, can be examined. How $\hat{R}(N, s_n)$ varies with alternative choices of values for N can also be studied.

Andreatta and Kaufman (1983) illustrated the use of this method of estimation in an application to North Sea data as treated by Smith and Ward (1981). Smith and Ward estimated the amount of recoverable oil remaining to be discovered in the North Sea by maximum likelihood methods using a discretized version of SS with seven magnitude classes along with a record of discoveries to 1977. A seven dimensional grid search resulted in estimates shown in the second column of Table 2.

Tables 1 and 2 (from Andreatta and Kaufman, 1983) display a partitioning of deposit (field) magnitudes into seven classes and the results of using estimators of the form in Eqs. (26) and (28) anchored on Smith and Ward estimates. The Smith and Ward (1981) estimates are closely matched, implying that for this data, approximately unbiased SS estimators are also approximate maximum likelihood estimators.

Moment Matching Estimation

When no credible point estimates of deposit population parameters, such as the number of deposits in the population or the sum of deposit magnitudes, are available and when observed data consist solely of deposit magnitudes in order of discovery, the method outlined above is not applicable.

Assuming as before that discovery is modeled as SS, Gordon (1983) suggested that estimation of population parameters be done by a procedure that invokes the following heuristic. Split a complete sample $s_n = \{A_1, \ldots, A_n\}$ of discovery magnitudes into an "early" part $s_m = \{1, 2, \ldots, m\}$ and a "late" part $s_{n:m} = \{m+1, \ldots, n\}$. That is, s_m contains magnitudes of the first n discoveries and $s_{n:m}$ contains magnitudes of the last $n - m$ discoveries.

Suppose for the moment that inclusion probabilities $\pi_k(n)$, $k = 1, 2, \ldots, N$ and $n = 1, 2, \ldots, N$ are known. Then, using s_m,

$$\sum_{j=1}^{m} A_j / \pi_j(m) \qquad (29)$$

Table 1. Field size classification scheme (size measured in million barrels).

Class	Class Interval	Representative Magnitude A_i	Frequency of Historical Discoveries
1	0–50	25	26
2	50–100	75	15
3	100–200	150	15
4	200–400	300	19
5	400–800	600	16
6	800–1,600	1,200	4
7	1,600–3,200	2,400	4

and using s_n,

$$\sum_{k=1}^{n} A_k / \pi_k(n) \qquad (30)$$

are both unbiased estimators of R [see Eq. (23)]. In addition, subsequent to observing s_m, $s_{n:m}$ is a successive sample drawn from $A - s_m = \{A_{m+1}, \ldots, A_N\}$ so that if we define $\pi_i(n - m|s_m)$, $i = m+1, \ldots, N$ as the probability that the deposit labeled i appears in a sample of size $n - m$ drawn successively from $A - s_m$,

$$\sum_{\ell=m+1}^{n} A_\ell / \pi_\ell(n - m|s_m) \qquad (31)$$

is an unbiased estimator of $R - (A_1 + \ldots + A_m)$. Consequently, Eq. (31) plus the sum of elements in s_m also estimates R unbiasedly.

The hitch in using Eqs. (29), (30), or (31) is that the inclusion probabilities are not known. If, however, each of the inclusion probabilities is approximated as in Eq. (19), with

$$\hat{\pi}_j(m) = 1 - \exp\{-\alpha A_j\} \qquad j = 1, 2, \ldots, m \quad (32)$$

$$\hat{\pi}_k(n) = 1 - \exp\{-(\alpha + \beta)A_j\} \quad k = 1, 2, \ldots, n \quad (33)$$

and

$$\hat{\pi}_\ell(n - m|s_m) = -\exp\{-\beta A_i\} \qquad \ell = m = 1, \ldots, n \quad (34)$$

and estimators of the same parameter are set equal, we have two equations in two unknowns α and β: using Eqs. (29) and (30) with Eq. (32) and (33) and using Eqs. (33) and (31) with Eq. (32) and (34), we have

$$\sum_{k=1}^{n} A_k / [1 - \exp\{-(\alpha + \beta)A_k\}]$$

$$= \sum_{j=1}^{m} A_j / (1 - \exp\{-\alpha A_j\}) \qquad (35a)$$

Table 2. Comparison of estimates for North Sea data: Smith and Ward method versus anchored estimates.

Class	Smith–Ward Estimates	Estimates of Number in Each Size Class using Eqs. (27) and (28)[a]					
$A_1 = $ 25	203	[203][b]	200.4	194.7	191.7	201.2	193.6
$A_2 = $ 75	44	44.5	[44]	42.9	42.3	44.2	42.7
$A_3 = $ 150	26	26.8	26.5	[26]	25.7	26.6	25.9
$A_4 = $ 300	23	23.5	23.4	23.1	[23]	23.5	23.1
$A_5 = $ 600	16	16.6	16.6	16.5	16.5	16.6	16.5
$A_6 = $ 1,200	4	4.0	4.0	4.0	4.0	4.0	4.0
$A_7 = $ 2,400	4	4.0	4.0	4.0	4.0	4.0	4.0
Total Fields	320	322.4	318.9	311.3	307.3	[320]	309.9
Total sum of field sizes	43,175	43,868	43,674	43,250	43,031	43,733	[43,175]
λ	—	0.005482	0.005559	0.005735	0.005831	0.005535	0.005767

[a]Double precision calculations rounded to one decimal place; λ reported to six decimal places, calculated to twelve.
[b]Square bracketed entry is the anchor chosen to implement Eqs. (27) and (28). In each column there is one bracketed number and it is the anchor for the estimates of the other numbers in the column.

$$\sum_{j=1}^{m} A_j/(1 - \exp\{-\alpha A_j\})$$
$$= \sum_{j=1}^{m} A_j + \sum_{\ell=m+1}^{n} A_\ell/(1 - \exp\{-\beta A_\ell\}) \quad (35b)$$

A solution $(\hat{\alpha}, \hat{\beta})$ to the pair of Eqs. (35a) and (35b) provides numerical approximations to the inclusion probabilities in Eqs. (32), (33), and (34). The approximation $\hat{\pi}_k(n) \equiv 1 - \exp\{(\hat{\alpha} + \hat{\beta})A_k\}$ *to* $\pi_k(n)$ in Eq. (30) yields an estimate

$$\hat{R} = \sum_{k=1}^{n} A_k/\hat{\pi}_k(n) \quad (36)$$

of R that is based solely on observed data. A large sample approximation to the variance of the estimator $\hat{R}$ of R is

$$\text{Var}\{\hat{R}\} \cong \sum_{k=1}^{n} A_k^2[1 - \hat{\pi}_k(n)]/\hat{\pi}_k^2(n) \quad (37)$$

Replacement of A_k in Eq. (36) with $g(A_k)$ as in Eq. (23) allows estimation of other features of A; for example $g(A_k) \equiv 1$ yields an estimate N of the number of deposits or elements in the population from which the sample was drawn.[2]

This method was applied to the combined western Gulf of Mexico Miocene–Pliocene deposit data as described in Drew et al. (1982). These authors used a modified form of Arps and Roberts' model to estimate the ultimate number of oil and gas fields in magnitude (size) classes (classes 9 to 19 in Table 3) subsequent to 2,700 wildcats. Western Gulf data were partitioned by Drew and co-workers to conform to two major geologic trends: Miocene–Pliocene and Pleistocene.

Only the former data set is examined here. Cumulative discoveries in each of the 5-year period beginning with 1955 (end of year) and ending with 1975, plus discoveries in the first half of 1976, appear in Table 3. (These data were recovered from figures 11–15 of USGS Professional Paper 1252 by interpreting graphical displays and thus may not be exact.) Field discoveries of magnitude < 0.76 BOE are not recorded by Drew et al. (1982) and are thus omitted in our analysis as well.

Several external factors may possibly influence the time pattern of discoveries in the offshore Gulf. First, nearly all acreage blocks drilled are owned by the U.S. Government. They were sold in about 30 Federal lease sales beginning in 1954. In addition, early exploratory drilling was limited to shallow depths. As deep water drilling technology improved, drilling expanded to deeper waters and to parts of trends inaccessible earlier in the exploration history of the Gulf. Finally, the rapid increase in natural gas prices in the 1970s stimulated a burst of drilling—for Pleistocene deposits in particular—formerly considered unprofitable.

The impact of these external factors on the time pattern of discoveries are ignored in our analysis. The data are assumed to have been generated by successive sampling. Moment-matched estimates are compared with those of numbers of fields in place presented by Drew et al. (1982).

Table 4 displays three different splits of the data into "early" discovery and "late" discovery fields. The first split defines "early" as discoveries made to the end of 1960, the second as discoveries made to the end of 1965, and the third as discoveries made to the end of 1970. Table 5 shows the effects of using alternative definitions of "early" on the computation of estimates, and a comparison of moment-matching estimates of the form in Eqs. (35a) and (35b) with Arps–Roberts type estimates.

An "early" sample composed of discoveries made between 1954 and 1965 matches Drew et al. (1982) estimates reasonably well. However, as the size of the "early" sample increases, the estimates of the ultimate number of fields in each magnitude class other than classes 17, 18, and 19 (the largest three magnitude classes) decrease. For this data, moment matching estimates of the ultimate number of fields in smaller magnitude classes appear to be quite sensitive to

---

[2]Gordon (1983) demonstrated that with certain regularity conditions on magnitudes of elements of A in force, an equation system similar to Eqs. (35a) and (35b) has a unique, consistent solution as $N \to \infty$ with m/N and n/N fixed. Barouch et al. (1983) showed that the mean $\bar{A}_m$ of the early sample greater than the mean $\bar{A}_n$ of the entire sample is sufficient to guarantee existence of a solution to Eqs. (35a) and (35b) and that the number of solutions must be odd.

Table 3. Cumulative discoveries in western Gulf Miocene–Pliocene trend[a].

Class	(10⁶ BOE) Magnitude	1954–1955	1954–1960	1954–1965	1954–1970	1954–1975	1954–1976
9	1.14	0	4	7	16	20	24
10	2.28	6	7	11	14	23	27
11	4.56	0	5	12	17	24	27
12	9.12	3	9	15	20	32	25
13	18.24	7	15	22	35	46	50
14	36.48	9	15	21	28	36	42
15	72.96	9	19	27	35	43	43
16	145.92	7	8	19	24	25	25
17	191.84	6	12	16	17	17	17
18	583.68	2	5	5	5	5	5
19	1,166.36	2	2	2	2	2	2

[a]From Drew et al. (1982).

Table 4. Gulf Miocene–Pliocene trend data split into early and late discovery periods.

Class	Early 1954–1960	Late 1961–1976	Early 1954–1965	Late 1966–1976	Early 1954–1970	Late 1971–1976
9	4	20	7	17	16	8
10	7	20	11	16	14	13
11	5	22	12	15	17	10
12	9	26	15	10	10	15
13	15	35	22	28	35	15
14	15	27	21	21	28	14
15	19	24	27	16	35	8
16	8	17	19	6	24	1
17	12	5	16	17	1	0
18	5	0	5	0	5	0
19	2	0	2	0	2	0
Total	101	196	157	140	213	84

Table 5. Effects of different partitions of western Gulf Miocene–Pliocene and comparison to USGS estimates.

Class	(10⁶ BOE) Magnitude	Moment Matching Estimates of Ultimate Numbers of Fields in East Class Size			Arps-Roberts[a]	
		Early Period 1954–1960	Early Period 1954–1965	Early Period 1954–1970	Number Fields Found By 1,300 Wildcats	Estimated Number In Each Class
9	1.14	1,296	838	654	24	889
10	2.28	736	478	375	27	539
11	4.56	375	246	194	27	327
12	9.12	252	168	135	35	198
13	18.24	193	134	111	50	120
14	36.48	93	69	60	42	66
15	72.96	61	51	47	43	51
16	145.92	27	26	25	25	26
17	291.84	17	17	17	17	17
18	583.68	5	5	5	5	5·
19	1,166.36	2	2	2	2	2
Total magnitude (10⁹ BOE)		32.9	27.4	25.3		
Standard deviation of estimate of total magnitude		1.0	0.62	0.46		

[a]From Drew et al. (1982).

how the sample is split. As Drew and co-workers point out, the observable upturn in discovery rates in 1970–1975 for fields in classes smaller than class 16 may possibly be due to low, regulated prices for natural gas in the 1950s and 1960s and to a corresponding delay in exploration and development of smaller fields until increased price incentives appeared in the 1970s. Leasing constraints add to a confounding of the pattern of exploration that might be expected to unfold in the absence of a regulated market. In addition, assuming SS as a model of discovery, a shift toward increased discovery rates for small fields with few or no large discoveries signals depletion of the initial population and may possibly reduce estimates of ultimate numbers of fields in smaller magnitude classes. This appears to be the case for these data.

CONCLUDING REMARKS

The examples presented in this paper demonstrate how the simplest form of successive sampling models can be used to estimate "field size" distributions. Elaborations of SS like those suggested in Eq. (1) that account for special features of particular plays or deposit populations are possible and warrant further study and testing.

We have not discussed here projections of patterns of future discoveries. However, if discovery is modeled as SS, an estimate $\hat{A}$ of A is computed from a sample s_n, and $\hat{A}$ is adopted as representative of A, then SS of the unobserved portion of $\hat{A}$ dictates a pattern of as yet unobserved discoveries *in order of occurrence*. How this type of forecasting is done is a story for another day.

APPENDIX

The variance of $\bar{n}(w)$ with A-R* in force can be calculated by the following elementary argument. We let $I_k(w) = 1$ if the deposit labeled k appears among discoveries made by the first w wells and $I_k(w) = 0$ otherwise. Similarly we let $I_{jk}(w) = 1$ (given that $j \neq k$) if *both* deposits j and k appear among discoveries made by the first w wells and $I_{jk}(w) = 0$ otherwise. The probability that any particular deposit is discovered by the first w wells is $1 - (1-p)^w \equiv \pi(w)$, so $\text{Var}[\bar{I}_k(w)] = \pi(w)[1 - \pi(w)]$.

The covariance of $\bar{I}_k(w)$ and $\bar{I}_j(w)$ is $E[\bar{I}_{jk}(w)] - \pi^2(w)$. To evaluate $E[\bar{I}_{jk}(w)]$ use the facts that the probability that *neither* deposit j nor deposit k are discovered by the first w wells is $(1-2p)^w$ and that $P\{j \text{ or } k \text{ or both are discovered by } w \text{ wells}\} = P\{\bar{I}_j(w) = 1\} + P\{\bar{I}_k(w) = 1\} - P\{\bar{I}_{jk}(w) = 1\}$. These two facts imply that

$$1 - (1-2p)^w = 2 - 2(1-p)^w - P\{\bar{I}_{jk}(w) = 1\}$$

Since $E\{\bar{I}_{jk}(w)\} = P\{\bar{I}_{jk}(W) = 1\}$,

$$E\{\bar{I}_{jk}(w) = 1\} = 1 - 2(1-p)^w + (1-2p)^w$$

$$= \pi(w) - (1-p)^w + (1-2p)^w$$

Hence,

$$\text{Cov}\{\bar{I}_j(w), \bar{I}_k(w)\} = P\{\bar{I}_{jk}(w) = 1\} - \pi^2(w)$$

$$= \pi(w)[1 - \pi(w)] - (1-p)^w + (1-2p)^w$$

The variance of $\bar{n}(w)$ is thus

$$\text{Var}\{\bar{n}(w)\} = \sum_{k=1}^{N} \text{Var}\{\bar{I}_k(w)\} + 2 \sum_{k=1}^{N} \sum_{j>k} \text{Cov}\{\bar{I}_j(w), \bar{I}_k(w)\}$$

$$= N\pi(w)[1 - \pi(w)] + N(N-1)[\pi(w)(1 - \pi(w)) - (1-p)^w + (1-2p)^w]$$

When p is small and w is large,

$$(1-p)^w - (1-2p)^w \cong e^{-wp}(1 - e^{-wp}) \cong \pi(w)[1 - \pi(w)]$$

and then

$$\text{Var}\{\bar{n}(w)\} \cong N\pi(w)[1 - \pi(w)]$$

REFERENCES CITED

Adelman, M. A., J. C. Houghton, G. M. Kaufman, and M. B. Zimmerman, 1983, Energy Resources in an uncertain future: coal, gas, oil, and uranium supply forecasting: Cambridge, Massachusetts, Ballinger Publishing Company, 434 p.

Andreatta, G., and G. M. Kaufman, 1983, Estimation of finite population properties when sampling is without replacement and proportional to magnitude: Journal of the American Statistics Associations, MIT-EL 82-027 Working Paper.

Arps, J. K., and T. G. Roberts, 1958, Economics of drilling for Cretaceous oil and gas on the east flank of the Denver–Julesberg basin: AAPG Bulletin, v. 42, n. 11, p. 2549–2566.

Barouch, E., S. Chow, G. M. Kaufman, and T. Wright, in press, Properties of successive sample moment estimators: Studies in Applied Mathematics, Cambridge, Massachusetts Institute of Technology.

Bloomfield, P., et al., 1979, Volume and area of oil fields and their impact on the order of discovery: Resource Estimation and Validation Project, Department of Statistics and Geology, Princeton University, 53 p.

Cochran, W. G., 1953, Sampling Techniques: New York, John Wiley.

Drew, L. J., J. H. Scheunemeyer, and D.H. Root, 1980, Petroleum-resource appraisal and discovery rate forecasting in partially explored regions, part A: an application to the Denver basin: U.S. Geological Survey Professional Paper 1138-A, B, C, p. A1–A11.

Drew, L. J., J. H. Schuenemeyer, and W. J. Bawiec, 1982, Estimation of the future rates of oil and gas discoveries in the western Gulf of Mexico: U.S. Geological Survey Professional Paper 1252, 57 p.

Gordon, L., 1981, Successive sampling in large finite populations: Annals of Statistics, v. 11, n. 2, p. 702–706.

———, in press, Estimation for large successive samples with unknown inclusion probabilities: Annals of Statistics.

Hájek, J., 1981, Sampling from a finite population: New York, Marcel Decker, Inc., 247 p.

Holst, L., 1973, Some limit theorems with applications in sampling theory: Annals of Statistics, v. 1, p. 644–658.

Horwitz, D. G., and D. J. Thompson, 1952, A generalization of sampling without replacement from a finite universe: Journal of the American Statistical Association, v. 47, p. 663–685.

Kaufman, G. M., Y. Balcer, and D. Kruyt, 1975, A probabilistic model of oil and gas discovery, *in* J. Haun, ed., Estimating the volume of undiscovered oil and gas resources: AAPG Studies in Geology Series, n. 1, p. 113–142.

Kaufman, G. M., and J. Wang, 1980, Model mis-specification and the Princeton study of volume and area of oil fields and their impact on the order of discovery: MIT Energy Laboratory Working Paper No. MIT-EL 80-003WP, 16 p.

Murthy, M. N., 1957, Ordered and unordered estimators in sampling without replacement: Sankhyā, v. 18, p. 378–390.

Root, D. H., and J. H. Schunemeyer, 1980, Petroleum-resource appraisal and discovery rate forecasting in partially explored regions: forecasting in partially explored regions: Mathematical foundations, U.S. Geological Survey Professional Paper 1138-B, p. B1–B11.

Rosén, B., 1972, Asymptotic theory for successive sampling with varying probabilities without replacement, I and II: Annals Mathematical Statistics, v. 43, p. 373–397, 748–776.

Smith, J. L., and G. L. Ward, 1981, Maximum likelihood estimates of the size distribution of North Sea oil fields: Mathematical Geology, v. 13, n. 5.

Wang, C. C., and P. J. Lee, 1983a, Probabilistic formulation of a method for the evaluation of petroleum resources: Mathematical Geology, v. 15, n. 1, p. 163–181.

—— , 1983b, Conditional analysis for petroleum resource evaluations: Mathematical Geology, v. 15, n. 2, p. 349–361.

Comparison of Predrilling Predictions with Postdrilling Outcomes, Using Shell's Prospect Appraisal System

D. Sluijk and J. R. Parker
Shell Internationale Petroleum Maatschappij B.V.,
The Hague, Netherlands

Since 1975 Shell Internationale Petroleum Maatschappij has used a Monte Carlo simulation model for worldwide prospect appraisal. The input parameters to this model describing charge (oil and gas available for trapping and retention), structure, reservoir, and retention (seal characteristics) are given in the form of probability distributions. For the estimation of charge and retention, the model follows a scheme of Bayesian update and uses equations derived from calibration studies, that is, statistical analysis of extensive data sets with a worldwide distribution. Comparison of predrilling predictions with postdrilling results suggests that the underlying calibration procedure is sound. It also demonstrates the importance of assessing geologic uncertainty in a quantitative manner. Geologists appear to have been successful in describing the geologic setting of prospects with respect to hydrocarbon charge and retention (the calibrated parts of the system); however, serious overestimation has occurred with respect to reservoir parameters, trap existence, and related factors (the uncalibrated parts of the system).

INTRODUCTION

The uncertainty about the expected outcome of exploration opportunities is generally expressed as a probability distribution derived from a prospect appraisal system (based on subjective, statistical, deterministic, or a combination of methods). Such approaches do not automatically guarantee that the predictions are realistic, in other words, that they correctly describe the uncertain situation being appraised. Therefore, after actual outcomes are available, they should be compared with the corresponding predictions to assess the validity of the system.

This study reports on a comparison made between predrilling predictions and postdrilling outcomes, using Shell's prospect appraisal system.

SHELL'S PROSPECT APPRAISAL SYSTEM

In a prospect appraisal system, one would ideally wish to construct a coherent picture of the processes of hydrocarbon generation, migration, accumulation, and retention as a completely deterministic model, by integrating all the relevant geologic concepts and research findings with the clear objective of quantitatively predicting hydrocarbon occurrences. However, despite many advances in our understanding of these processes, this ideal situation is rarely realized. Although a given model may be logically consistent and based on the best of present knowledge, it is likely to be incomplete and only partially correct. The reason for this is that there are variables that are either unknown or cannot be measured with sufficient precision and processes that cannot be accommodated in the model. Hence, a somewhat different, more pragmatic approach has been adopted by Shell, which uses a deliberately simplified Monte Carlo simulation model, the input parameters of which are generally available in an early exploration phase. The input parameters to this model describing charge (oil and gas available for trapping and retention), structure, reservoir, and retention (seal characteristics) are given in the form of probability distributions that reflect the input uncertainty.

For the estimation of charge and retention, the model uses equations derived from calibration studies, which are statistical analyses of extensive data sets having a worldwide distribution (so-called learning sets). Such calibration studies test the importance (or irrelevance) of the various hypotheses implied by the model and define the range of uncertainty resulting from the incompleteness of the model, the simplification of the parameters describing the model, the varying levels of input information, and the confidence in that information. The results from the calibration studies are then used to update worldwide prior distributions by a Bayesian process using information specific to the prospect being appraised.

The principles of the Shell system have been described by Sluijk and Nederlof (1984) and the statistical background by Nederlof (1981), but the calibration data and specific results from using the system remain largely proprietary information.

TESTING THE VALIDITY OF THE SYSTEM

Because of the probabilistic nature of a prospect appraisal system, any test of the validity of the system that compares prediction with outcome cannot be made using a single result. All that can be done with a single result is to see whether it falls within the predicted range. Meaningful comparisons can only be made with a fair sample of actual outcomes that can be tested statistically. This study analyses a set of 165 prospects, which have been appraised since 1975 using the Shell prospect appraisal system and subsequently conclusively tested by drilling.

The purpose of the analysis was to ascertain whether the probabilistic predictions truly represented the state of uncertainty about the prospects before drilling. Such uncertainty can be misrepresented in various ways: it can be too optimistic or too pessimistic (location bias) and too certain or too uncertain (dispersion bias). We also assessed the ability of the system to predict the rank order of the actual outcomes (forecasting efficiency).

LOCATION AND DISPERSION BIAS

Location bias can be tested by comparing the sum of the expectations (the expectation being the mean of the probability distribution for an individual prospect, including both success and failure cases) with the sum of the actual outcomes after drilling. If we compare oil-in-place for all 165 conclusively tested prospects, we find that the sum of the expectations is greater than the sum of the actual outcomes by slightly more than one standard deviation of the sum of the expectations. Although not significant in a statistical sense, this points to appreciable overestimation because the standard deviation of the sum of the expectations is large (Table 1).

A further test for location bias considers the probability, as read from a cumulative probability distribution for each prospect, that a volume equal to or greater than the volume actually discovered would be found. Such after drilling probabilities for all 165 prospects should be evenly spread along a probability axis if no location bias exists: in other words, the average probability should be 0.5. The actual average probability is 0.7, about one standard deviation higher. Although again not significant in a statistical sense, this again suggests considerable overoptimism in the predictions.

A breakdown of the sources of this overoptimism shows a small but consistent overestimation of reservoir parameters:

Table 1. Statistics of predictions and outcomes.

Sample Size	165
A. Sum of the STOIIP[a] expectations (million bbls)	19932
Standard deviation of A	10905
B. Sum of the Actual STOIIP	7121
C. Separation of A and B	12811
In terms of standard deviation of A	1.17
[a]STOIIP stock tank oil initally in place.	

gross reservoir by 21%, net: gross ratio by 25%, porosity by 14%, and the chance of the presence of closure and of reservoir by 16%. The multiplicative nature of the calculations using these parameters explains much of the overoptimism, which is a factor of approximately 2–2.5. Certainly this is geologically, if not statistically, significant. It is worth noting that this overoptimism relates to the uncalibrated, purely subjective part of the prospect appraisal system.

There are two principal sources of uncertainty in a probabilistic prediction: (1) the uncertain input by the geologist, which is based on a subjective assessment of the likelihood that particular geologic conditions will occur; and (2) the prospect appraisal system itself, because of the imperfection in the calibration between geologic factors and hydrocarbon volumes. The overoptimism in this analysis appears to be mainly the result of the subjective overestimation of geologic parameters by the geologist. This, in turn, suggests that the system uncertainty reasonably reflects the state of nature. Thus, we emphasize the importance of an independent expert who can give guidance on subjective judgment and can act as an outside technical auditor during the course of a prospect appraisal interview.

To test for dispersion bias, we compared the probability of a successful outcome, as predicted by the system, with the actual success ratio. We did this by dividing the 165 prospects into subsets, each subset containing those prospects with a similar predicted probability of a successful outcome; thus, each group of prospects had a predicted success ratio. The achieved success ratio was calculated from the actual drilling results of this subset of prospects. Statistical testing showed reasonable agreement between the predicted and observed success ratios. Hence, we concluded that no appreciable dispersion bias exists and that the predictions were neither too certain nor too uncertain.

RANKING ABILITY

The ability of the prospect appraisal system to predict the rank order of the actual outcomes (forecasting efficiency) is shown on Figure 1. The vertical axis represents the cumulative volume of oil in place discovered in the 165 prospects selected for testing; the horizontal axis represents the hypothetical sequence numbers of the wells testing the prospects. These wells can be ordered in a large variety of ways (in fact, there are 165! hypothetical sequences possible). The diagonal line is the average of all possible hypothetical ranking orders and represents the average cumulative plot if the selected prospects had been drilled in random order (0% forecasting efficiency). The upper curve is a plot of the actual

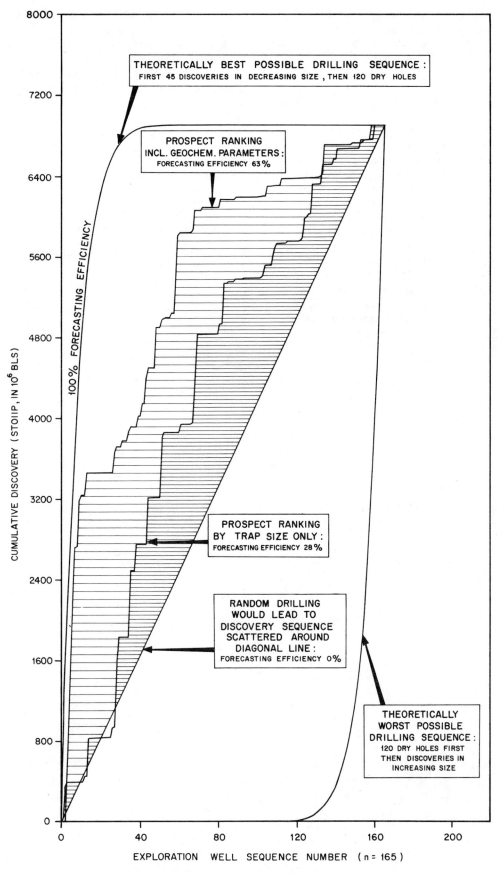

Figure 1. Forecasting efficiency using Shell's prospect appraisal system (see text for further details). STOIIP, stock tank oil initially in place.

outcomes (in hindsight) in decreasing size and is thus the theoretical best possible ranking (+ 100% forecasting efficiency). Conversely, the lower curve is a plot of the actual outcomes (in hindsight) in increasing size and is thus the theoretically worst possible ranking (− 100% forecasting efficiency).

The better the ability of the system in predicting the actual rank order of outcomes, the closer the cumulative plot of the outcomes in decreasing predicted size will approach the cumulative plot of the outcomes in decreasing actual size. The forecasting efficiency is defined as that percentage of the area between the random line and the upper curve that lies below the predrilling line. If we rank prospects according to the predrilling trap size estimate only, then the forecasting efficiency is only 28%. However, if we use predrilling predictions from the complete prospect appraisal system (including hydrocarbon charge and retention parameters), then the forecasting efficiency is 63%, which represents a significant improvement. It should be stressed that for the purposes of this study, the sequences in which the wells are plotted are purely hypothetical and are not the actual time sequences in which the prospects, from many different ventures throughout the world, were drilled.

ACKNOWLEDGMENTS

Our thanks are extended to Shell Internationale Petroleum Maatschappij for permission to publish this paper and to our colleagues, M. H. Nederlof and J. W. A. Bodenhausen, for their contribution to the study.

REFERENCES CITED

Nederlof, M. H., 1981, Calibrated computer simulation as a tool for exploration prospect assessment: U.N. ESCAP CCOP Technical Publication 10, p. 122–138.
Sluijk, D., and M. H. Nederlof, 1984, Worldwide geological experience as a systematic basis for prospect appraisal, in G. Demaison and R. J. Murris, eds., Petroleum Geochemistry and Basin Evaluation: AAPG Memoir 35, p. 15–26.

Stratigraphic Aspects of Petroleum Resource Assessment

Gregory Ulmishek
Argonne National Laboratory, Energy and Environmental Systems Division
Argonne, Illinois

Four major factors control petroleum richness of a region: source rock, reservoir rock, seal, and trap. Assessment of undiscovered resources of oil and gas in poorly known regions should be based on comparative analysis of these factors in a forecast region and in a well-explored analog area. Three of these factors mainly reflect stratigraphic, rather than tectonic, conditions, and the fourth, the trap factor, includes both stratigraphic and tectonic aspects. The predominance of stratigraphic information in the factors indicates that the main unit used for petroleum resource assessment done by comparative geologic analysis should be a stratigraphic unit. Such a proposed unit, which is called an independent petroliferous system (IPS), is understood here as a body of rocks separated from surrounding rocks by regional barriers to lateral and vertical migration of fluids, including oil and gas. Stratigraphically, an IPS is essentially homogeneous. It includes source rocks, reservoir rocks, traps, and a regional seal, and thus, it is a suitable unit for comparative analysis of the factors and petroleum genetic studies. For oil and gas resource assessment in poorly known regions, an IPS has certain advantages over a basin or play as an assessment unit. The concept of an IPS can also be used in statistical methods of resource appraisal and can increase reliability of these results.

INTRODUCTION

The purpose of this article is to consider the use of a stratigraphic versus tectonic approach to the assessment of undiscovered resources of oil and gas, mainly in frontier areas or poorly known regions. A poorly known region is understood here to be a region for which information is restricted chiefly to regional geologic and/or geophysical studies. Commercial productivity may be proven, but the main exploration plays are not established. It is assumed, however, that the major characteristics of facies composition, thickness of sedimentary cover, and structural style of a region are known or may be inferred with sufficient reliability. Exploration in such regions requires large investments and is characterized by a high risk factor; thus reliable appraisal of expected resources becomes especially important. Some of the concepts discussed here also have application in more maturely explored regions.

All methods involved in resource assessment of poorly known regions are invariably based on geologic comparisons. A geologic analog with known productivity is used to predict the productivity of the forecast objective. The volumetric method is the most frequently used method. The geochemical balance (or volume–genetic) method is based on geochemical material balance calculations of the amount of

generated hydrocarbons (Semenovich et al., 1977.) This and the play analysis method deal primarily with data obtained in a forecast region itself, but when these data are scarce, both methods have to depend on external analogs.

Wide use of the volumetric method as a consistent approach for resource appraisal began with the work of Weeks (1949) who proposed the volumetric yield, or specific resources per unit of rock volume, as a "yardstick" for the measurement of a basin's richness. In spite of a long history of application, however, the volumetric method and other comparative geologic methods are not sufficiently formalized to permit a truly objective assessment. The major deficiency of these methods lies in the absence of a common understanding for the terms *analogy* and *similarity*, which results in widely variable criteria for comparative analysis. Consequently, very different assessments of a region can be made depending on the chosen analog and the approach to the analytical procedure. We address these problems here.

This article is essentially conceptual, although the proposed ideas are based on my study of a variety of basins, mainly in Russia and China. Some aspects of the approach discussed in this article were used for the assessment of petroleum resources of the Timan–Pechora basin and the Barents–Northern Kara Shelf (Ulmishek, 1982) and gave satisfactory results. Successful application of the concept is now hampered by the absence of a file of analogs derived from maturely explored basins. At present, the compilation of such a file is in progress.

FACTORS CONTROLLING A REGION'S PETROLEUM RICHNESS

It is evident that any comparative geologic method of resource assessment should be based on the successive comparison of factors that control the richness of a petroliferous region. Many such factors have been identified in works of different authors. Probably the most complete single list, which included 29 factors, was presented by Weeks (1975). Unfortunately, although most of these factors are connected in some way to a region's richness, they are too general to be of much use. For example, such factors as "a favorable historic and physical relationship between source rock and reservoir facies" or "an optimum degree of mobility during sedimentation" are very indefinite and permit different evaluations. Terms such as "favorable" and "optimum" may have a highly variable sense in different geologic situations and are almost useless for comparative analysis, especially in frontier areas. Another group of Weeks' factors describes geologic processes rather than their results ("a high degree of sorting or screening of muds from the sands"), and others describe features of paleogeographic environments, such as the type and position of sources of clastic material or the rate of bacterial decomposition in sediments. Each factor from this group requires a large amount of data as well as difficult and often uncertain reconstructions. The significance or, at least, the universality of some factors listed by Weeks (1975) is doubtful. For example, broad lateral variability in the rate of deposition does not seem to be a feature controlling oil and gas richness. Many very rich basins, such as the West Siberian and Volga–Ural basins in Russia, are characterized by moderate lateral changes in the rate of deposition during most of their geologic histories (Dikenshtein et al., 1982; Rudkevich, 1974). And finally, the last group of factors ("favorable results from, and significance of, any exploratory drilling" and others) is not geologic, but rather reflects the exploration status of a region and has no significance for assessment of a region's richness by comparative geologic analysis.

Factors that can be combined to create a reliable basis for the comparative analysis for assessment of undiscovered oil and gas resources in poorly known basins must meet a few major requirements. First, they must reflect major conditions for generation and accumulation of petroleum and preservation of pools. Second, these factors must not require complicated and often uncertain geologic reconstructions but should be easily recognizable in sedimentary sections in frontier areas. And third, the factors must be measurable or at least comparable in qualitative terms of better or worse. Probably only four factors meet all these requirements. These are (1) character of potential source rocks and their maturation, (2) presence of reservoir rocks and their quality, (3) presence and quality of regional seals, and (4) presence of traps and their abundance and size.

These four factors do not exhaust all the complexities of geologic systems; they do, however, reflect the most important conditions controlling the richness of a basin's fill. For example, if regional geophysical studies and drilling of the first few wells in a region show that it contains mature, highly bituminous rocks, satisfactory reservoir rocks covered by a good evaporite or shaly seal, and supposedly adequate structural and stratigraphic traps, then the assessment of the region's richness will be high whatever known or unknown geologic processes have resulted in this favorable combination. All four factors are characteristics of the sedimentary column or the present-day structural pattern of frontier regions and are relatively easily recognizable with limited information. They are measurable in qualitative and often in quantitative terms and are thus readily comparable with the same factors in a geologically analogous region.

Systematic comparative analysis of the four factors in frontier and analog regions is facilitated if each factor is subdivided into a few major types (Table 1). Then the comparison of any factor that fits the same type in both regions can be done with maximum simplicity. The best analog fits in the same type with respect to all four major factors observed in the frontier region, and thus provides easy geologic comparability. The selection of such an analog is feasible if the subdivision reflects only the most basic characteristics of the factors and consists of only a limited number of types.

Subdivision of organic matter into humic and sapropelic types shown in Table 1 reflects different hydrocarbon products generated by these types. Further subdivision of sapropelic organic matter into dispersed and concentrated types is related to the very high oil-generating potential of formations that are commonly relatively thin but extremely enriched with bitumen. These formations, usually deposited in anoxic environments, are known in almost all prolific petroliferous regions, such as in the Middle East, West Siberia, California, and the North Sea. A similar subdivision of source rocks containing humic organic matter is not essential. Because of the small amount of sorption and the high migration ability of gas, which is the predominant hydrocarbon product of this organic matter, the total amount of humic organic matter in a section is more important than its concentration in separate formations.

Subdivision of the trap factor into two types, intensely and slightly deformed, is conditional; a definite boundary between these two types can not be established. Although an intermediate degree of deformation is common, a clear difference between almost undeformed regions and regions containing numerous well-expressed structures is evident. Thus, almost any forecast region can be classified in terms of degree of deformation without significant difficulties and an appropriate analog can be chosen.

A positive or negative connection between the type of reservoir rocks and the richness of a region is not necessarily valid (Halbouty et al., 1970). More often, a large volumetric yield is provided by reservoir rocks of high capacity (massive or multistrata). There are contradictory cases, however, such as the Volga–Ural region in the Soviet Union where large reserves are concentrated in a few permeable horizons of high quality (Aliyev et al., 1983). The bulk of giant oil reserves (14.3 billion bbl) in the Romanshkino field, for example, is found in two pay zones formed by an alternation of sandstone, siltstone, and shale beds having a total thickness

Table 1. Subdivision of factors controlling petroleum richness of a region.

Factor	Type	Characteristics of a Type
Source rock	Humic organic matter	Mainly terrestrial, including coal-bearing rocks. Dry gas is the major hydrocarbon product.
	Sapropelic organic matter, dispersed	Marine and lacustrine rocks. The content of organic matter is usually close to the clarke level and seldom reaches 2–3% in separate samples. Significant admixture of humic organic matter is common.
	Sapropelic organic matter, concentrated	Marine and lacustrine rocks, sometimes in relatively thin formations. Average concentration of exclusively sapropelic organic matter commonly exceeds 4–5% and reaches 20% or more in individual samples.
Trap	Intensely deformed	Structural (including halokinetic) and combination traps are abundant and predominate.
	Slightly deformed	Structural traps are rare; stratigraphic (including paleogeomorphic) traps predominate.
Reservoir rock	Massive	Usually thick carbonates (including reefs). Reservoir properties are significantly determined by cavernous porosity and fracturing, although matrix porosity can be important. Thick sandstone formations with laterally nonpersistent shales may also fit this type.
	Stratified	One to a few, usually sandstone beds in a relatively narrow stratigraphic interval. Intergranular porosity predominates but leaching and fracturing sometime play significant roles. "Blanket," often biostromal, carbonate reservoirs also fit this type.
	Multistrata	Numerous sandstones in thick clastic formations often of paralic or deltaic origin. Intergranular porosity predominates.
Seal	Perfect	Almost absolutely impermeable. These are salt and/or anhydrites, over-pressured shales, other thick (hundreds of feet) plastic shale formations, and evidently permafrost.
	Imperfect	Partly permeable, especially in zones of tectonic faulting and fracturing. Can be represented by different compacted shales, dense carbonates, marls, etc.

of only 100–200 ft (Muslimov, 1979). The subdivision of reservoir rock into stratified and multistrata types (Table 1) is conditional, and intermediate cases are common. In each particular case, however, classification of reservoir rocks into one of these three types and subsequent comparison with an analog does not seem to be difficult.

The presence and quality of a regional seal plays a significant role in the formation and preservation of oil and gas fields. Such seals control the hydrodynamic regime in underlying permeable formations and, hence, strongly influence the migration and accumulation of hydrocarbons and preserve oil and gas pools from destruction. Although many oil and gas pools in rich petroliferous regions are directly controlled by local seals (which are often thin), these covers are effective only in the presence of a regional seal that isolates the hydrodynamic system. In Table 1 regional seals are subdivided into perfect and imperfect types. The group of perfect seals includes formations that are characterized by plasticity in subsurface conditions. These formations include salt, anhydrite, overpressured shales, and plastic shales, usually having the essential participation of smectite minerals. Such a unique and poorly studied seal as permafrost can also be included in this group. The connection between giant oil and gas fields and evaporite seals has been emphasized repeatedly in the geologic literature; the presence of such seals is even considered as

one of the major factors controlling the formation of giant fields (Halbouty et al., 1970).

The imperfect type of regional seals mainly includes dense shales (often intercalated with siltstone beds) and dense carbonates. These rocks can not be considered as perfect seals because they lose their impermeability in zones of faulting and fracturing. If a regional seal is absent, it is unrealistic to expect significant petroleum potential in this region, although separate pools under local covers may be found.

The subdivision of factors shown in Table 1 can serve as a reliable basis for comparative analysis in assessing a region's richness. In the best case, the analog chosen should fit all four major factors permitting easy geologic comparison.

INDEPENDENT PETROLIFEROUS SYSTEM (IPS): A NEW UNIT FOR RESOURCE ASSESSMENT

It is evident that three of the four major factors controlling a region's petroleum richness (source, reservoir, and seal) contain much more stratigraphic than tectonic information. The fourth, the trap factor, tends to reflect both stratigraphy and tectonics depending on the type of trap. It seems reasonable, therefore, that a unit chosen for comparative assessment of petroleum resources should be more related to the stratigraphy of an area than to the tectonics. The analysis of factors of richness in such a unit will be an easier task than the analysis of these factors in any tectonic unit that is "heterogeneous" from a stratigraphic point of view. Because the four listed factors reflect the conditions for successive processes of generation, accumulation, and preservation of oil and gas, such a unit must meet two major requirements: (1) it must be a confined system in which these processes take place independently from surrounding rocks, and (2) it must be the simplest of these systems, to provide maximum internal geologic uniformity and to permit sufficient depth of analysis. Such an assessment unit is here called an independent petroliferous system (IPS), which is defined as a continuous body of rocks separated from surrounding rocks by regional barriers to lateral and vertical migration of liquids and gases (including hydrocarbons) and within which the processes of generation, accumulation, and preservation of oil and gas are essentially independent from those occurring in surrounding rocks.

Laterally, an IPS is bounded by structural barriers or barriers of permeability. Vertically, IPSs may be separated from each other by perfect seals, such as salt or thick shales. An IPS may be partly breached at the ground surface or open downward if it is underlain (without a separating seal) by a suite of rocks that does not contain its own source of hydrocarbons. In the latter case, oil and gas pools can occur, as a rule, only at the very top part of the underlying suite. Pools in Paleozoic buried hills of eastern China serve as an example of this (Tang Zhi, 1982).

It is evident that absolute isolation of one hydrodynamic system from another is rare in natural conditions. Some leakage of fluids and gas through faults, mud volcanoes, and other hydrodynamic windows can occasionally occur. In the presence of a perfect seal, however, this leakage is restricted, and migration paths between systems are ephemeral because these paths are quickly dammed by plastic material that forms the seal.

It is not always easy to distinguish an IPS in a poorly known region, and good judgment in determining IPS boundaries is required. When information is insufficient, some subjectivity is unavoidable as is true for any other assessment unit, such as a basin or a play. It seems likely, however, that possible mistakes will be neither common nor significant. Evidently, long-distance lateral migration of petroleum is the exception rather than the rule. In many basins where source rocks have been reliably mapped, such as the North Sea (Ziegler, 1980), the Timan–Pechora basin in the Soviet Union (Ulmishek, 1982), and many located in China (Fan Pu et al., 1980), oil fields are present only inside the boundary of mature bituminous facies. The expressions of rarely observed extremely long-distance lateral migrations in nature are the giant accumulations of heavy oil and bitumen along basin (and IPS) peripheries. Thus, lateral boundaries of IPSs in a poorly known region could be established to the first approximation on the basis of regional structural analysis. The upper and lower boundaries of an IPS determined by perfect seals are usually revealed by the first wells, and often these seals can be traced laterally by regional seismic surveys.

The most important task in developing the proposed approach is the determination, for analog purposes, of IPSs in well-explored basins. At present, 40–50 well-explored basins worldwide certainly contain not less than 150–200 IPSs. These could provide an excellent file of analogs with most combinations of factor types. Volumetric yields of the well-studied IPSs could serve as a basis for the evaluation of undiscovered resources of the forecast IPSs.

COMPARISON OF IPS, PLAY, AND BASIN AS ASSESSMENT UNITS

An IPS is purely an assessment unit; its application for other purposes is limited. As an assessment unit, however, it has significant advantages over two other such units that are widely used in practice: the play, or petroleum zone, and the basin. Probably the earliest definition of a play (petroleum zone) was published by Bois (1975, p. 87) who understood it to be "... a continuous portion of sedimentary volume which contains pools showing the following common characteristics: (1) reservoirs within the same productive sequence occur throughout the zone; (2) hydrocarbons are of similar chemical composition; and (3) traps are of the same type, or belong to a small number of types." Most geologists generally follow this definition with some changes. Baker et al. (1984, p. 426), for example, substituted the requirement of similar chemical composition of hydrocarbons with "presumably the same hydrocarbon source" and emphasized the areal continuity of a play.

The assessment of undiscovered oil and gas resources of a play (play analysis) is undoubtedly very important for the decision-making process in an oil company, but as a method for resource assessment of large regions, especially poorly studied regions, it is deficient. The concept of a play assumes the existence of a group of identified or inferred prospects that form the play. It is clear, however, that identification, or even reliable inference, of a prospect requires a great amount of detailed data that are always absent in the poorly known region. This can be seen in the play analysis formula (Procter et al., 1982) in which undiscovered resources are estimated by multiplication of (1) area of closure, (2) reservoir thickness, (3) reservoir fraction, (4) porosity, (5) hydrocarbon saturation, (6) gas fraction, (7) recovery factor, and (8) gas volume factor. All or most of these factors are completely unknown in frontier areas. External analogs in this case are rarely of help because there are no firm criteria for choosing a reliable analog for such widely variable characteristics as reservoir fraction, porosity, and area of closure.

The summation of petroleum resources in separate plays for the appraisal of a large region or basin likewise seems unacceptable. Tens or even hundreds of plays may exist in such a region, and prediction of many of them, especially in the early stages of exploration, is impractical. Resource assessment by play analysis in frontier areas is basically a process of going from the details to the general picture. It is analogous, for example, to beginning the mapping of a completely unknown region at a scale of 1:10,000. An opposite process, going from the general picture to the details, is usually more productive in any scientific investigation. Thus, play analysis appears to be an adequate method for studying the distribution of petroleum resources during the gradual growth of a data base, rather than being a method for their quantification in a large region.

A play does not meet the first requirement for an assessment unit because it is not an independent system with respect to the generation and migration of petroleum. A requirement for the same hydrocarbon source in the definition of a play does not have any significance for the play analysis method and can not be proved or disproved by the study of a play itself. Basically, the genetic aspects of productivity are not a fundamental part of the play analysis method; that is why this method can be applied with equal success to a whole play, to any part of it, or to a single prospect. An attempt to introduce genetic aspects into play analysis (Bishop et al., 1983) demonstrates that large numbers of highly uncertain reconstructions outside the play should be included in the consideration. Such reconstructions are very difficult even in maturely explored regions. As for frontiers, the evaluation of participating factors, such as amount of hydrocarbons spilled from downdip traps, gas volumes lost by diffusion, amount of dispersed oil along secondary migration pathways, and many others, even in more explored regions, are no more than guesses, and multiplication of guesses produces nothing but uncertainty.

A play in its strict sense is chiefly an exploration concept. Its definition as "a practical meaningful planning unit around which an integrated exploration program can be constructed" (Miller, 1982, p. 55) aptly expresses its role. It seems likely that the results of play analysis will be more valid if the play is considered as part of an IPS. Richness of the IPS and quality of the major factors controlling its productivity can serve as a regional framework for play analysis. For example, decisions that are critical for play analysis (Baker et al., 1984), such as the assessment of risk and the truncation of the field size distribution curve where the large fields are plotted, can be strongly supported by an assessment of the IPS's richness.

In practice, many authors studying frontier regions where particular prospects are difficult to identify often understand a play as a stratigraphic unit rather than a group of prospects. In play analysis assessment of oil and gas resources of the Arctic Wildlife Range (Mast et al., 1980), for example, a play is understood as a suite of regionally developed and stratigraphically confined reservoir rocks bounded by shales. Kent (1980) writes about a "Jurassic play of the North Sea" that includes Middle Jurassic sandstones overlain by the Upper Jurassic regional seal. H. D. Klemme (personal communication, 1984) has even proposed the term *megaplay*, which is a large stratigraphic sequence that may be productive in a few basic types of traps worldwide. This list of examples can be enlarged, but my point here is that during early stages of exploration the practical meaning of a play sometimes closely approaches the concept of an IPS.

The second of the assessment units and the one in most common usage is the basin. A clear understanding of this term and definitive criteria for distinguishing basins, however, are very imprecise. Commonly a basin is a synonym

for a structural depression or trough. Basins can be very large, with areas measuring millions of square miles, or very small, with areas of only a few hundred square miles. They can be structural or sedimentary, be confined to one regional tectonic structure, or include many of them. One basin can include several others, as do, for example, West Texas, Midland, and Delaware basins. This chaotic picture can also be aggravated by numerous subbasins that are completely indefinable. Commonly a basin is considered to be the structural opposite of a platform, although both can be equally rich in oil and gas. Nevertheless, a basin is a convenient practical unit for exploration, but even an ideal, well-defined basin is not a convenient unit for resource assessment. Although basins are independent systems, they are not the simplest of these systems, and thus they do not meet the second requirement of an assessment unit. A basin usually contains several IPSs, and only rarely does an IPS coincide with a small, simple basin.

A basin is essentially a tectonic unit. This is illustrated by all known basin classifications (~ 20) being invariably based on tectonic principles (Sokolov, 1980). The resulting problem is that basins with different qualities of major assessment factors are "lumped" into the same tectonic group. For example, there appears to be little value in comparing the Zagros foredeep in the Middle East with any foredeep basins of the Alpine System in Europe. Productivity of the first is related to massive carbonate reservoirs of great capacity (Murris, 1980), whereas productive zones in the second are thin clastic beds (Paraschiv and Olteanu, 1970; Burshtar and Lvov, 1979). The first has a perfect evaporite seal, the second does not. The first contains excellent source rocks; the quality of those in the second is poor. In spite of tectonic similarity, one basin can not be considered as an analog for the other.

The North Sea basin and the West Siberian basin in the Soviet Union are another demonstrative example of two basins with tectonic similarities, yet with striking differences in the major features of productivity. In all tectonic classifications, both basins fit perfectly into the same type: a rift structure overlain by a sedimentary sag (Klemme, 1977, 1980; Bally and Snelson, 1980; Sokolov, 1980; Kingston et al., 1983b). Productivity of the North Sea basin, however, is strongly controlled by the rift system itself, (Ziegler, 1980), whereas in the West Siberian basin oil production is on large, gentle arches in the sag sequence (Kontorovich et al., 1975). Many of these arches are situated hundreds of miles from the major rift system and are not directly controlled by the rift structure. Major oil fields have been found in this basin in Neocomian deltaic deposits, whereas the Triassic rift sequence contains no oil fields. Thus again, for the assessment of undiscovered resources, one basin can in no way be considered as an analog for the other.

It would be logical to assume that if the tectonic type of a basin controls its richness, different basin types would significantly differ from each other in the level of their productivity. But experience shows that tectonic basin classifications are limited in their effectiveness of rating the richness of basins. In fact, ranges of volumetric yields for different basin types are barely distinguishable from one another and widely overlap. This can be seen, for example, in the works of Klemme (1980) and Bally and Snelson (1980). Most authors of classification charts do not try to assess quantitatively the richness of various types of basins and correctly imply that the major difference between these basin types results from the different distribution of plays (Kingston et al. 1983a; Sokolov, 1980).

The essentially tectonic concept of a basin does not include stratigraphic information adequate for resource assessment. As a result, attempts have been made to subdivide basins vertically into smaller assessment units such as megasequences (Porter and McCrossan, 1975), or cycles (Kingston et al., 1983b). All these units, however, reflect tectonic phases, or stages, in the development of a basin and thus have much more tectonic than stratigraphic information. Such units are convenient for tectonic analysis, but are often unsuitable for resource assessment. This is well demonstrated in an article by Kingston et al. (1983b), which is probably the most exhaustive analysis ever published of the dynamics of basin development based on the plate tectonic concept. In this article a cycle is considered to be identical to a basin and is the sedimentary reflection of a significant tectonic event confined by regional unconformities at the bottom and top. However, these unconformities separating tectonic cycles unite, rather than separate, productive sequences in the regressive top of the lower cycle and transgressive bottom of the upper one (Figure 1). Such unconformities are common migration paths that supply oil and gas to any suitable trap both upward and downward. Separate analysis of adjoining cycles artificially separates the resource assessment from the genetic aspects of petroleum productivity. On the contrary, the concept of an IPS unites permeable parts of adjoining cycles (despite probable structural differences between them) into one natural reservoir complex underlain and overlain by impermeable seals. An unconformity at the base of the Miocene Lagunillas Formation in the Maracaibo basin, which controls the giant oil accumulations here, is an actual example of relationships that are shown schematically in Figure 1.

A detailed analysis of tectonic–sedimentary cycles separated by unconformities (facies cycle wedges) performed by White (1980) concentrated on the spatial relationship between oil and gas pools, source rocks, and reservoir rocks (potential plays) inside a cycle. In his study, major petroleum resources appeared to be located in the lower and middle (transgressive) parts of a typical facies cycle wedge. Oil and gas reserves in the upper regressive part of the wedge are much smaller, although this part is thicker and contains the best reservoir rocks. In only a few cases did the anhydrite seal provide substantial pools at the top of the cycle. The result of this study may seem surprising, because upward migration from source rocks in the middle part of the cycle is usually more facile than migration downward. Evidently, most of the oil and gas that was generated in the organic-rich middle part of the cycle and that migrated upward had crossed the top unconformity. This oil and gas could accumulate in basal course facies of the overlying cycle under the regional seal of the cycle's middle portion. Thus, White (1980) clearly demonstrated the difference between a cycle and IPS. The independent resource assessments of these two cycles would separate the source for oil and gas from potentially

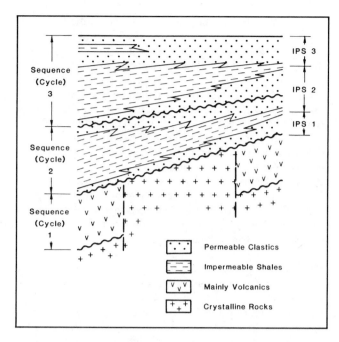

Figure 1. Interrelationship between sequences, or cycles, and IPSs in a sedimentary section. Major structural unconformities separate sequences, whereas IPSs are separated by regional impermeable seals.

productive reservoirs. Using the concept of an IPS, on the contrary, one would consider the regressive part of the lower cycle and the transgressive part of the upper cycle as a permeable part of one IPS overlain and underlain by impermeable seals.

The advantages of an IPS as a major assessment unit can be clearly seen from the above discussions. Compared with a play, an IPS is a regional unit suitable for the application of a petroleum genetic analysis. Furthermore, the assessment of an IPS in a poorly known region can avoid uncertain (considering the restricted data base) reconstructions such as timing, paths or secondary migration, and the interplay between oil and gas accumulation processes, whereas these uncertain reconstructions are necessary for assessment of a play.

If source rock studies show that petroleum was generated, it can be expected that this petroleum is preserved somewhere inside the isolated system of the IPS, although not all of the plays would receive hydrocarbons. Conditions of this preservation are generally reflected by the quality of the regional seal, the availability of traps, and the maximum formational paleotemperature. That is why directions of secondary migration inside an IPS in a poorly known region are not considered a major factor that can significantly influence petroleum resource assessment. If sealed reservoirs, rich source rocks, and traps are present in the IPS, oil will reach at least part of these reservoirs regardless of migration paths. Even if conventional reservoir rocks are insufficient in the IPS, they can be created in rich source rocks, such as bituminous cherty shales, by the fracturing that accompanies oil and gas generation. The Jurassic Bazhenov Formation of

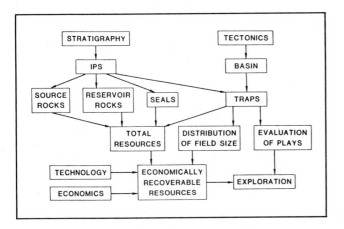

Figure 2. Interplay between stratigraphic and tectonic approaches in petroleum resource assessment and development of exploration strategy.

West Siberia (Skorobogatov and Krasnov, 1984) and, evidently, the Miocene Monterey Formation of California (Isaacs, 1984) are good examples of this process.

Compared with a basin, the IPS is a smaller unit with a higher uniformity of stratigraphic factors controlling richness: source rocks, reservoir rocks, seals, and often traps. Choosing an analog for the IPS is easier, the analogy itself is more comprehensive, and the quality of comparative analysis is higher.

The concept of an IPS as a major assessment unit does not reject the importance of both basin study and play analysis in the process of evaluating of a frontier region. The advantages of basin analysis lie not in the appraisal of a basin's richness but rather in the location of major structural plays and evaluation of field size distribution. The latter is important for the assessment of economic factors of exploration, especially in remote regions. The relationship of field size distribution to basin types is due to the major predominance of structural traps among giant oil and gas fields (Klemme, 1973, 1974). The prediction of such traps is a matter of comparative tectonic analysis. Play analysis in poorly known regions becomes a useful approach to the evaluation of expected exploration success when more detailed data are available for that part of the region.

Figure 2 is an attempt to show in simplified form the interplay between stratigraphy and tectonics in the procedure of resource appraisal. The assessment of total resources in a basin is, to a great extent, a result of the stratigraphic approach through comparative analysis of the major factors of productivity, source rocks, reservoir rocks, seals, and traps in each of the basin's IPSs. The tectonic approach using comparative basin analysis within a well-developed classification scheme gives expected field size distribution on the basis of an IPS's richness and locates major structural plays. The field size distribution greatly contributes to the appraisal of economically recoverable resources. The evaluation of plays usually becomes possible during late stages of exploration and greatly enhances the exploration success.

STRATIGRAPHIC APPROACH TO STATISTICAL RESOURCE ANALYSIS

The stratigraphic approach can also be successfully applied to the statistical method for resource appraisal and, in some cases, can significantly increase its reliability. This method gives good results in maturely explored regions when the efficiency of exploration begins to decline. It is well known that plotting values of reserve additions against the number of wells drilled (or footage drilled) gives a broken line that can be approximated by a curve. Significant deviations upward from the curve of segments of this broken line commonly reflect discoveries followed by successful exploration of new plays or stratigraphic complexes. Petroleum resources of yet unexplored complexes can be underestimated if they are not analyzed separately.

A simple example of this situation is provided by a rift structure covered by a sag, shown in Figure 3A. Production from the upper sag complex, or IPS, was established at an early stage of exploration, but discoveries from the rift itself came more recently. Analysis of the rate of reserve growth for the basin as a whole shows decreasing efficiency of exploration (Figure 3B). Only small amounts of oil are yet to be found. Dividing resource growth and drilling by depth gives approximately the same result (Figure 3C). This is understandable because exploration at greater depths was directed mainly at continuation of known plays in the same upper sag complex where major resources had already been found. Separate statistical analysis of each complex (or IPS), however, reveals different trends for exploration efficiency (Figure 3D). Assessment of undiscovered resources in the lower complex in this case can not be done by the statistical method; the lower complex is instead a subject for comparative geologic analysis.

CONCLUSIONS

The comparative geologic analysis is the only possible approach to the assessment of undiscovered oil and gas resources in poorly known regions, because data for the application of other methods in these regions are always insufficient. The comparative geologic analysis, however, will be successful only if (1) the criteria of comparison are adequate and (2) the units being compared are sufficiently homogeneous and similar to one another with respect to criteria of comparison. A basically stratigraphic unit, an independent petroliferous system (IPS), has certain advantages for comparative analysis of the main factors controlling generation, accumulation, and preservation of oil and gas.

The three units—an IPS, a basin, and a play—do not exclude but rather supplement each other in a study of a frontier region. A comparative analysis of IPSs is most suitable for the assessment of petroleum richness. A comparative basin study can provide the best conceptual model concerning the expected field size distribution. A play analysis is a useful exploration tool when more data become available.

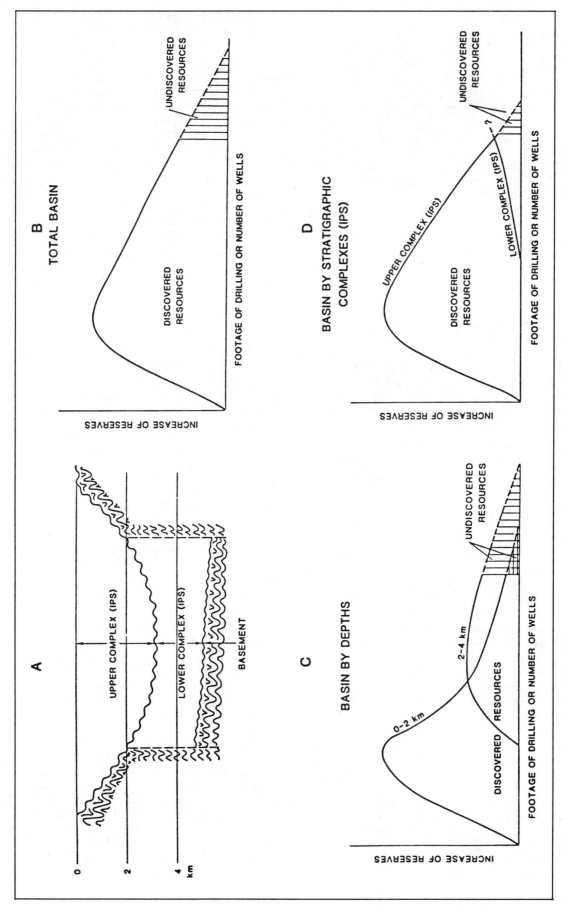

Figure 3. Petroleum resource assessment by the statistical method: general structural style of a basin (A), and statistical curves for the whole sedimentary cover (B), for the sedimentary cover divided into depth intervals (C), and for the sedimentary cover divided into stratigraphic complexes, or IPSs (D).

Assessment of undiscovered oil and gas resources is a difficult task. Satisfactory results can be obtained only on the basis of all our knowledge and by using all suitable methods. Each of these methods has its own advantages as well as disadvantages. Tectonic analysis is only one of these methods, albeit a very important one. Undoubtedly, tectonics controls subsidence and sedimentation, but in particular cases, the same tectonic regime may result in a great variety of lithologic, geochemical, hydrogeologic, and other characteristics. These characteristics are equally, if not more, important to the assessment of undiscovered oil and gas resources than are the tectonic features of a region. The concept of an IPS may become a useful approach to the sequential analysis of stratigraphic factors controlling oil and gas productivity.

ACKNOWLEDGMENTS

The work was sponsored by the U.S. Geological Survey, World Energy Resources Program. The author is grateful to C. D. Masters and A. B. Coury for their valuable comments and editorial assistance. The author also thanks his colleagues in the World Energy Resources Program for helpful discussions of the ideas contained in this article.

REFERENCES CITED

Aliyev, I. M., et al., 1983, Petroliferous provinces of the USSR [Neftegazonosnye provintsii SSSR], G. K. Dikenshtein et al., eds.: Nedra, Moscow, 271 p.

Baker, R. A., H. M. Gehman, W. R. James, and D. A. White, 1984, Geologic field number and size assessments of oil and gas plays: AAPG Bulletin, v. 68, p. 426-432.

Bally, A. W., and S. Snelson, 1980, Realms of subsidence, in A. D. Miall, ed., Facts and principles of world petroleum occurrence: Canadian Society of Petroleum Geologists Memoir 6, p. 9–75.

Bishop, R. S., H. M. Gehman, and A. Young, 1983, Concepts for estimating hydrocarbon accumulation and dispersion: AAPG Bulletin, v. 67, p. 337–348.

Bois, C., 1975, Petroleum-zone concept and the similarity analysis—contribution to resource appraisal, in J. D. Haun, ed., Methods of estimating the volume of undiscovered oil and gas resources: AAPG Studies in Geology Series, n. 1, p. 87–89.

Burshtar, M. S., and M. S. Lvov, 1979, Geography and geology of oil and gas of the USSR and foreign countries [Geografiya i geologiya nefti i gaza SSSR i zarubezhnykh stran]: Nedra, Moscow, 364 p.

Dikenshtein, G. K., S. P. Maksimov, and T. D. Ivanova, 1982, Tectonics of petroliferous provinces and regions of the USSR [Tektonika neftegazonosnykh provintsiy i oblastey SSSR]: Nedra, Moscow, 222 p.

Fan Pu et al., 1980, Formation and migration of continental oil and gas in China: Scientia Sinica, v. 23, p. 1288–1295 and 1417–1427.

Halbouty, M. T., A. A. Meyerhoff, R. E. King, R. H. Dott, Sr., H. D. Klemme, and T. Shabad, 1970, World's giant oil and gas fields, geologic factors affecting their formation, and basin classification, in M. T. Halbouty, ed., Geology of giant petroleum fields: AAPG Memoir 14, p. 502–555.

Isaacs, C. M., 1984, The Monterey—key to offshore California boom: Oil and Gas Journal, Jan. 9, p. 75–81.

Kent, P. E., 1980, The North Sea—evolution of a major oil and gas play, in A. P. Maill, ed., Facts and principles of world petroleum occurrence: Canadian Society of Petroleum Geologists Memoir 6, p. 633–652.

Kingston, D. R., C. P. Dishroon, and P. Williams, 1983a, Hydrocarbon plays and global basin classification: AAPG Bulletin, v. 67, p. 2194–2198.

——— , 1983b, Global basin classification system: AAPG Bulletin, v. 67, p. 2175–2193.

Klemme, H. D., 1973 and 1974, Structure-related traps expected to dominate world-reserve statistics: Oil and Gas Journal, Dec. 31, p. 112–118, and Jan. 7, p. 97–103.

——— , 1977, World oil and gas reserves from analysis of giant fields and petroleum basins (province), in R. F. Meyer, ed., The future supply of nature-made petroleum and gas: New York, Pergamon Press, p. 217–260.

——— , 1980, Petroleum basins—classifications and characteristics: Journal of Petroleum Geology, v. 3, p. 183–207.

Kontorovich, A. E., et al., 1975, Geology of oil and gas of West Siberia [Geologiya nefti i gaza Zapadnoy Sibiri]: Nedra, Moscow, 679 p.

Mast, R. F., R. H. McMullin, K. J. Bird, and W. P. Brosge, 1980, Resource appraisal of undiscovered oil and gas resources of the William O. Douglas Arctic Wildlife Range: U.S. Geological Survey Open-File Report 80-916, 34 p.

Miller, B. M., 1982, Application of exploration play-analysis techniques to the assessment of conventional petroleum resources by the USGS: Journal of Petroleum Technology, v. 34, p. 55–64.

Murris, R. J., 1980, Hydrocarbon habitat of the Middle East, in A. D. Maill, ed., Facts and principles of world petroleum occurrence: Canadian Society of Petroleum Geologists, Memoir 6, p. 765–800.

Muslimov, R. K., 1979, Influence of geologic framework on the effectivity of exploitation of the Romashkino field [Vliyaniye osobennostey geologicheskogo stroyeniya na effektivnost razrabotki Romashkinskogo mestorozhdeniya]: Kazan, USSR, Kazan University, 211 p.

Paraschiv, D., and G. Olteanu, 1970, Oil fields in Mio-Pliocene zone of eastern Carpathian (District of Ploiesti), in M. T. Halbouty, ed., Geology of giant petroleum fields: AAPG Memoir 14, p. 399–427.

Porter, J. W., and R. G. McCrossan, 1975, Basin consanguinity in petroleum resource estimation, in J. D. Haun, ed., Methods of estimating the volume of undiscovered oil and gas resources: AAPG Studies in Geology Series, n. 1, p. 50–75.

Procter, R. M., P. J. Lee, and G. C. Taylor, 1982, Methodology of petroleum resource evaluation: Geological Survey of Canada, 52 p.

Rudkevich, M. Y., 1974, Paleotectonic criteria of oil and gas productivity [Paleotektonicheskiye kriterii neftegazonosnosti]: Nedra, Moscow, 183 p.

Semenovich, V. V., et al., 1977, Methods used in the U.S.S.R. for estimating potential petroleum resources, in R. F. Meyer, ed., The future supply of nature-made petroleum and gas: New York, Pergamon Press, p. 139–153.

Skorobogatov, V. A., and S. G. Krasnov, 1984, Some criteria of oil potential of the Bazhenov Formation of West Siberia: Geologiya nefti i gaza, n. 3, p. 15–19.

Sokolov, B. A., 1980, Evolution and productivity of sedimentary basins [Evolutsiya i neftegazonosnost osadochnykh basseynov]: Nauka, Moscow, 243 p.

Tang Zhi, 1982, Tectonic features of oil and gas basins in eastern part of China: AAPG Bulletin, v. 66, p. 509–521.

Ulmishek, G., 1982, Petroleum geology and resource assessment of the Timan–Pechora basin, USSR, and the adjacent Barents–Northern Kara shelf: Argonne National Laboratory Report ANL/EES-TM-199, 195 p.

Weeks, L. G., 1949, Highlights on 1948 development in foreign petroleum fields: AAPG Bulletin, v. 33, p. 1029–1124.

——— , 1975, Potential petroleum resources—classification, estimation, and status, in J. D. Haun, ed., Methods of estimating the volume of undiscovered oil and gas resources: AAPG Studies in Geology Series, n. 1, p. 31-49.

White, D. A., 1980, Assessing oil and gas plays in facies-cycles wedges: AAPG Bulletin, v. 64, p. 1158–1178.

Ziegler, P. A., 1980, Northwest European basin: geology and hydrocarbon provinces, in A. D. Miall, ed., Facts and principles of world petroleum occurrence: Canadian Society of Petroleum Geologists Memoir 6, p. 653–706.

U.S. Geological Survey Quantitative Petroleum Resource Appraisal Methodologies

Robert A. Crovelli
U.S. Geological Survey
Denver, Colorado

A number of methodologies are described that provide estimates of volumes of undiscovered hydrocarbons. These include the direct assessment of undiscovered volumes of oil and gas and statistical methodologies that extrapolate the expected volumes to be found by future drilling from past performances. The latter include finding-rate studies expressed as decline curves and as cumulative resource curves, both of which provide a measure of undiscovered volumes to be found with additional drilling. Another technique described is the application of probability theory to the aggregate of a number of probability distributions under different dependency assumptions and the comparison of these results to those obtained by the use of Monte Carlo aggregation techniques. In the assessment of undiscovered amounts of oil and gas, where the estimates are based on little direct knowledge, the probability theory aggregation methodologies look very promising. The assessments of the United States, of the entire world, and of wilderness lands are discussed.

INTRODUCTION

A variety of statistical and probabilistic methodologies are used by the U.S. Geological Survey (USGS) to make oil and gas resource assessments. The purpose of this paper is to explain some of the basic quantitative methodologies that we have used to make basin assessments and to aggregate basin probability distributions to generate probabilistic estimates of a collection of basins. We apply a statistical methodology in the extrapolation of historical trends and use a probabilistic methodology in direct assessment methods. We discuss comparative aggregation methodologies using an example from the last national assessment. Several computer graphics routines are illustrated with examples.

EXTRAPOLATION OF HISTORICAL TRENDS

Finding-rate studies use statistical procedures to predict future discoveries by extrapolation of past performances. The most commonly used historical statistics are the discovered volume of hydrocarbons related to the exploratory footage drilled, or to the number of exploratory wells, or to time. Hubbert (1967), Moore (1966), and Davis (1958) used techniques of this category.

Decline Curve Models

Finding-rate studies provide an analytical tool when adequate drilling and discovery information is available. These studies are often used to describe the relationship between discovered amounts of hydrocarbons and exploratory footage drill. Curves fitted to these historical data allow for extrapolation from which undiscovered resources can be determined, assuming a continuation of existing exploration trends and successes.

One of the basic methods of estimating the future discovery rate for oil and gas fields in a maturely explored province is the use of a decline curve. There is a long history of the use of these methods that permit the extrapolation of the trend of the past into the future, and they are useful as long as no major unforeseen or unexpected event disrupts this general trend.

A good model should have a reasonably high index of determination. The index of determination (a measure of how well a model fits past data) cannot be the only criterion on which models are compared if the purpose of a model is to describe the future behavior of a process. A model could have a high index of determination but be inappropriate for modeling the future if it is inconsistent with fundamental characteristics of the process. More importantly, a good prediction model needs to be consistent with the properties

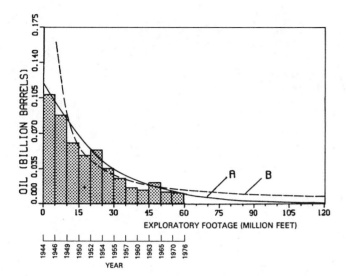

Figure 1. Decline curves consisting of exponential curve A and hyperbolic curve B for volumes of oil found plotted against exploratory footage. Data are for the Illinois basin. From Dolton et al. (1981).

that are known about the future behavior of the process, as in the case of decline curves that must approach zero. Therefore, it is possible for a model with a relatively low index of determination to be a much better model for describing the future if it is more consistent with what is known about the future behavior of the process.

We considered a number of statistical models for describing finding rates. Regression analysis was used to estimate those parameters of a model to be determined by the historical data. The exponential and hyperbolic decline curves fit by regression analysis to historical data show the best promise for finding-rate projection capability.

The model of the exponential decline curve is

$$y = ae^{bx}$$

and the model of the hyperbolic decline curve is

$$y = \frac{a}{x^b}$$

where a and b, in both cases, are estimated by the historical data using regression analysis. The variable y represents the volume of hydrocarbons, and x is the exploratory effort measured in footage drilled, number of wells, or time.

An example of a finding-rate study showing extrapolation of both exponential and hyperbolic decline curves based on exploratory footage drilled is illustrated in Figure 1. The historic data is for the Illinois basin from 1944 to 1976. The estimated amount of undiscovered recoverable oil to be found in the next 60 million ft of exploratory drilling is 0.038 billion bbl for the exponential curve, and 0.115 billion bbl for the hyperbolic curve. This method was used by the USGS in recent petroleum assessment of the United States (Dolton et al., 1981).

A second example of a finding-rate study using extrapolation of exponential and hyperbolic decline curves based on exploratory wells or drill holes is illustrated in

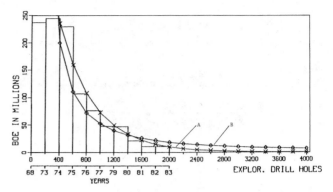

Figure 2. Decline curves consisting of exponential curve A and hyperbolic curve B for bbl of oil equivalent found plotted against number of exploratory drill holes. Data are for the northern Michigan Niagaran pinnacle reefs.

Figure 2. The historical data (collected by Dan Gill) for the northern Michigan Niagaran pinnacle reefs is from 1968 to 1983. The estimated amount of undiscovered recoverable bbl of oil equivalent (BOE) to be found in the next 2000 exploratory drill holes is 21.8 million BOE for the exponential curve and 106.8 million BOE for the hyperbolic curve. The historical data and additional resources discovered with each future drilling increment of 200 wells are shown in Table 1.

Cumulative Curve Model

Another type of curve that is used for the extrapolation of historical trends is the cumulative resource curve, which gives estimates of the remaining resources consistent with those of the decline curve. The cumulative resource curve uses the same data as the decline curve model but relates the cumulative discovered volume of hydrocarbons to the exploratory effort, which may be expressed as footage drilled, number of wells, or time. The extrapolation of estimated cumulative resources based on an exponential decline curve is discussed below.

The undiscovered cumulative resource (Z) of m future increments is given by the formula

$$Z = \frac{ae^{bd(n+1)} - ae^{bd(n+m+1)}}{1 - e^{bd}}$$

where d is the increment size and n is the number of past increments.

The total cumulative resource (W) of $n + m$ increments is given by

$$W = H + Z$$

where H is the historical cumulative resource of n increments.

The ultimate undiscovered cumulative resource (U) is given by the formula

$$U = \frac{ae^{bd(n+1)}}{1 - e^{bd}}$$

The ultimate total cumulative resource (V) is given by

$$V = H + U$$

Table 1. Extrapolation of historical trends for the northern Michigan Niagaran pinnacle reefs trend showing number of barrels of oil equivalent (BOE) found or to be found by the number of wells drilled or to be drilled, based on an exponential decline curve.[a]

Number of Wells Drilled Past Effort	Million BOE		Number of Wells Drilled Future Effort	Million BOE		
	Discovered Resource	Discovered Cumulative		Undiscovered Resource	Undiscovered Cumulative	Total Cumulative
200	237	237	2200	7.168	7.168	1040.2
400	244	481	2400	4.864	12.033	1045.0
600	229	710	2600	3.301	15.334	1048.3
800	107	817	2800	2.240	17.574	1050.6
1000	77	894	3000	1.520	19.093	1052.1
1200	48	942	3200	1.031	20.125	1053.1
1400	47	989	3400	0.700	20.825	1053.8
1600	22	1011	3600	0.475	21.300	1054.3
1800	11	1022	3800	0.322	21.622	1054.6
2000	11	1033	4000	0.219	21.841	1054.8

[a]Historical cumulative = 1033.00 million BOE; ultimate undiscovered cumulative = 22.302 million BOE; ultimate total cumulative = 1055.3 million BOE.

The cumulative resource curve for the northern Michigan Niagaran pinnacle reefs is displayed in Figure 3. The data are presented in Table 1.

DIRECT ASSESSMENT METHODOLOGY

A probabilistic methodology used by the USGS for direct assessment is described in detail in Crovelli (1984b). The probabilistic methodology for estimation of undiscovered, recoverable oil and gas resources in a basin or other assessment area is briefly discussed as follows.

Geologic and statistical methodology is used to assess the quantity of undiscovered petroleum potential in a basin; it consists of the same techniques described by Dolton et al. (1981) for domestic petroleum assessment. The first stage is the information or data gathering stage; the types of information and data compiled relate to the following: (1) petroleum geology, (2) exploration history, (3) seismic exploration, (4) production and reserves statistics, and (5) analog basins or formations. The second stage involves the analysis of the data by one or more of the following methods: (1) extrapolation of historical trends (finding rates), (2) areal or volumetric yield analysis, and (3) geochemical material balance analysis. The third stage is direct subjective estimation of a petroleum resource in an assessment area. A team of geologists uses a Delphi type of approach, which involves the following steps: (1) presentation and discussion of the geologic and geophysical data and interpretation; (2) presentation and discussion of the data analyses made in stage 2; (3) probabilistic assessments made independently by each team member; (4) reviews of all estimates by team and revision of estimates by individuals if desired; (5) possible iteration of steps 1–4; and (6) averaging of individual resource estimates.

The resource assessment procedure results in probabilistic estimation of two uncertain events: (1) the presence of assessed hydrocarbon, which results in an estimate of the probability that the resource is present; and (2) if present, its quantity, which involves conditional probability and thus results in conditional resource estimates. More explicitly, the probabilistic estimates are as follows:

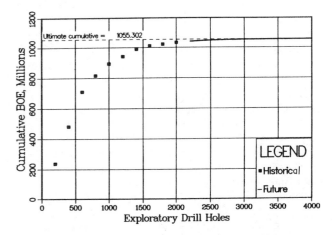

Figure 3. Cumulative resource curve for the northern Michigan Niagaran pinnacle reefs. Ultimate resource to be found, including that already discovered, 1055.302 million bbl of recoverable oil equivalent.

1. An estimate of the probability of occurrence of the particular hydrocarbon called the marginal probability, that is, a subjective probability of the condition that the resource is actually present in conventionally recoverable quantities.
2. Conditional on that specific resource being present, three estimates of its probable quantity are
   a. a low resource estimate, the 95th fractile, corresponding to a 95% probability of more than that amount;
   b. a high resource estimate, the 5th fractile, corresponding to a 5% probability of more than that amount; and
   c. a modal ("mostly likely") estimate of the quantity of resource associated with the greatest likelihood of occurrence.

The conditional low, high, and modal estimates of undiscovered recoverable resource are used to determine a conditional probability distribution of the quantity of undiscovered recoverable resource for a basin. The

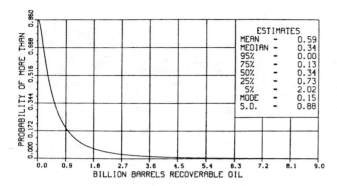

Figure 4. Probability distribution of undiscovered recoverable oil for the Montana overthrust belt province derived from subjective probability estimation procedures. Estimates are mean, median, mode, standard deviation (S.D.), and fractiles that correspond to the percentages listed.

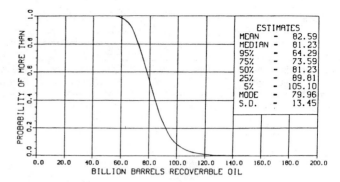

Figure 5. Monte Carlo aggregate probability distribution of undiscovered recoverable total oil for the United States, assuming independent province assessments. Estimates are mean, median, mode, standard deviation (S.D.), and fractiles that correspond to the percentages listed. From Dolton et al. (1981).

conditional probability distribution represents the judgmental probability distribution of the quantity of undiscovered recoverable resource conditional on the resource being present. (This distribution is also referred to as the "unrisked" distribution.) The lognormal distribution is used as a probability model for the conditional probability distribution in a basin. One approach to fitting the lognormal distribution is as follows. Because each pair of values among the three conditional estimates determines a lognormal distribution, there are three possible lognormal distributions. The fitted lognormal distribution with the largest standard deviation is chosen. The marginal probability of the resource for a basin is applied to its corresponding conditional probability distribution to produce the unconditional probability distribution of the quantity of undiscovered recoverable resource. (This distribution is also referred to as the "risked" distribution.) The unconditional mean (μ_u) and variance (σ_u^2) are related to the conditional mean (μ_c) and variance (σ_c^2) through the marginal probability (p) as follows:

$$\mu_u = p\mu_c$$

$$\sigma_u^2 = p\sigma_c^2 + p(1 - p)\mu_c^2$$

The conditional and unconditional probability distributions are described by graphs of their complementary cumulative distribution functions that give the probability of more than a specific amount of resource. As an example, the unconditional probability distribution of the quantity of undiscovered recoverable oil for the Montana overthrust belt province is displayed in Figure 4.

COMPARATIVE AGGREGATION METHODOLOGIES

Aggregation techniques are applied to the unconditional probability distributions for a region composed of several basins to generate probabilistic estimates of the total undiscovered recoverable resource in the region.

A Monte Carlo aggregation computer program was used by Dolton et al. (1981) to generate the approximate probability distribution of the quantity of undiscovered recoverable total oil for the United States (Figure 5). The United States was divided into 137 provinces that were aggregated; the Montana overthrust belt province in Figure 4 was one of those provinces. The number of Monte Carlo samplings was 5000. For each individual sampling, a random value was selected from each of the 137 oil probability distributions, and the 137 random values were then added to produce a random value for total oil; this procedure was repeated 5000 times, taking considerable computer time.

Since the national petroleum assessment was completed by the USGS in 1981, some alternative aggregation methodologies have been investigated that are mathematically tractable and inexpensive and that utilize a minimum of computer programming and running time. The Monte Carlo simulation technique that had been used did not meet all of these conditions, especially when the total gas for the United States was aggregated. In this case 274 probability distributions were aggregated because there were two components (nonassociated gas and associated/dissolved gas) per province.

One approach was simply to approximate an aggregate distribution with a lognormal distribution. Since the means and variances of the province probability distributions were known, they were added, respectively, to obtain the aggregate mean and variance under the assumption of independence. A lognormal distribution was fitted to these first two central moments. The lognormal aggregate probability distribution of undiscovered recoverable total oil for the 137 provinces of the United States is displayed in Figure 6. The lognormal approximation approach compares quite favorably with the Monte Carlo technique. Regarding the fractiles, the 95% estimate was slightly lower and the 5% slightly higher for the approximating lognormal, but the differences are too small to be of importance, particularly when we consider that we are aggregating estimates of undiscovered amounts.

As another alternative, and for purposes of experiment and study, the total oil for the United States was again aggregated

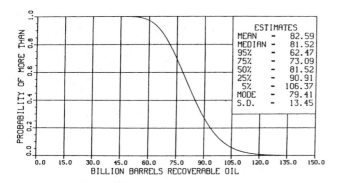

Figure 6. Lognormal aggregate probability distribution of undiscovered recoverable total oil for the United States, assuming independent province assessments. Estimates are mean, median, mode, standard deviation (S.D.), and fractiles that correspond to the percentages listed.

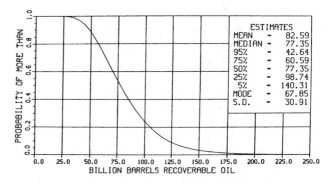

Figure 7. Lognormal aggregate probability distribution of undiscovered recoverable total oil for the United States, assuming dependent province assessments within regions, but independent regions. Estimates are mean, median, mode, standard deviation (S.D.), and fractiles that correspond to the percentages listed.

by fitting a lognormal curve, but this time under a dependence assumption. Dependence is defined in this study in terms of how sets of the first two central moments are combined. In Dolton et al. (1981) the 137 provinces of the United States were partitioned into 15 regions. In the present study, to get a partial dependency, we made the province assessments within regions dependent, while the 15 regions are treated independently. Under the dependence assumption within regions, the means and standard deviations of the provinces were added, respectively. Therefore, dependence here means that the data are approximately perfectly positively correlated. Then, under the independence assumption among regions, the means and variances of the regional distributions so derived were added, respectively. Since the dependency within regions is overstated, and the dependency among regions is understated, these assumptions introduced compensatory effects. A lognormal distribution was fitted to the final aggregate mean and variance. The lognormal aggregate probability distribution of undiscovered recoverable total oil for the United States, assuming dependent province assessments within independent regions, is shown in Figure 7. The mean estimate remained the same as in the independent case (Figure 6), but the width of the interval estimate formed using the 95th and 5th fractiles increased with the partial dependency. The dependent aggregation methodology used in this study is similar to that of Ford and McLaren (1980), whose techniques were employed by the Department of Energy in its regional and national assessment of uranium resources. Their method is ideal for aggregating a large number of probability distributions with dependency.

WORLD ASSESSMENT

The USGS assessed the sedimentary basins of the world for undiscovered, conventionally recoverable, crude oil resources. Probabilistic methods were applied to each of the basin assessments to produce estimates in the form of unconditional probability distributions. The methodology is described in detail in Crovelli (1984c). The basin probability distributions were then aggregated to produce resource

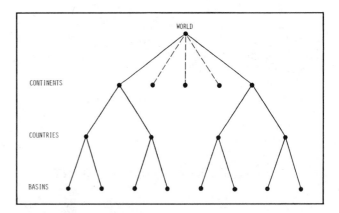

Figure 8. Generalized tree diagram for aggregations of undiscovered recoverable oil resources in the world.

estimates by region, by country, by continental area, and for the entire world. The aggregation methodology is described in detail in Crovelli (1985).

The aggregations were approximated by three-parameter lognormal distributions that were produced by combining the first three central moments of the basin distributions assuming dependency among the basin assessments. An assumption was made that the basin assessments were dependent because the same team of geologists assessed all of the basins. Dependence is defined in this study in terms of how sets of the first three central moments are combined. Under the dependence assumption, each of the following functions of the moments are additive: mean, standard deviation, and cube root of the third moment. Therefore, dependence here means that the data are approximately perfectly positively correlated. A three-parameter lognormal distribution was fitted to the three moments for each aggregation in the study as an approximate aggregate distribution. Each aggregate distribution of crude oil potential was described graphically by a complementary cumulative distribution function.

A large tree diagram was used to organize the basins and the required aggregations. A simplification of this diagram is given in Figure 8. The lowest level of the tree diagram was the basin level. Many basins were grouped into regions of countries; China, for example, was divided into offshore and onshore (onshore had five branching regions, one of which had another level of two basins). Hence, the second level up the tree was the region, and the third level was the country. Since the countries were grouped into continental areas, the fourth level was the continental area. The top of the tree was the world level. The aggregations began at the base of the tree diagram, aggregating all of the basins in each region, each country, and each continental area; finally, the continental areas were aggregated to generate probabilistic estimates for the entire world. The world was divided into eight continental areas: North America, South America, Europe (excluding Soviet Union), Soviet Union, Africa, Middle East, Asia (excluding Soviet Union)/Australia/Oceania, and Antarctica.

Members of the USGS reviewed the probability distributions generated from the computer aggregations, and as a result, a few of the probabilistic estimates were slightly modified. Estimates of undiscovered, conventionally recoverable oil resources in the form of the 95th and 5th fractiles and the modal value were reported by Masters et al. (1983) by country, by continental area, and for the entire world; a summary by continental area and for the entire world is given in Table 2. The interval estimate for the entire world yields a 90% probability that the value lies between 321 and 1417 billion bbl of oil, with a most likely (modal) value being 550 billion bbl of oil.

WILDERNESS LANDS ASSESSMENT

Probabilistic methodology was developed for a quantitative assessment of the petroleum potential of wilderness lands in the Western United States; it is described in detail in Crovelli (1983). The objective was to estimate, in terms of probability distributions, the quantity of undiscovered recoverable conventional crude oil and natural gas on wilderness lands in each of 11 individual western states and the total oil and gas resources in those States: Arizona, California, Colorado, Idaho, Montana, Nevada, New Mexico, Oregon, Utah, Washington, and Wyoming.

We decided for this study to utilize petroleum resource estimates derived from the USGS Circular 860 study (Dolton et al., 1981; Varnes et al., 1982). These estimates were made at the geologic province level and were denoted in the wilderness lands report as follows: P(B) is the marginal probability for the province, F_{95} is the conditional 95th fractile, and F_5 is the conditional 5th fractile.

The approach was to determine what fraction of the province resources is in the wilderness lands. All of the wilderness lands were partitioned into 183 "clusters" of wilderness tracts that occur within geologically similar areas. The area fraction f_a of each cluster was calculated by dividing the cluster area by the province area. Each cluster was rated as to its relative potential compared with the average quantity of resource per unit area in the province. The rating consisted of a range of values from minimum, R_{min}, to maximum, R_{max}. Each cluster was assigned a conditional marginal probability P(A|B), which is a subjective probability of the resource being present in a cluster, given

Table 2. Estimates by continental areas of undiscovered recoverable resources of conventional crude oil (billion bbl).

	Undiscovered Recoverable Resources Probability Range		
	F_{95}[a]	Mode	F_5
North America	104	163	322
South America	20	33	69
Europe (excluding Soviet Union)	13	20	49
Soviet Union	59	107	343
Africa	28	46	105
Middle East	72	125	337
Asia (excluding Soviet Union)/ Australia/Oceania	33	58	176
Antarctica	0	0	19
World	321	550	1417

[a]F_{95} denotes 95th factile; the probability of more than the amount F_{95} is 95%. F_5 is defined similarly.

that the resource is present in the province. An example of the data format for oil that uses only two wilderness clusters in Colorado is shown in Table 3. The province number is the assigned Resource Appraisal Group (RAG) code number of that geologic province. The ratings approach allowed the geologists to take into account the facts that (1) petroleum resources were not evenly distributed throughout geologic provinces; (2) there had been little or no production of oil and gas in the wilderness lands; (3) some states containing wilderness lands had been explored heavily, while others had not; and (4) past exploration and production results in the 11 states provided valuable information to an assessment of the wilderness lands.

The information from the data forms was used as input for a computer program, which was designed on the basis of probability theory. For each cluster, the program produced (1) the marginal probability, (2) the conditional probability distribution, (3) the unconditional probability distribution, and (4) the first three unconditional central moments.

The estimates of undiscovered recoverable petroleum resources in all the wilderness clusters in a given state were aggregated. An assumption was made that the cluster estimates in a state were dependent. This assumption was made because of the known geologic dependency of the clusters and because the same geologist(s) assessed all of the clusters in a state. Under this assumption, the sets of three moments for the clusters in a state were combined, respectively, to produce the first three moments of a probability distribution for that state. A three-parameter lognormal distribution was fitted to the three moments for the state as an approximate aggregate distribution.

The estimates of undiscovered recoverable petroleum resources of the wilderness lands in the 11 western states were aggregated, one aggregation for total oil and one for total gas. An assumption was made that the 11 distributions within an aggregation were independent. This assumption was made because of the wide geographic separation of wilderness clusters from state to state, and because most of the states were assessed by different geologists. Since the

Table 3. Data format for assessing petroleum (oil) resources of wilderness lands in state of Colorado.

| | Wilderness Cluster | | | | Province | | | |
| | | | | | | | Conditional Fractiles (billion bbl) | |
Cluster Number	Conditional Marginal Probability $P(A|B)$	Area Fraction f_a	Ratings or Richness Factors R_{min}	R_{max}	RAG[a] Province Number	Marginal Probability $P(B)$	Low F_{95}	High F_5
1	1.0	0.00124	2	4	85	1.0	0.24	3.22
2	0.8	0.00215	0.5	1	85	1.0	0.24	3.22

[a]RAG = Resource Appraisal Group.

Table 4. Petroleum potential in wilderness lands by state in western United States.

| | Undiscovered Recoverable Oil (million bbl) | | | Undiscovered Recoverable Gas (billion cu ft) | | |
State	Low[a] F_{95}	High F_5	Mean	Low F_{95}	High F_5	Mean
Arizona	0.06	11.24	2.91	0.16	31.37	8.11
California	2.69	30.12	11.60	1.37	13.64	5.37
Colorado	3.20	25.54	9.16	31.11	320.16	128.28
Idaho	77.56	427.38	206.79	850.15	3351.45	1841.25
Montana	1.70	105.98	29.55	623.86	8049.33	3031.33
Nevada	3.03	349.21	92.11	16.28	1423.76	383.35
New Mexico	2.43	20.16	7.59	52.67	273.92	122.13
Oregon	.04	13.19	3.42	5.01	242.25	69.80
Utah	12.45	391.14	120.83	34.66	829.84	270.22
Washington	.03	4.74	1.24	1.35	71.86	20.43
Wyoming	116.49	755.08	348.51	1506.08	7692.67	3848.69
Total	555.29	1490.39	833.71	5536.14	16638.60	9728.96

[a]F_{95} denotes the 95th fractile; the probability of more than the amount F_{95} is 95%. F_5 is defined similarly. Fractile values are not additive.

dependency within a state was overstated and the dependency among states was understated, the respective assumptions introduced compensatory effects and thus the means were not affected. Under the independence assumption, the sets of three moments for the 11 states were combined respectively, to produce the first three moments of the probability distribution for the wilderness lands. A three-parameter lognormal distribution was fitted to the three moments for the wilderness lands as an approximate aggregate distribution. The results of this analysis are given in Table 4. A study of the petroleum potential of designated versus proposed wilderness lands in the western United States can be found in Crovelli (1984a).

CONCLUSIONS

A number of methodologies have been used by the USGS for petroleum resource appraisal. Extrapolation of historical trends involves a statistical methodology that requires historical data from past performance in a maturely explored province. Finding-rate studies using decline curves and cumulative resource curves are such extrapolations, and they produce estimates of volumes of undiscovered hydrocarbons to be found with additional drilling.

The direct subjective assessment methodology is a probabilistic procedure for estimation of undiscovered petroleum resources. This approach is applicable in the assessment of both mature and frontier areas. In the case of mature areas various statistical methodologies are first applied and form part of the basis for the direct assessment methodology.

An analytical approach to the aggregation of distributions can result in savings in cost and time as compared with the Monte Carlo technique, especially when the number of distributions to be aggregated is very large or the aggregation scheme is complex. When evaluating the probability theory aggregation methodology, we should keep in mind that the individual assessment distributions are only rough subjective approximations. Therefore, the lognormal aggregate for the United States is quite adequate compared to the Monte Carlo aggregate. Probability theory aggregation allows various dependency assumptions. In the case of partial dependency there was a significant increase in the standard deviation for the total U.S. oil distribution that may be desirable. The world and wilderness lands assessments illustrate the flexibility of the analytical aggregation technique.

ACKNOWLEDGMENT

The author wishes to thank Richard H. Balay for computer programming.

REFERENCES

Crovelli, R. A., 1983, Probabilistic methodology for petroleum resource appraisal of wilderness lands, *in* B. M. Miller, ed., Petroleum Potential of wilderness lands in western United States: U.S. Geological Survey Circular 902-A-P, p. 1–5.

———, 1984a, Petroleum potential of designated versus proposed wilderness lands in the western United States: U.S. Geological Survey Open-File Report 84-363, 55 p.

———, 1984b, U.S. Geological Survey probabilistic methodology for oil and gas resource appraisal of the United States: Journal of the International Association for Mathematical Geology, v. 16, n. 8, p. 797–808.

———, 1984c, Procedures for petroleum resource assessment used by the U.S. Geological Survey—statistical and probabilistic methodology, *in* C. D. Masters, ed., Petroleum resource assessment: International Union of Geological Sciences, n. 17, p. 24–38.

———, 1985, Comparative study of aggregations under different dependency assumptions for assessment of undiscovered recoverable oil resources in the world: Journal of the International Association for Mathematical Geology, v. 17, n. 4, p. 367–374.

Davis, W., 1958, Future productive capacity and probable reserves of the U.S.: Oil and Gas Journal, February 24, p. 105–119.

Dolton, G. L., K. H. Carlson, R. R. Charpentier, A. B. Coury, R. A. Crovelli, S. E. Frezon, A. S. Klan, J. H. Lister, R. H. McMullin, R. S. Pike, R. B. Powers, E. W. Scott, and K. L. Varnes, 1981, Estimates of undiscovered recoverable conventional resources of oil and gas in the United States: U.S. Geological Survey Circular 860, 87 p.

Ford, C. E., and R. A. McLaren, 1980, Methods for obtaining distributions of uranium occurrence from estimates of geologic features: Oak Ridge, Tennessee, K/CSD-13, Union Carbide, Computer Services Division, 121 p. (1 microfiche).

Hubbert, M. K., 1967, Degree of advancement of petroleum exploration in United States: AAPG Bulletin, v. 51, n. 11, p. 2207–2227.

Masters, C. D., D. H. Root, and W. D. Dietzman, 1983, Distribution and quantitative assessment of world crude-oil reserves and resources: U.S. Geological Survey Open-File Report 83-728, 11 p.

Moore, C. L. 1966, Projections of U.S. petroleum supply to 1980, with annex entitled The Gompertz curve for analyzing and projecting the historic supply patterns of exhaustible natural resources: Washington, D.C., U.S. Department of the Interior, Office of Oil and Gas, 42 p.

Varnes, K. L., R. R. Charpentier, G. L. Dolton, 1982, Conditional estimates and marginal probabilities for undiscovered recoverable oil and gas resources by province—statistical background data for U.S. Geological Survey Circular 860: U.S. Geological Survey Open-File Report 82-666-A, 29 p.

Limitations of Geological Consensus Estimates of Undiscovered Petroleum Resources

L. F. Ivanhoe
Novum Corporation
Santa Barbara, California

The geological consensus is a useful method for estimating undiscovered oil and gas resources because the optimists balance the pessimists. Such group averages, however, are limited by the participants' local expertise. Estimates of undiscovered oil and gas by petroleum geologists and engineers are much more conservative for areas where the experts have been humbled by dry holes and more optimistic wherever they have no direct experience. It is impossible to judge any expert's degree of optimism or pessimism until his estimates are compared with those of a group of his peers. All group appraisals should be plotted on graphs, wherein each individual's estimates are identified by a unique symbol to show the range of opinions. Estimates for virgin basins where no one has first-hand experience will normally be on the optimistic side.

INTRODUCTION

The geological consensus is an accepted method for quickly appraising any area's undiscovered petroleum resources. The procedure averages the estimates made by a committee of experts and then presents the results as a group opinion. Unfortunately, such estimates are limited by the local expertise of the individuals making the appraisals, and the range of opinions that exist between the optimists and pessimists is usually lost in the process. A careful study that identified and preserved all of the experts' considered and independent estimates was conducted in South America. Conclusions from this project are pertinent to resource appraisals everywhere.

COLOMBIAN STUDY

In 1982, a department of the Colombian government contracted with an independent exploration consultant to make a special geological appraisal of the nation's remaining petroleum potential, in order to check previous estimates for planning purposes. The consultant was given a free hand. He

An abbreviated version of this paper was presented at the May, 1983, Annual Meeting of the Pacific Section of the American Association of Petroleum Geologists.

established procedures, conducted all interviews, synthesized the data, and wrote the final report.

The consultant decided to utilize the competent local petroleum expertise of senior geologists and engineers working for various oil companies operating in Colombia to obtain the best possible geological consensus. Twelve experts were recommended as the nation's best for the study by the helpful exploration manager of the national oil company, Ecopetrol. These were all solid professionals with advanced degrees from high-ranking American, European, and Colombian universities, and a combined total of more than 200 years of local experience working for oil companies in Colombia. It would be hard to find a more qualified group of experts anywhere. Their names and companies, however, remain anonymous for this study, because none of them wanted their estimates to be revealed in the local newspapers. Consequently, each expert was interviewed confidentially in his own office by the consultant, and no one other than the consultant knew who the other experts were or what their opinions were, even within the same company. All of them were working on responsible operating jobs and were familiar with the petroleum geology of Colombia, but not with statistical methods for estimating future undiscovered oil or gas. The consultant acted as a roving moderator to ensure that all of the experts used common factors and units, but he tried not to influence any individual's estimates.

77

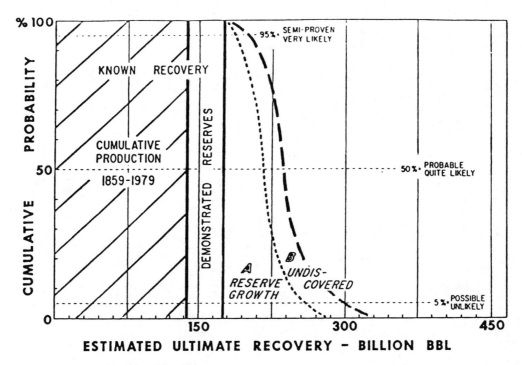

Figure 1. Graph showing probability distribution for estimated ultimate recovery (EUR) of oil and natural gas liquids in the United States. Known recovery includes cumulative production plus demonstrated reserves. About 65% of the estimated oil still to be discovered will be from reserve growth around known fields (A) and 35% from still undiscovered fields (B). (After Nehring, 1981, figure 5-1.)

PROCEDURES

Conventional probability curve procedures were used. A good example is shown for oil and gas in the United States (Nehring, 1981) in Figures 1 and 2[1]. Nehring's graphs were modified to prepare hypothetical graphs for Colombia's oil and gas to show the Colombian experts the final objectives for the project (Figure 3 and 4).

A one-page questionnaire and a table were prepared. This table was designed to show the country's prospective regions divided into 18 basins, sub-basins, or areas, depending on local geological or operating problems. Pertinent, up-to-date statistics, including kilometers of seismic lines, exploratory wells drilled, reserves discovered, and cumulative production, were obtained from Ecopetrol's company reports (Chona and Bendeck, 1982) and listed for each basin or area. Nine columns were used in the table for estimates of oil, gas, and heavy oil at the 95%, 50%, and 5% probability levels. Definitions for probability terms used in this study are as follows: 95% is semiproven or very likely; 50% is probable or likely (median); and 5% is possible or unlikely. (A 95% probability means that there is a 95% chance that there will be more, and a 5% chance that there will be less, than the estimate.) Definitions for petroleum terms are as follows: light oil (>15° API) includes condensate and natural gas liquids; gas includes both associated and nonassociated gas; and heavy oil (tar) (<15° API) will not flow without heat treatment and is limited to what might be produced within

25 years. The reason for using this terminology stems from the particular geography of oil and gas in Colombia. In 1982, Colombia had produced oil since 1921 and imported over 25% of its needs of 190,000 BOPD. There is only a limited local market for natural gas. Steam recovery of heavy oil is a recent development in a few old oil fields. Heavy oil (tar) resources may be enormous (e.g., Venezuela's Orinoco tar belt) but are often completely uneconomical and must be considered as inactive reserves if located in a remote area with no infrastructure. Consequently, for this project and planning purposes, heavy oil (tar) was separated from (light) oil and a practical time limit of 25 years was set for the heavy oil recovery.

Not all experts filled out the forms completely for all basins. After the first replies were plotted on graphs, the consultant returned to each respondent to show him how his estimates compared with the others and to clarify any ambiguous points. A common misunderstanding occurred when the respondents estimated oil-in-place rather than the requested recoverable oil. Several experts subsequently modified their appraisals of this item. After these semifinal results were plotted, the consultant returned a last time to discuss the problems with those few experts whose estimates, particularly for gas, seemed erratic. Few corrections were made at this point because most individuals had strong feelings that their estimates were realistic. All of the estimates were posted on the final probability graphs, but a few of the highest erratics were slightly reduced by the consultant on the basis of visual inspection to bring them more into line with the other estimates (Figure 5, 6, 7, and 8). Failure to do this would have unduly skewed the final

---

[1]All figures in this paper are taken from Ivanhoe (1984).

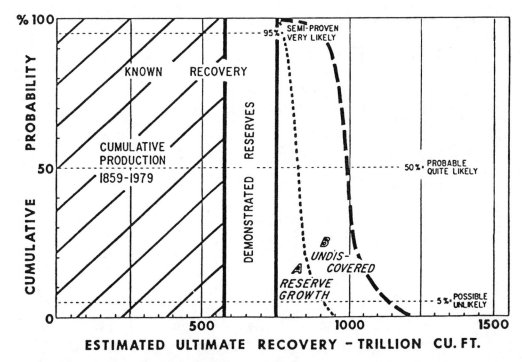

Figure 2. Graph of probability distribution showing EUR of natural gas in the United States. Cumulative production plus demonstrated reserves are indicated. Approximately 20% of the estimated gas still to be discovered will be from reserve growth around known fields (A) and 80% from still undiscovered fields (B). (After Nehring, 1981, figure 5-2.)

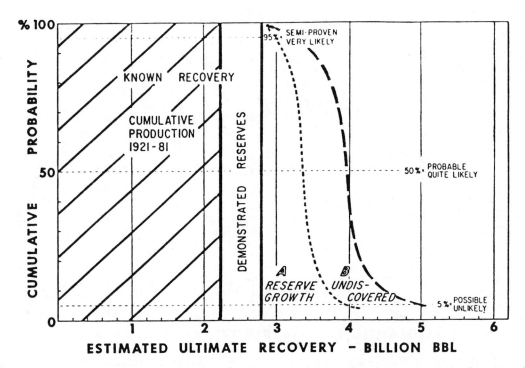

Figure 3. Graph of probability distribution showing hypothetical EUR of light oil (> 15° API) in Colombia. Actual cumulative production plus demonstrated reserves are indicated. The hypothetical EUR of oil still to be discovered (1.2 billion bbl at 50% probability) and the relative amounts from reserve growth (A) at 50% versus that from new fields (B) at 50% are unknown and were drawn schematically to show the methodology and objectives of the study to the Colombian experts. (Adapted from Figure 1.)

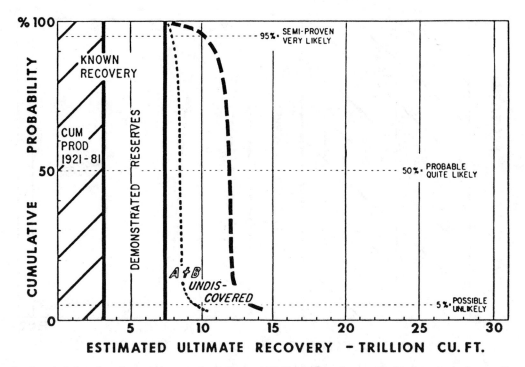

Figure 4. Graph of probability distribution showing hypothetical EUR of natural gas in Colombia. Actual cumulative production plus demonstrated reserves are indicated. The hypothetical EUR of gas still to be discovered (5 trillion cu ft at 50% probability) and the relative amount from reserve growth (A) at 20% versus new fields (B) at 80% are unknown and shown schematically to explain the methodology and objectives of the study to the Colombian experts. (Adapted from Figure 2.)

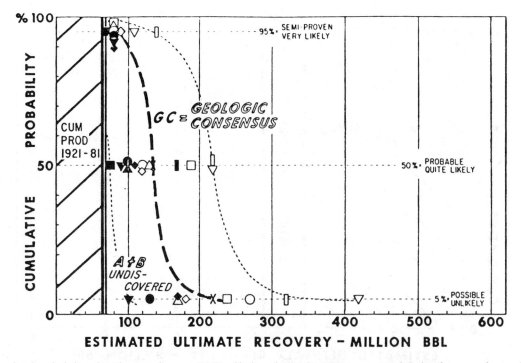

Figure 5. Graph of probability distribution showing EUR of light oil (> 15° API) in a typical Colombian producing basin. Unique symbols identify the various experts' estimates at the 95%, 50%, and 5% probability levels. Light dashed lines are drawn through the high and low estimates and a heavy dashed line through the computed average estimate (X). No distinction was made between reserve growth (A) around known fields and still undiscovered fields (B), although both will occur; consequently any new additions could be from either A or B.

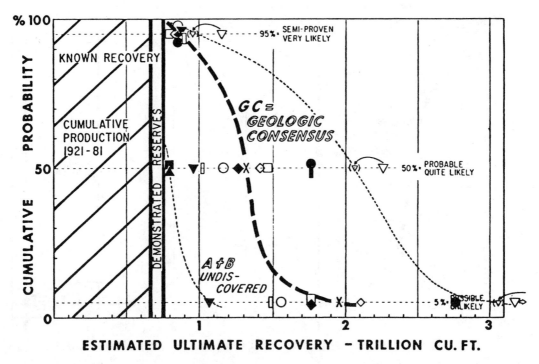

Figure 6. Graph of probability distribution showing EUR of natural gas in a typical Colombian producing basin. Extra-high erratics were adjusted downward as indicated by small arrows to avoid skewing the average. (See Figure 5 legend for further details.)

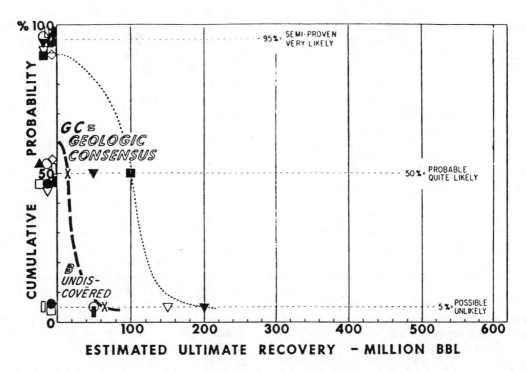

Figure 7. Graph of probability distribution showing EUR of light oil (> 15° API) in a typical Colombian nonproducing basin. Unique symbols identify the various expert's estimates at the 95%, 50%, and 5% probability levels. A light dashed line is drawn through the high estimates, whereas the low estimate is at EUR = 0 for all probabilities from 0 to 100%. A heavy dashed line is drawn through the average estimate (X). There can be no reserve additions (A) in a virgin basin, consequently any new oil will be from presently undiscovered fields (B).

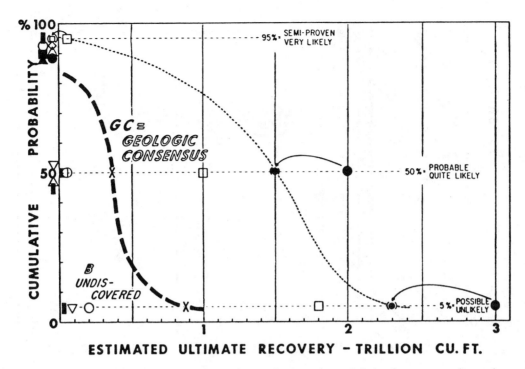

Figure 8. Graph of probability distribution showing EUR of natural gas in a typical Colombian nonproducing basin. Extra-high erratics were adjusted downward as indicated by small arrows to avoid skewing the average. (See Figure 7 legend for further details.)

averages. Each expert's estimates were identified by a unique symbol on all of the 36 (18 light oil and 18 gas) final probability graphs. To maintain confidentiality, only the consultant and each individual knew his symbol. Heavy oil (tar) was not graphed because of the lack of quantitative replies.

RESULTS

The pattern of the experts' views is thought provoking. No person's estimate was consistently the most pessimistic or optimistic for prospects in all basins, nor was their education, nationality, or job of any significance, with two exceptions:

1. Greater operating experience within a basin almost always made an individual more conservative in that specific area.
2. Petroleum engineers replied with "no opinion" on estimates of areas more often than did geologists.

Both of these tendencies might have been anticipated. Whenever explorationists know the local problems, they are more pessimistic about finding new oil; there is nothing like a string of dry holes to dampen enthusiasm. The jobs of petroleum engineers tend to make them more cautious than exploration geologists. Both disciplines are needed for a balanced consensus.

At the most used median value (50% probability level), each individual's optimism varied considerably between basins, but 90% of the total group's estimates fell routinely

within a range of 0–200% of the presently known oil and gas (discovered between 1920 and 1981) in each producing area. This can be considered as an acceptable range for order-of-magnitude ("ball-park") resource appraisals by experienced industrial personnel. When plotted on a graph, the acceptable estimates can be readily distinguished from the rare, very optimistic erratic opinions that had to be slightly reduced to avoid skewing the final averages.

CONCLUSIONS

The final averaged geological consensus from the 12 industrial experts provided a realistic evaluation of Colombia's remaining petroleum potential—the optimists balanced the pessimists. Appraisals by one individual for any basin are always suspect because there is no way to establish the degree of optimism or pessimism until such an expert's estimate is compared with those of a group of peers. A minimum of six independent opinions from qualified experts is probably required to produce a truly valid geological consensus. Several interviews are needed to draw opinions from cooperative but busy chief geologists and engineers of operating oil companies, who rarely fill out complicated questionnaires that are mailed to them. The confidential interview procedure allows each person to give a candid opinion, which is rare in any committee where members tend to agree with their superiors. Considerable range in the various geological estimates should be tolerated, but erratics must be recognized and properly adjusted. Final results are only as good as the geological expertise of the persons involved; electronic computers will not improve their assumptions. Any published group resource estimates should be supported by indicating the years and nature of local oil and gas expertise for the geological views being averaged. Finally, we should always remember that there is no sure way to predict the unknown, thus all persons involved should be humble with respect to their consensus.

REFERENCES CITED

Chona, F., and J. Bendeck, 1982, Estadisticas de la industria petrolera colombiana—a Diciembre 31, 1981, Cuarta edicion: Empresa Colombiana de Petroleo, Vicepresidencia de produccion, Coordinacion de produccion, CDP-046-82, Apartado Aereo 5938, Bogota, Colombia, p. 106.

Ivanhoe, L. F., 1984, Advantages and limitations of geological consensus estimates of undiscovered petroleum resources, in Oil and gas fields, Proceedings of the 27th International Geological Congress: Utrecht, The Netherlands, VNU Science Press, v. 13, p. 277–285.

Nehring, R., 1981, The discovery of significant oil and gas fields in the United States: Report R-2654/1-USGS/DOE, Rand Corp., Santa Monica, California, p. 236.

Field Size Distribution Related to Basin Characteristics

H. D. Klemme
Geo Basins Ltd.,
Bondville, Vermont

Economic appraisal of frontier and developing basins requires input from an analysis of the distribution of the most likely field size. An analysis of 65 of the world's basins containing 85% of the world's discovered and produced petroleum (given as barrels of oil equivalent, or BOE) has concentrated on the larger fields in each basin. Significant variations in the percentage of a given basin's oil and gas (given as BOE) that is contained in its five largest fields (which contain an average of 40–50% of the world's reserves) appear to be fundamentally related to the architectural form or morphology (a major element in basin classification) and size (volume of sediment in cubic miles) of the basins. Second-order variations in the morphology/size relationships of a basin's field size distribution by percent-rank appear to be modified by the nature of the sedimentary basin fill, the diversity and variation in the type of "plays," and the propensity of many basins to overlap or resemble other basin types (hybridism). With few exceptions, variations in the relative richness (BOE recovery factor per cubic mile of basin sedimentary volume) of maturely developed basins of the same class do not alter the percent-rank field size distribution in the five largest fields.

Previous modeling of field size distribution has attempted to consider basins in general through one all-encompassing model. This study tends to support more recent studies suggesting that several distribution patterns should be considered. These apparently consistent variations in the percent-rank of field size distribution might also serve as an additional check on any appraisals or reserve estimates of frontier or developing basins.

INTRODUCTION

Any evaluation of potential hydrocarbons in either frontier or developing basins can be enhanced by estimates that provide a breakdown of potential petroleum resources by field size distribution. Conversely, if, as current opinion indicates, half of the world's petroleum resources have been discovered, any patterns found in field size distribution may serve as analogs suggesting limits to the magnitude of a basin's potential petroleum resource.

Probabilistic models have been proposed (Kaufman et al., 1975), size-rank relationship analogs have been constructed (Jeffries, 1975; Ivanhoe, 1976; Coustau, 1979; Perrodon, 1980), and field size distributions have been analyzed on the basis of the percent-rank of a basin's total reserves that an individual field represents (Klemme, 1975; Roy and Ross, 1980), all in an attempt to refine our knowledge of field size relationships. In this study, we use the percentage of a basin's oil and gas (given as barrels of oil equivalent, or BOE) that is contained in its five largest fields to obtain the percent-rank.

To separate the shape of the field size distribution from the absolute size of the oil and gas fields, the recoverable BOE in a field (in this study, the five largest fields) is given as a percentage of the recoverable BOE in the basin rather than of the actual BOE.

Recent compilation and analysis of the large sample represented by the United States (Nehring, 1981) tends to confirm that there is a variation in field size distribution patterns when fields are related to their tectonic genesis through basin classification. This study includes an analysis of 65 producing basins (Figure 1 and Table 1) representing greater than 85% of the world's discovered petroleum reserves (BOE).

In the 65 basins considered in this study, an average of 40% of a basin's present reserves (50% if the Middle East Arabian–Iranian basin is excluded) are in its five largest fields. Reserves in the five largest fields in individual basins range from 10 to 90%, a variation of considerable significance, which is the departure point for this paper.

Some small differences in the estimated sizes of reserves

Figure 1. Numbers indicate major producing basins used in the analysis of field size distribution in this study.

Table 1. Key to basins outlined in Figure 1 arranged by basin type. (See Figure 1 for locations and Figure 3 for explanation of nomenclature.)

I. CRATON INTERIOR BASINS	**III. CONTINENTAL RIFTED BASINS**
1. Denver (1.4)*	A. Craton and Accreted Zone Rift
2. Michigan (1.6)	33. Reconcavo (1.2)
3. Illinois (5.5)	34. Suez (4.2)
4. Williston (4.6)	35. Cambay–Bombay (2.8)
II. CONTINENTAL MULTICYCLE BASINS	36. Gippsland (5.0)
	37. Aquitaine? (2.6)
A. Craton Margin (Composite) Basins	38. Viking (25.0)
	39. Central Graben (10.0)
6. Fort Worth (1.4)	40. Dnieper–Donetz (11.0)
7. Big Horn (2.7)	41. Sirte (26.0)
8. Wind River (0.8)	(Cooper sub-basin of Great Artesian basin)
9. Arkhoma (1.8)	
10. Piceance–Uinta (1.5)	B. Rifted Convergent Margin
11. Powder River (2.6)	42. Santa Maria (b) (1.05)
12. Oriente (2.4)	43. Central Sumatra (a) (10.0)
13. Sichuan (4.0)	44. Los Angeles (b) (9.0)
14. Green River–Overthrust (7.3)	45. Vienna (c) (1.3)
15. Greater Anadarko (42.0)	46. Cook Inlet (b) (2.3)
16. Greater Permian (47.0)	47. Ventura–Santa Barbara (b) (4.0)
17. Alberta (27.0)	48. Middle Magdalena (c) (1.5)
18. Volga–Ural (77.0)	49. Java Sea (a) (1.8)
	50. South Sumatra (a) (2.1)
B. Craton-Accreted Margin (Complex) Basins	51. North Sumatra (a) (3.0)
	52. Baku–Kura (c) (13.0)
5. San Juan (4.5)	53. Maracaibo (c) (39.0)
19. Paris (0.16)	54. Greater San Joaquin–Sacramento (b) (14.6)
20. Erg Oriental (19.0)	
21. Erg Occidental (14.0)	C. Rifted Passive Margin
22. Southern North Sea (24.0)	55. Cuanza (0.13)
23. Timan–Pechora (6.6)	56. Sergipe–Alagoas (0.35)
24. Great Artesian (1.0)	57. Campos (0.8)
25. West Siberian (156.0)	58. Congo (2.0)
	59. Gabon (1.3)
C. Crustal Collision Zone–Convergent Plate Margin	60. Greater North West Shelf (3.0)
	IV. DELTA BASINS
Closed	
26. East Venezuela–Trinidad (18.0)	61. Mahakan (3.4)
27. Greater Middle Caspian–Turkomen (40.0)	62. Greater Mackenzie (>2.7)
28. Middle East Arabian–Iranian (712.0)	64. Niger (27.0)
	65. Mississippi (50.0)
Open	
29. Tampico (13.0)	
30. Vera Cruz–Reforma–Campeche (38.0)	
31. North Borneo–Sarawak (7.5)	
32. Greater Gulf Coast (66.0)	

*Numbers in parentheses indicate present discovered BOE × 10^9, that is, the cumulative production and present recoverable reserves.

among many fields or in the total discovered reserves among basins were encountered during the compilation of an extensive data base (which is not included here). The amount of error (20%) arising from these discrepancies, however, does not seem seriously to affect the quality of the main conclusions. Data for the fields and present total reserves in basins of the United States were taken from Nehring (1981), and the potential resource data for basins of the United States were taken from USGS Open-File Report 81-192 (1981). Resources for many other areas outside the United States were estimated from various sources, including *Oil and Gas Journal, World Oil,* Nehring (1978), *Petroconsultants,* Perrodon (1980), and, other existing literature.

FIELD SIZE DISTRIBUTION

From the study of the five largest fields in each basin, two factors appear to be most important in variations of field size distribution: basin size and basin morphology, the latter of which is used as part of the classification scheme of basins. The tectonic framework for basin classification is shown in Figure 2, and cross sections of basin types (Klemme, 1980, modified and revised) are shown in Figure 3. The major criteria for basin classification include basin morphology, morphologic history, and location on either a crustal Precambrian craton (i.e., upper Proterozoic petroleum economic basement) or a crustal accreted zone (i.e., petroleum economic basement unknown). Figures 4–12 show

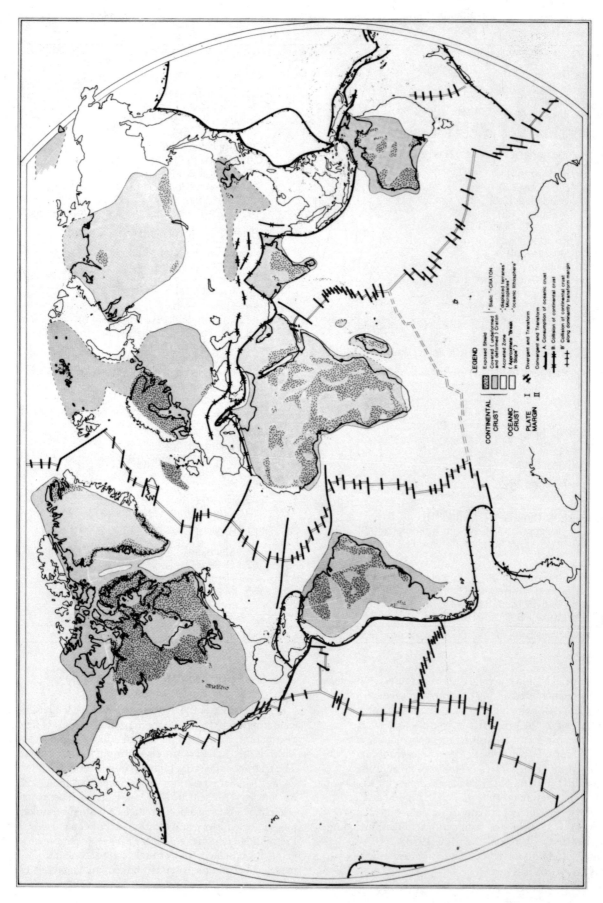

Figure 2. Tectonic framework for basin classification (see Figure 3).

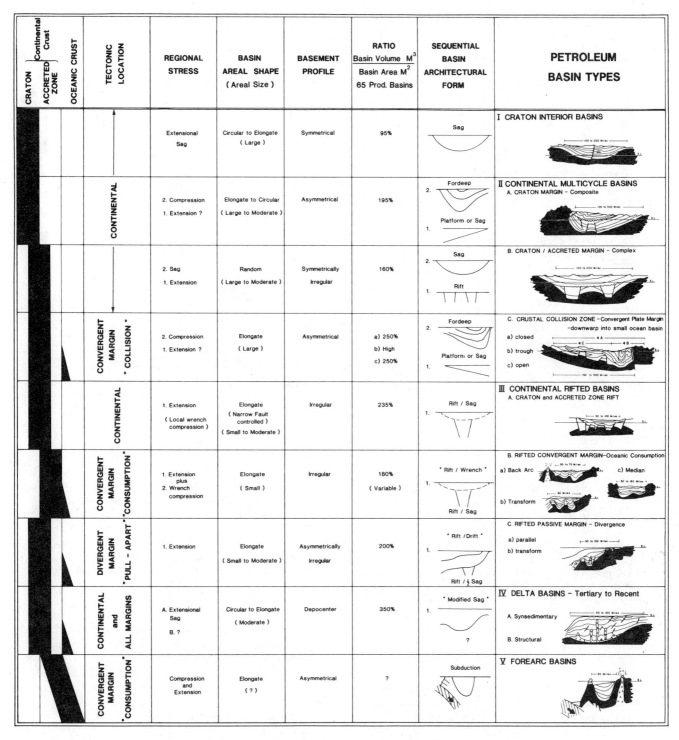

Figure 3. Petroleum basin classification scheme.

plots of the percentages of the total cumulative production and present reserves from the five largest fields of various types of basins, in relationship to basin size. You will note that most variations can be accounted for by basin size, expressed as the volume of sediment in cubic miles, and by basin type, which is based on morphology and tectonic genesis (Figure 3).

A basin's size sets limits on the opportunity to find fields. The smaller the basin, the fewer traps available; the larger the basin, the more space for the formation of traps. In general, however, most traps capable of becoming fields have proportionately greater limitations on spatial dimensions than do the variations of basin size. This tendency is best displayed by some of the larger basin types, such as rifts (Type

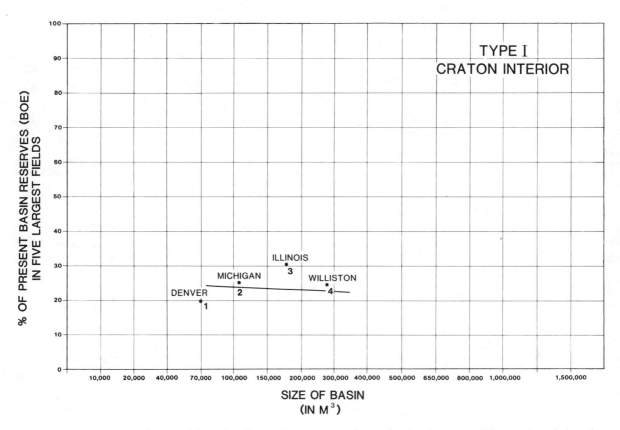

Figure 4. Type I craton interior basin field size distribution by percent-rank as related to basin size. The numbered plots shown here and in Figures 5–12 are the 65 basins examined in this study, grouped by basin type.

IIIA), rifted convergent margins (Type IIIB), possibly rifted passive margins (Type IIIC), and craton margins (composite) (Type IIA). With increasing basin size, there is a decreasing proportion of these basins' total reserves in the five largest fields. The cratonic interior (Type I), plate margin (collision) (Type IICA), and delta (Type IV) basins generally have a pattern in which the five largest fields possess a lower percentage of present total reserves. It appears, however, that there is a greater percentage decrease in reserves in the five largest fields in the larger basins than in the smaller ones. The lines in Figures 4–12, as well as in Figure 13, were chosen visually as the most likely percent-rank for a particular basin class, and consideration was made for those basins that are immaturely developed.

Researchers have often noted that a considerable variation in field size distribution exists between the California basins and the onshore Mississippi Delta–Gulf Coast province (Jeffries, 1975; Klemme, 1975, 1977; Nehring, 1981). These two extreme basin types (Type IIIB rifted convergent margin and Type IV deltas, respectively) represent different basin "habitats" (Coustau, 1979; Perrodon, 1980): the former is called a *concentrated* habitat (e.g. Ventura basin) and the latter a *dispersed* habitat (e.g., Mississippi Delta). Mature Type IIIB rifted convergent margin basins presently average 75% of the basin's total reserves in their five largest fields, whereas moderately mature to mature Type IV delta basins with synsedimentary traps have only ~10% of their total reserves in their five largest fields.

Other basin types (Figure 13) can be arranged between these two extreme examples. Rifted passive margins (Type IIIC) and rift (Type IIIA) basins are generally similar to Type IIIB rifted convergent margin basins. In contrast, Type IIA cratonic margin (composite) basins have a slightly smaller portion of their reserves in the five largest fields, and Type I cratonic interior basins have a much smaller portion and show less dependency on basin size.

Table 2 shows how the geological character of basins and basin type relate to the field size distribution, habitat, and basement profile. Basins with irregular basement profiles and extensional modes of deformation have the greatest amount of reserves in the five largest fields (normal to concentrated habitat in the smaller basins). Although Type IIIB basins occur on convergent margins, where extension and wrench faulting appear to predominate, the commonly oblique nature of convergence apparently results in considerable tensional deformation in a regional framework of convergent compression. Basins with symmetrical basement profiles and depocenters with synsedimentary structures possess the smallest percentage of the basin's total reserves in the five largest fields. These latter basins display a dispersed habitat of field size distribution. Extension–compression basins include Type IIA craton margin (composite) and Type IICa plate margin (collision) basins; they display an asymmetrical basement profile resulting from late stage compression. The percent-rank of the five largest fields in these basins appears to lie between the extremes outlined above.

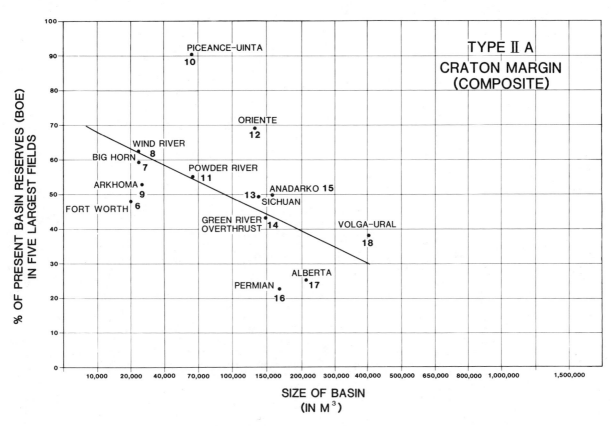

Figure 5. Type IIA craton margin (composite) basin field size distribution (see Figure 4 for details).

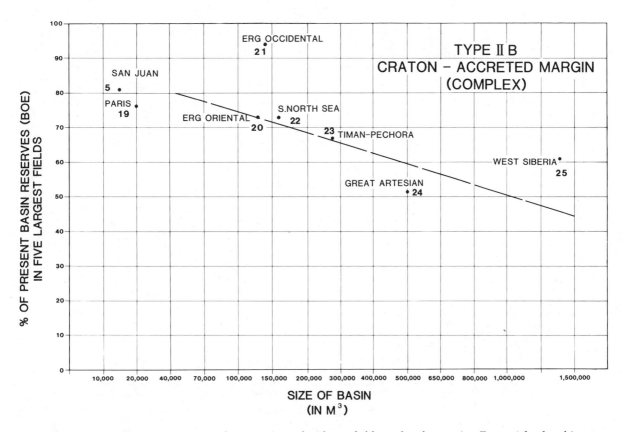

Figure 6. Type IIB craton-accreted margin (complex) basin field size distribution (see Figure 4 for details).

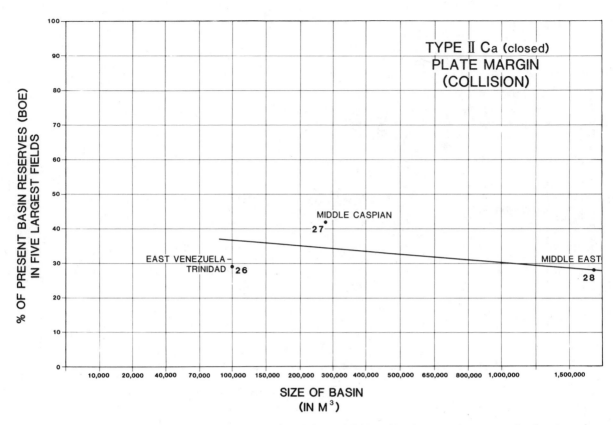

Figure 7. Type IICa plate margin (collision) (closed) basin field size distribution (see Figure 4 for details).

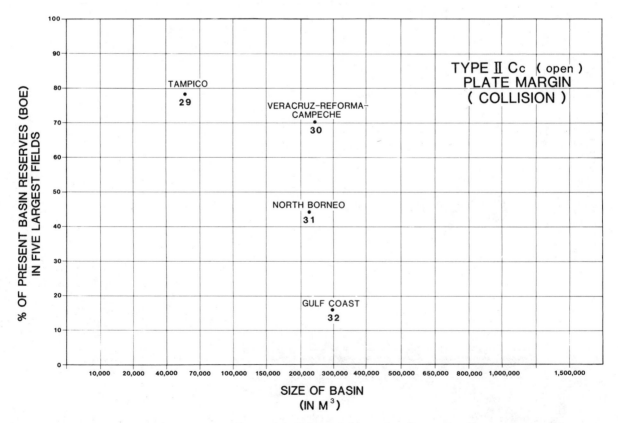

Figure 8. Type IICc plate margin (collision) (open) basin field size distribution (see Figure 4 for details).

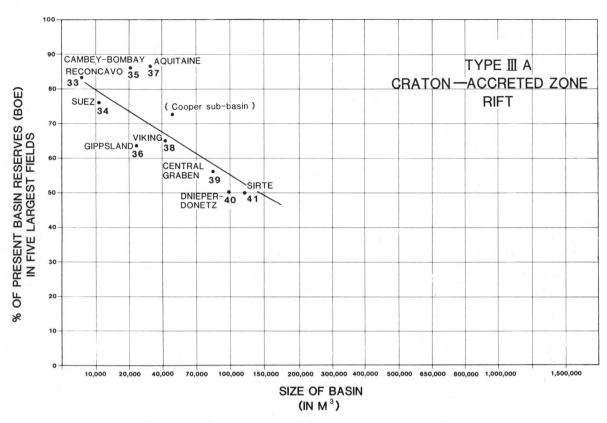

Figure 9. Type IIIA craton-accreted zone rift basin field size distribution (see Figure 4 for details).

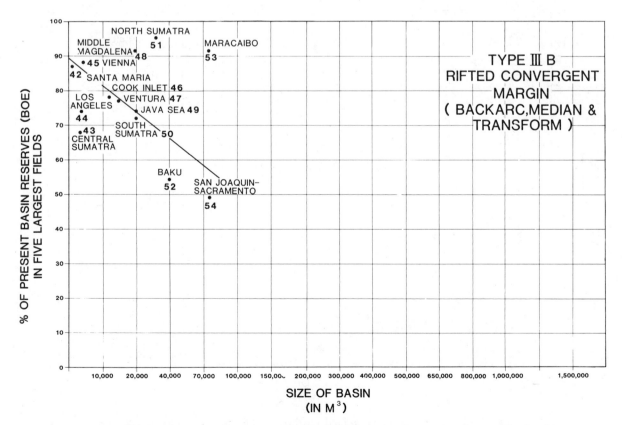

Figure 10. Type IIIB rifted convergent margin basin field size distribution (see Figure 4 for details).

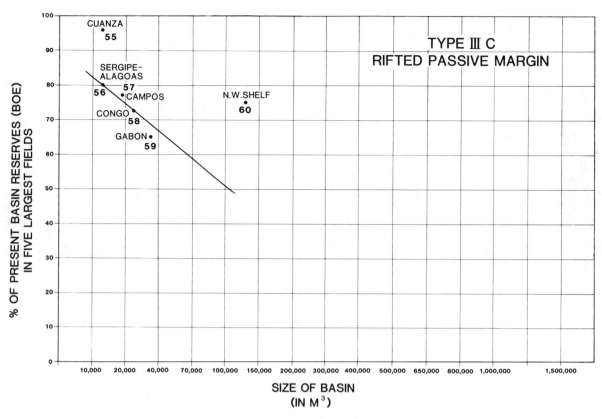

Figure 11. Type IIIC rifted passive margin basin field size distribution (see Figure 4 for details).

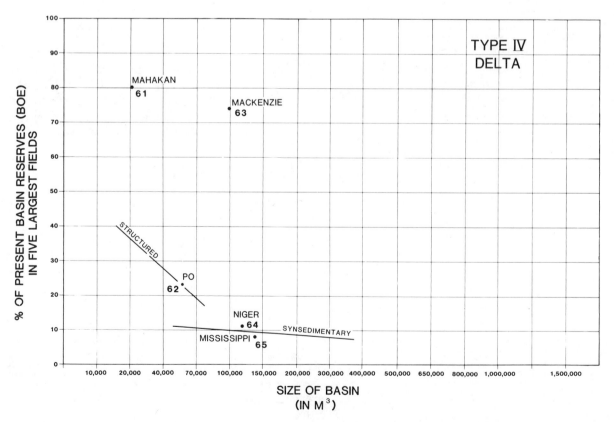

Figure 12. Type IV delta basin field size distribution (see Figure 4 for details).

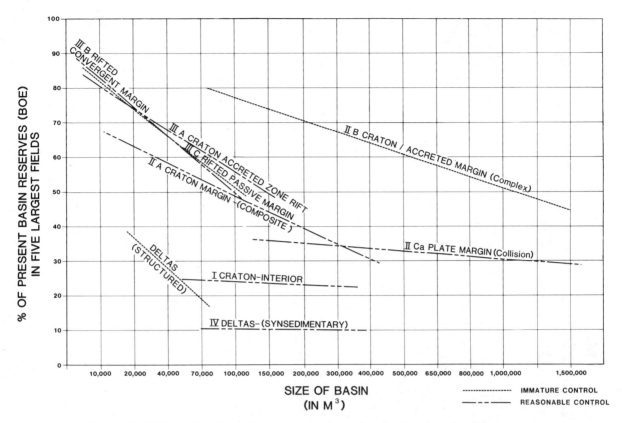

Figure 13. Field size distribution by percent-rank as related to basin size for all basin types.

Percentages of the present basin discovered oil and gas (BOE) in the five largest fields as modified by basin size are shown in Figure 13. Because we can expect to find additional reserves in most basins, it is likely that the estimated ultimate recovery in the basins studied would lower the percentage of the total reserves in the largest five fields by 7–12%, depending on the maturity of the basin, basin type, basin size, and possible additional reserves in the large fields.

The largest field in the Type IIIA rift, Type IIIB rifted convergent margin, and possibly Type IIIC rifted passive margin basins contains between 15 and 80% of the basins' present total reserves. If modifying factors (mainly basin size and development maturity) are also considered, the range is closer to 20–45% and the average is ~25% (Figure 14). In Type IIA cratonic margin (composite) basins the largest field contains an average of 19% of the basin's total discovered oil and gas, in Type IICa plate margin (collision) basins ~9%, in Type I cratonic interior basins 8%, and in mature synsedimentary structured Type IV delta basins 2–3%. The fifth largest field for all basins contains 2–6% of the basins' present total reserves and averages 1.5–3% of projected future resources. Figure 14 shows the averages of the ultimate reserves (including cumulative production, present reserves, and best guessed future discoveries) in each basin type and thus gives a more representative picture of the basins' ultimate field size distribution than does the present distribution. Those fields that are smaller than the five largest fields in any basin are more likely to assume a log normal distribution.

FIELD SIZE DISTRIBUTION: MODIFIERS

Several factors directly or indirectly modify the fraction of a basin's total reserves that is in the five largest fields.

Basin Development Maturity
A basin's maturity affects the percentage of a basin's total reserves attributable to both the total of the five largest fields and the percent-rank of the individual fields. Tables 3 and 4 indicate the time it takes to discover a quantity of reserves equal to 75 and 100% of that quantity eventually attributable to a basin's five largest fields. State-of-the-art advances in exploration techniques, increased hydrocarbon demand, and in particular, the move to offshore basins have considerably reduced both the amount of drilling and the time required to find the five largest fields (Table 4). Immature basins have a greater percentage of reserves in their five largest fields (because large fields are commonly found at an early stage of development) and often appear anomalous on percent-rank/basin size plots. As a basin matures, the percentage of its total reserves in the first or largest fields is usually reduced.

Basin Fill
A basin sedimentary fill can range from almost entirely clastic (e.g., Type IV deltas, many Type IIIB rifted convergent margins, some Type IIIA rifts, and some Type IICc plate margin [open]), through mixed clastic and

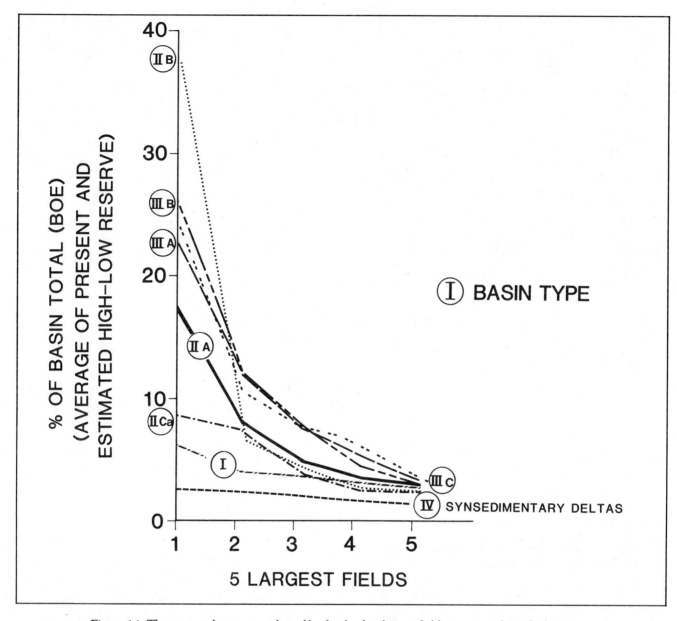

Figure 14. The averaged percent-rank profiles for the five largest fields in a given basin by basin type.

carbonate (e.g., Type IIA craton margin [composite] and Type IICa plate margin [collision]), to predominantly carbonate (e.g., some Type IICc's). (See Figure 3 for explanation of basin types.)

A deltaic clastic fill tends to lower the percentage of a basin's reserves in the largest five fields; examples include Denver basin (Figure 4), Fort Worth and Arkhoma (Figure 5), Gabon (Africa) (Figure 11), Baku (Soviet Union), and San Joaquin–Sacramento (Figure 10), North Borneo (Malaysia) and Gulf Coast (Figure 8), and the Mississippi, Niger, and Po deltas (Figure 12). In turn, it appears that the influence of deltaic sedimentary fill is modified by the type and magnitude of structural development, (e.g., tectonically structured versus synsedimentary structured deltas) (see Figure 12). An interesting example of structured deltaic fill is the anomalous Maracaibo basin of Venezuela (Figure 10),

which displays a concentrated habitat due to the crosswarp or arching of an Eocene delta resulting in an extensive structural stratigraphic trap for the many pools in the Bolivar coastal field.

In contrast, although not as clearly demonstrated as most deltaic environments, basins of any given type having a predominance of carbonate contain reservoirs that commonly have a higher proportion of basin reserves in the five largest fields; that is, they fall above the lines indicating basin types shown in Figures 13 and 14. Perhaps the best example of this can be provided by Type IICc plate margin (open) basins: the Tampico and Vera Cruz–Reforma–Campeche basins which contain a predominantly carbonate basin fill, have a higher proportion of the basins' BOE in the five largest fields than the predominantly deltaic North Borneo and Gulf Coast basins (Figure 8).

Table 2. Percent-rank field size distribution and "habitat" related to basin types and geologic character (morphology, regional stress, and basement profile).

Regional Stress	Extension	Sag Over Extension	Wrench Extension	2. Compression 1. Extension	Sag	Synsedimentary Depocenter
Basin	III A Craton-Accreted Zone Rift III C Rifted Passive Margin	II B Craton/Accreted Margin (complex)	III B Convergent Margin Rift	II A Craton Margin (composite) III Ca Plate Margin (collision)	I Craton Interior	IV Delta
Basement Profile	Irregular			Asymmetric	Symmetric	Depocenter
Percent of total basin reserves in 5 largest fields	High			Low		
Habitat	Concentrated to Normal			Dispersed		
Discovered Hydrocarbons (this study)	26% (26%)*			74% (30%)*		
Number of Basins	33			32		
*Excluding Middle East basin.						

Table 3. Time required to find 75 and 100% of the present reserves in a basin's five largest fields (all basin types).

	Average Years to Find 75 and 100% of the Reserves Attributable to the Five Largest Fields*	
	75%	100%
1880–1900	27	41.6
1900–1910	19	31.2
1910–1920	16	40.1
1920–1930	23	30
1930–1940	15	28
1940–1950	10	14
1950–1960	4.6	9
1960–1970	5.8	7.3
1970–1980	3.3?	6?
*From the earliest discovery date of any of the five largest fields.		

Petroleum Plays: Diversity and Variation of Play Types

The more petroleum plays (multiple source sequences and multiple trap and reservoir types) a basin contains, the lower the percent-rank field size distribution. We can see this in two ways: (1) larger basins more often have a multiplay character, and (2) certain basin types are prone to contain multiple plays (e.g., Type IICa crustal collision zones [closed]) while others are dominated by a single play (e.g., Type IIB craton/accreted margins [complex], small Type IIIA rifts and Type IIIB convergent margin rifts). In the case of craton interior basins, the accumulations often appear to result from combination structural–stratigraphic traps that are simply numerous and dispersed, whereas deltas display a wider (horizontal and vertical) stratigraphic range of multiple source and reservoir units.

An increase in the diversity of trap types (flowage, reef, stratigraphic, or wrench/compression fold traps) in any basin tends to result in below average percent-rank field size distribution (i.e., below the average plotted in Figure 13). In contrast, a dominance of arch, tilted fault block, or other similar subsidence-related features (relatively uplifted blocks) tends to result in above average percent-rank field size distribution (i.e., above the average plotted in Figure 13).

Variation in Individual Basin Character

Basins can be classified into more than one type, because individual basins can have characteristics that are similar to other basin types. Few basins, however, are completely similar, thus gradations in similarity occur. This gradation of character from one typical basin type to another is not unexpected; it simply emphasizes the difficulty of classifying basins. When morphological gradation in basin type occurs within an individual basin, the field size distribution commonly shows a similar "skew." Examples include the following basins: Illinois, gradation from Type I to IIA; Greater Permian, Type IIA to I; Volga–Ural, (Soviet Union), Type IIA to IIB; Piceance–Uinta, Type IIA to IIB; Greater Anadarko, Type IIA to IIB; and Alberta, (Canada), Type IIA, but mainly Paleozoic and more like Type I. Clearly the class gradation phenomenon occurs most commonly in the cratonic and continental basin categories.

The large arches with large fields on the edge of some Type IIA craton margin (composite) basins (e.g., Panhandle-Hugoton arch of Greater Anadarko basin, Tartar Arch of Volga–Ural basin, Douglas Creek arch of Piceance–Uinta basin, and the Casper Arch of the Powder River basin) resemble the large arches that develope in the Type IIB craton-accreted margin (complex) basins. When these arches are present, Type IIA basins tend toward a distribution

Table 4. Time required to find 75 and 100% of the present reserves of the five largest fields by basin type.

Basin	Average Discovery Time of Five Largest Fields					
	Pre-1940		1940–60		1960–80	
	75%	100%	75%	100%	75%	100%
Type I Cratonic Interior	7	36	10	11		
Type IIA Craton Margin (Composite)	17	37	7	11		
Type IIB Craton (Accreted) Margin (Complex)			1	14	2	7
Type IIIA Rift			8	19*	5	11*
Type IIC Collision IICa Closed	28	45*				
IICc Open	28	45*			5	7
Type IIIC Rifted Passive Margin			5	16	5	7
Type IIIB Rifted Convergent Margin-Backarc and Transverse	15	35*	7	32*		
Median	12	43*			2	12
Type IV Delta	10	53*			3	8

*Includes several basins with production onshore and offshore where onshore development was followed by extended gap until offshore portion of basin was developed.

pattern more like Type IIB basins, that is, they have a higher proportion of reserves in the five largest fields. Large regional arches and the consolidation of pools in the Type I Illinois basin causes it to resemble a Type IIA basin, whereas the circular shaped, more symmetrical profile of the Type IIA Greater Permian basin causes it to resemble a Type I basin. The field size distributions of these two basins are similar.

Richness: BOE Recovery per Volume of Sediment

The percent-rank profile, size-rank profile, and percent-rank of the largest five fields relative to basin size (Figure 13) do not vary directly with the relative richness of a basin. Two exceptions to this are the world's most prolific basins (three or four times richer than the next richest): the Los Angeles basin and the Central Sumatra (Indonesia) basin. These basins are Type IIIB rifted convergent margin basins, and they contain smaller reserves in their five largest fields for their size, than do other basins of this type. This suggests that some extremely rich basins may contain more reserves in fields smaller than the five largest (i.e., the statistcal "tail") than is normally the case. It is possible that the factor of richness, which has less influence on field size distribution at the high end (large fields) of a basin's distribution profile, may have more influence on the low or tail end (small fields) of the distribution profiles. This is because in any given size and class, extremely rich basins have more fields than do less rich basins.

Other Parameters

The geothermal regime, source rock, cap rock, trap type ("plumbing"), and greatest depth of a basin relate to its type and thus, through its morphology and genesis indirectly to its field size distribution. The quality of the individual elements of the basin's plumbing and their interrelationship (e.g., juxtaposition and timing) are among those factors that appear to influence relative richness of a basin. Regardless of the relative richness, however, the percent-rank distribution of a basin's five largest fields is related to basin type and basin size as modified by those factors discussed thus far in this section.

Habitat: Dispersed, Normal, or Concentrated

The size-rank relationship of the five largest fields in basins of a similar type may or may not vary according to basin size. Moderately mature or mature basins in Type I cratonic interiors, Type IV deltas, and possibly Type IICa plate margins (collision) show a dispersed pattern regardless of the basin's size. In contrast, in Type IIA craton margins (composite), Type IIB craton accreted margin (complex), Type IIIA rifts, Type IIIC rifted passive margins, and Type IIIB rifted convergent margins there is a tendency for smaller basins to display concentrated to normal patterns or habitats, whereas the larger basins in these types commonly display a dispersed habitat. As a result, these basins average an overall normal habitat.

It appears that basin size modified by fill, diversity and type of plays, and variations in individual basins from typical class character (hybridism) all influence the habitat (dispersion pattern) of the five largest fields. The smaller the basin, the more concentrated the size-rank, whereas the larger the basin or more deltaic the fill, the more dispersed the size-rank.

PREDICTIVE UTILITY

Improved, state-of-the-art exploration methods and techniques are creating a tendency toward relatively shorter exploration times for determining 75 and 100% of a basin's five largest fields (see Table 4). It is probably unreasonable, however, to suggest that within the first five to six years of a new basin's development (or longer if economics, politics, terrain, or climate interfere) a review of the five largest fields and their modifying factors (as outlined above), could be used to predict eventual basin reserves. This seems particularly unlikely when one realizes that any single appraisal technique is subject to limitations (Miller, 1979; White and Gehman, 1979). Consideration of the conclusions outlined above, however, can serve as an additional check on any appraisals or estimates made by other techniques.

Comparative analysis of the size distribution or profile of the five largest fields (Figure 14) may allow for the prediction of undiscovered fields in basins when the five largest fields have a profile that varies from its basin type profile as given in Figure 14. Gaps in the profiles of individual basins may suggest that other major field discoveries could occur. Nehring (1981) has recommended caution in this approach, citing the necessity for (1) relating the basin being appraised to its appropriate type, (2) considering the variation from typical basin type, and (3) considering other geological characteristics peculiar to the basin being examined (i.e., the major factors and modifiers of this study).

Certain basins (Piceance–Uinta, Powder River, and Greater Anadarko) show anomalous profiles for their type, which may be the result of a Type IIA craton margin (composite) basin resembling a Type IIB craton-accreted margin (complex) basin both tectonically and in field size distribution. These would not seem anomalous. Others in which a break between the largest and second largest fields exists are either immature basins or basins that extend offshore where an immature portion of the basin exists; these include Cambay–Bombay (India), Central Graben (United Kingdom and Norway), Santa Maria and Ventura (California), North Sumatra, and Java Sea (Indonesia). The break or gap in the profiles of these basins may be anomalous and may indicate that additional larger fields are a reasonable projection. Gaps appear to exist in such mature basins as the Middle East Arabian–Iranian and Russian Volga–Ural basins, as well as in most immature basins.

CONCLUSIONS

Field size distribution by percent-rank generally shows a relationship to basin classification as defined by geological character, including morphology and tectonic genesis, and to basin size. This relationship is imperfect, partly because of variable modifying factors such as basin fill (sedimentary character), diversity and character of play types, and hybridism (gradation between basin types).

More analysis of field size distribution and its relationship to basin geological character is warranted. This study strongly suggests that there are not one but two or more distributions of field size based on the geologic characteristics of petroleum basins or provinces. These patterns in field size distribution warrant consideration when any appraisals or resource estimates are made of frontier or partially mature basins.

ACKNOWLEDGMENTS

The writer wishes to thank Charles D. Masters, David H. Root, and Gregory Ulmishek for review and editing.

REFERENCES CITED

Coustau, H., 1979, Logique de distribution des tailles des champs dans les bassins: Petrole et Technique, v. 262, p. 23–30.

Ivanhoe, L. F., 1976, Evaluating prospective basins: Oil and Gas Journal, v. 74, n. 49, 50, 51.

Jeffries, F. S., 1975, Australian Oil Exploration—A great lottery: Australian Petroleum Exploration Association Journal, v. 15, pt. 2, p. 48–51.

Kaufman, G. M., Y. Balcer, and P. Kruyt, 1975, A probabilistic model of oil and gas discovery; in J. D. Haun, ed., Methods of Estimating the Volume of Undiscovered Oil and Gas Resources: AAPG Studies in Geology 1, p. 113–142.

Klemme, H. D., 1975, Giant oil fields related to their geologic setting—a possible guide to exploration: Bulletin of Canadian Petroleum Geologists, v. 23, p. 30–66.

—— , 1977, World oil and gas reserves from an analysis of giant fields and basins (provinces); in R. F. Meyer, ed., The future Supply of Nature Made Petroleum and Gas: New York, Pergamon Press, p. 217–260.

—— , 1980, Petroleum Basins—classifications and characteristics: Journal of Petroleum Geology, v. 3, n. 2, p. 187–207.

Miller, B. M., 1979, The evolution in the development of petroleum resource appraisal procedures in the U.S. Geologic Survey: Society of Petroleum Engineers Hydrocarbon Economics and Evaluation Symposium, Dallas, Texas.

Nehring, R., 1978, Giant oil fields and world oil resources: Report R-2284/CIA, Rand Corp., Santa Monica, California.

—— , 1981, The discovery of significant oil and gas fields in the United States: Report R-2654/1-USGS/DOE, Rand Corp., Santa Monica, California.

Perrodon, A., 1980, Geodynamique Petroliere—genese et repartition des gisements d'hydrocarbures; Bulletin Cent. Rech. Exploration–Production, ELF-AQUITAINE, Memoir 2, Masson S.A.

Roy, K. J., and W. A. Ross, 1980, Apportioning estimates of basins potential to fields; in A. D. Miall, ed., Facts and Principles of World Petroleum Occurrence: Canadian Society of Petroleum Geology, Memoir 6, p. 319–328.

White, D. A., and H. M. Gehman, 1979, Methods of estimating oil and gas resources, AAPG Bulletin, v. 63, n. 12, p. 2183.

Examination of the Creaming Methods of Assessment Applied to the Gippsland Basin, Offshore Australia

D. J. Forman and A. L. Hinde
Bureau of Mineral Resources, Geology and Geophysics
Canberra, Australia

The creaming method of estimating undiscovered petroleum resources depends on modeling the petroleum explorer's ability to find the large fields early. There are, however, a number of other creaming phenomena evident in the exploration statistics for the Central Deep of the Gippsland basin. Consideration of these suggests a new method for estimating undiscovered petroleum resources and allows the limitations of the creaming methods to be examined. The new method gives results similar to the old method. It involves simultaneous extrapolation of the predrill area of closure of the prospects and reserves per unit area, and it can be applied either to individual trap types or to all trap types within a region. In areas having a record of drilling but few discoveries, reserves per unit area can be determined from relationships among fields in a geologically similar region. Future success rates and the existence risk could be determined by geological risk analysis. The assessments probably underestimate the full potential of the Central Deep. Researchers should interpret them in the light of their limitations and of current knowledge of the region to determine the extent of the underestimation, and should see if there are any areas of significant potential that have been ignored. Any such areas could be assessed by other methods.

INTRODUCTION

Meisner and Demirmen (1981) described the creaming phenomenon as the diminishing effectiveness of exploration effort with advancing exploration. They considered that the phenomenon is best expressed as a decline in the field size, but is also expressed as a decline in the success rate. Meisner and Demirmen (1981) and Forman and Hinde (1985) described the creaming method of estimating undiscovered petroleum resources which depends on modeling the ability of the petroleum explorer to find the large fields early. Forman and Hinde modeled the tendency by fitting a straight line to the plot of log field size (log V) versus discovery number and used extrapolations to indicate the likely size of future discoveries. They also showed that the order in which the fields are discovered can be approximated by computer simulations using the hypothesis that the probability of each field's discovery is proportional to field size raised to a constant power, λ (lambda).

The value of λ is a measure of the explorer's ability to find the large fields early during exploration, and it is called the creaming factor. It has also been called the discoverability factor or coefficient (Schuenemeyer and Drew, 1983). The value of λ has a theoretical range from 0 to infinity, but typically is < 1. When $\lambda = \infty$, the fields will be discovered exactly in the order of their size. When $\lambda = 0$, the order of discovery of the fields will be random and independent of size. Hence, high values of λ indicate a strong creaming phenomenon.

The tendency to find the large fields early is only one of a number of results that petroleum explorers may want to achieve while attempting to locate the most profitable fields at the lowest cost. A number of other results are likely because the explorer looks at many geological and cost factors before deciding which prospect to drill next. These include area of closure of prospect, height of closure of prospect, richness and maturity of source rocks, thickness of reservoir, effectiveness of trap and seal, drainage area of trap, timing of trap development with respect to generation of hydrocarbons, nature of hydrocarbons, water depth, distance from shore or pipelines, and the depth to the primary objective. Our study of the Gippsland basin indicates that there are a number of creaming phenomena among these factors listed here.

Several important conclusions are drawn from the existence of these other creaming phenomena. Most

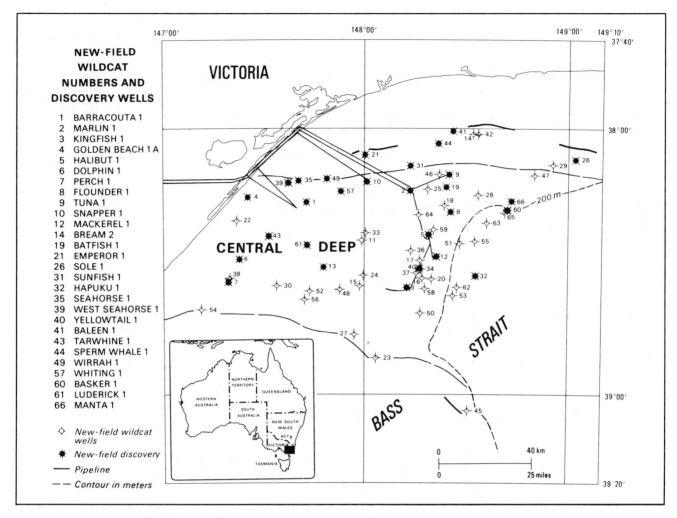

Figure 1. Location of new-field wildcat wells and new-field discoveries, Central Deep of Gippsland basin, Australia.

importantly, we suggest that undiscovered petroleum resources can be assessed by a variation of the creaming method that is based on simultaneous extrapolation of the predrill area of closure (A_p) and the reserves per unit area (V/A_p). The new method can be applied either to individual trap types or to areas with a record of drilling but with few discoveries by using data from a geologically similar area for the extrapolation of V/A_p.

Another important conclusion is that the overall decline of field size with discovery number and of predrill area of prospect with new-field wildcat number can be thought of as resulting from drilling the large prospects in the most favorable trap types early. Hence, the overall trend may not become evident until several prospects from several trap types have been drilled.

Finally, some creaming phenomena, such as are evident in water depth trends, form inclines rather than declines, and if significant enough could invalidate an assessment. Further research is needed to fully determine the strength of these effects.

The Central Deep of the Gippsland basin is a good location to examine the interplay of the various geological

and cost factors and to test the new method. Assessment by the new method can be checked against the results of the assessment done by projection of the plot of log field size versus discovery number.

CREAMING PHENOMENA EVIDENT IN PAST EXPLORATION TRENDS IN THE GIPPSLAND BASIN

The offshore part of the Central Deep of the Gippsland basin and its northern margin (Figure 1) occupy an area of about 4000 mi² (30,000 km²) to the 660 ft (200 m) bathymetric contour (Hocking, 1976). It is a fault-bounded graben containing as much as 40,000 ft (12,200 m) of upper Mesozoic to Cenozoic sediments. From 1965 to May 1984, 69 new-field wildcat wells were drilled in this area, with the resulting discovery of ~ 11 trillion ft³ (tcf) (310 billion m³) of sales gas and 3150 million bbl (500 million m³) of crude oil. A total of 82% of the gas and 95% of the oil were contained in 11 new-field discoveries identified by the first 13 new-field wildcat wells drilled.

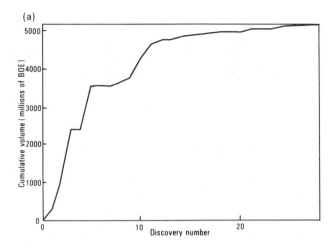

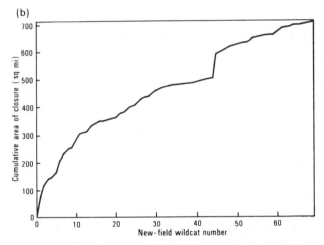

Figure 2. (a) Cumulative volumes of hydrocarbons discovered plotted in chronological order against discovery number and (b) cumulative area of closure of all prospects drilled plotted in chronological order against new-field wildcat number.

Postdrill Estimates of Field Size

The cumulative volumes of oil and gas in barrels of oil equivalent (BOE) discovered in the 28 fields in the Central Deep of the Gippsland basin are graphed against discovery number in Figure 2a. Identified sales gas resources were converted to crude oil equivalents on the basis of the approximate energy equivalence of 1000 ft^3 of sales gas to 0.182 bbl of crude oil. The slope of the graph is an indication of the size of the field discovered, with the steeper slopes indicating larger fields. The strong decline in the slope of the graph with discovery number indicates a strong tendency to find the large fields early.

We have calculated the correlation coefficient (r), r^2, and the z-statistic (Davies, 1961) to test if there is a correlation between log field size (log V) and discovery number. If z lies between ± 1.96, correlation is uncertain. If z is < -1.96, there is at least a 95% certainty that there is correlation between log V and discovery number, and that the data form a declining trend. Our calculated z-statistic is equal to -3.7,

and the square of the correlation coefficient (r^2) is equal to 0.4, indicating a good correlation.

A straight line appears to be a good fit to the plot of log V versus discovery number, and this conclusion is supported by a normal plot of the residuals, which fall close to a straight line. The straight line fitted to the data is described by the equation

$$\log V = A + BN_d \qquad (1)$$

where A is the intercept, B the slope of the fitted line, and N_d the discovery number of each field.

The value of lambda (λ) is a measure of the strength of the tendency to find the large fields early. The value of λ can be estimated by comparing the order in which the fields were actually discovered with the order obtained when discovery of the same fields is simulated in a computer program according to the rule that the probability of discovery of a field is proportional to field size raised to the power λ. We can also estimate λ using the method of maximum likelihood (Kendall and Stuart, 1967). This method finds the value of λ with maximum probability of producing the observed discovery order. The value of λ determined by the maximum likelihood method is 0.5.

Predrill Area of Prospect

The cumulative predrill area of closure (A_p) of all 69 structural and stratigraphic prospects drilled in the Central Deep and its northern margin is plotted against new-field wildcat number, in Figure 2b. There is a good correlation ($z = -4.5$ and $r^2 = 0.26$, see Table 1) between log A_p and new-field wildcat number, and a moderately strong creaming factor ($\lambda = 0.45$). The correlation and the creaming factor are strengthened and the standard deviation (σ_{res}) of the linear least-squares fit is reduced if the stratigraphic prospects are omitted from the data.

A straight line appears to be a good fit to the plot of log A_p versus new-field wildcat number, and this conclusion is supported by a normal plot of the residuals, which fall close to a straight line. The straight line is described by the equation

$$\log A_p = A' + B'N \qquad (2)$$

where A' is the intercept, B' the slope of the fitted line, and N the new-field wildcat number.

Ratio of Field Size to Area of Closure

The noted decline in field size (V) with discovery number is only partly due to a parallel decline in the area of closure (A_p) of the prospects drilled. Explorers will also seek to maximize such factors as height of closure of prospect, thickness of reservoir, effectiveness of trap and seal, drainage area of trap, richness and maturity of source rocks, timing of trap development with respect to generation of hydrocarbons, and nature of hydrocarbons. Their ability to maximize these factors is reflected in the decline in the ratio of field size to area of closure (V/A_p) with discovery number (Figure 3a).

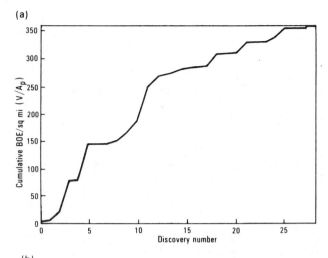

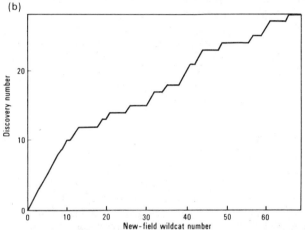

Figure 3. (a) Cumulative barrels of oil equivalent (BOE) per square mile plotted in chronological order against discovery number and (b) discovery number plotted against new-field wildcat number.

It can be shown by subtracting Eq. (2) from Eq. (1) that the relationship between log V/A_p and discovery number should be linear:

$$\log V/A_p = A - A' + (rB - B')N \qquad (3)$$

where r is a constant success rate. Alternatively Eq. (3) can be rewritten as

$$\log V = \log A_p + A - A' + (rB - B')N \qquad (4)$$

to show that the relationship between log V and log A_p should be nonlinear, except when $rB - B' = 0$. When $rB - B'$ does equal zero, a straight line can be fitted to the plot of log V versus log A_p, but it would have unit slope. We cannot conclusively prove a nonzero slope for the straight line described by Eq. (3) using Gippsland basin data because the square of the correlation coefficient is small ($r^2 = 0.09$) and the z-statistic ($z = -1.6$) does not quite reach the critical value (-1.96) of the 95% confidence region. However, the value of z does indicate an 89% chance of a nonzero slope.

Because of the creaming factor ($\lambda = 0.24$) in V/A_p, an explorer in 1984 looking for a field as large as the Kingfish oil field (Figure 1) located in 1967 by the fourth new-field wildcat well would have to drill a prospect with a closed area averaging about four times larger than the Kingfish structure.

Success Rate

Figure 3b shows the number of petroleum fields discovered plotted in chronological order against the number of new-field wildcat wells; the slope of the graph is the success rate. The initial success rate is high, with 12 discoveries in 14 new-field wildcat wells. The success rate then drops to an average of about 28% in a nearly straight-line trend for the next 56 wells. The decline is a creaming phenomenon brought about by early drilling of comparatively low risk prospects.

Trap Type and Stratigraphic Level of Main Objective

The hydrocarbons of the Gippsland basin have been sought in a variety of trap types: faulted anticline, paleotopographic high, anticline, fault, and stratigraphic. Within these traps, the reservoirs are composed of sandstone of the Cretaceous–Tertiary Latrobe Group. Seal is provided either by thick regional shales unconformably overlying the top of the Latrobe Group or by thinner shales interbedded within it.

The primary objectives of early exploration (Figure 4) were large, well-sealed faulted anticlines, paleotopographic highs, and anticlines at the top of the Latrobe Group (top porosity). The high initial success rate and the large sizes of the early discoveries prompted further drilling aimed at repetitions in smaller structures, higher risk fault and stratigraphic traps at the top of the Latrobe Group, and a wide variety of trap types deeper within the Latrobe Group (intra-Latrobe or intraporosity). Statistical analysis shows another creaming phenomenon. A tendency to identify hydrocarbons in the trap types with the greatest resources early is demonstrated by a negative correlation between the logarithm of the total petroleum resources identified in each trap type and the sequential discovery number of the trap type ($r^2 = 0.48$, $z = -2.09$).

Predrill Area of Prospect for Individual Trap Types

Because of the low success rate, within each trap type there is typically only a small number of fields compared to the number of prospects drilled. Therefore the exploration trends are best displayed by the plots of cumulative predrill area of closure (A_p) versus prospect number (Figure 5).

Values for the correlation coefficient (r), r^2, the z-statistic (z), and λ have been determined for the relationship between log A_p and prospect number for each trap type (Table 1). On the basis of the z values, we find a strong correlation between log area of closure (log A_p) and prospect number for anticlines and faulted anticlines at the top of the Latrobe Group, but weak or uncertain correlation for paleotopographic highs and faults at the top of the Latrobe Group. The analysis suggests that, where sufficient data are available, a straight line may be fitted to the data and may be projected to indicate the likely predrill areas of undrilled prospects.

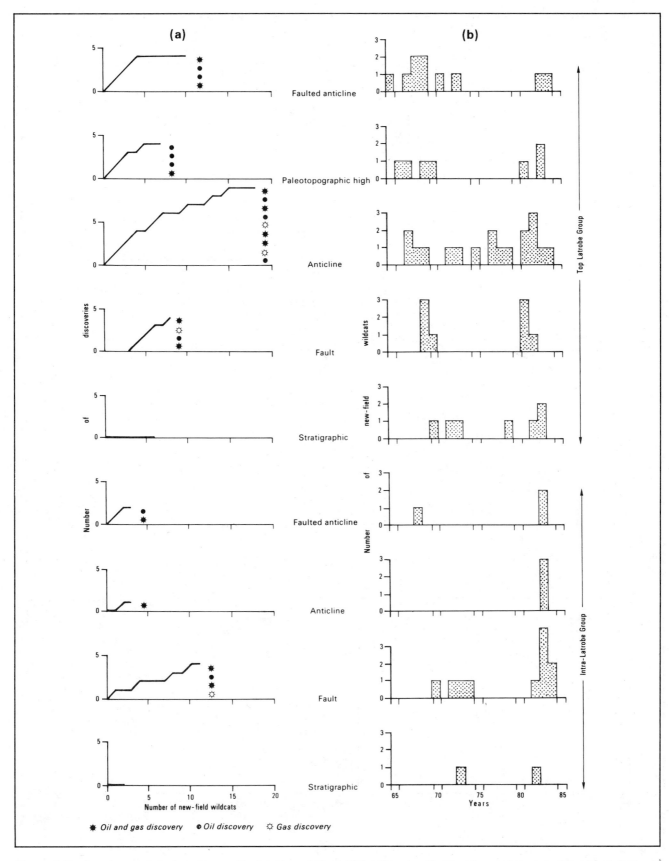

Figure 4. Exploration statistics by trap type for the Central Deep of the Gippsland basin: (a) cumulative number of new-field discoveries plotted against number of new-field wildcat wells, showing types of hydrocarbons, and (b) histograms showing totals of annual new-field wildcat wells drilled.

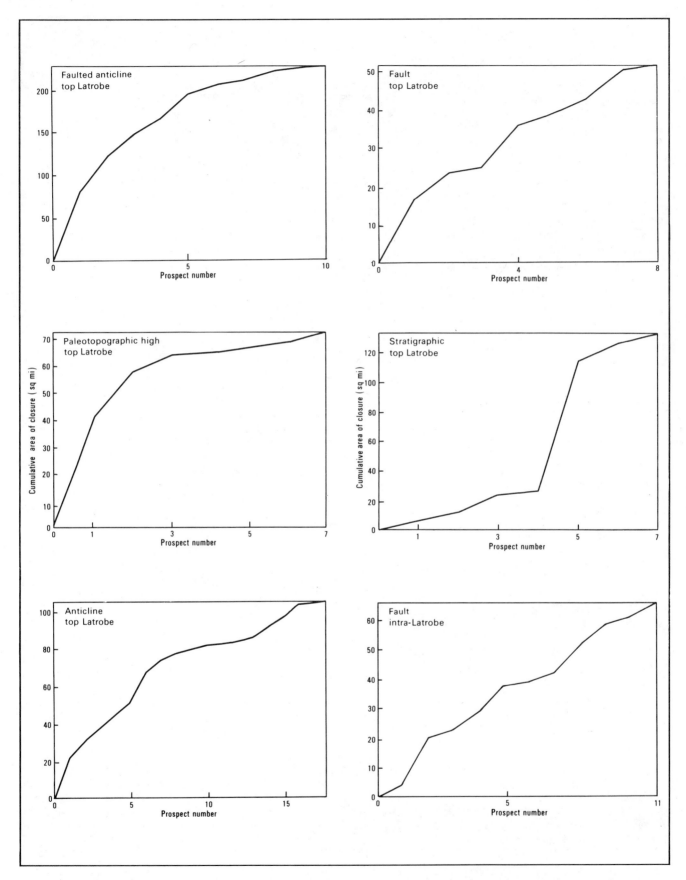

Figure 5. Cumulative area of closure (A_p) of prospects drilled plotted against prospect number for five trap types at top of the Latrobe Group and one trap type within the Latrobe Group.

Table 1. Values of r, r^2, z, and λ for plot of log A_p versus prospect number and average success rate for individual trap types and for all prospects combined.

Trap Types	r	r^2	z	λ	Average Success Rate	Comments
Faulted anticline, top Latrobe Group	−0.96	0.92	−5.1	3.3	0.4	
Paleotopographic high, top Latrobe Group	−0.69	0.48	−1.7	0.97	0.6	
Anticline, top Latrobe Group	−0.64	0.41	−2.9	0.69	0.5	
Fault, top Latrobe Group	−0.52	0.27	−1.3	0.86	0.5	
Stratigraphic, top Latrobe Group	0.23	0.05	0.46	−0.09	0	
Faulted anticline, intra-Latrobe Group	−0.95	0.9			0.7	Only 3 prospects
Anticline, intra-Latrobe Group	−0.8	0.65			0.3	Only 3 prospects
Fault, intra-Latrobe Group	−0.19	0.04	−0.6	0.27	0.4	
Stratigraphic, intra-Latrobe Group					0	Only 2 prospects
All prospects combined	−0.5	0.26	−4.5	0.45	0.4	

The tendency to drill the areally large structures early—the creaming factor (λ)—is highest for faulted anticlines and other structural traps at the top of the Latrobe Group (Table 1), which possibly suggests that only areally small structures are likely to be drilled in the future. At the same time, there is only a low creaming factor for the predrill areas of stratigraphic traps and intra-Latrobe fault traps, which instead suggests that areally large prospects of this type may still remain to be drilled. These could be the main targets of future exploration.

Water Depth

Figure 6 shows the plot of log water depth versus new-field wildcat number. There is a good correlation between the two variables with $z = 3.0$ and $r^2 = 0.13$. The data clearly indicate a preference for drilling the prospects in shallow water early, and the creaming factor, though inverse, is quite strong ($\lambda = -0.46$).

The preference for drilling a prospect in shallow water indicates that a prospect in deeper water would need to have larger potential resources before equal consideration would be given to drilling it. If the creaming factor for water depth is strong enough, it could increase the creaming factors for field size and area of prospect. Similar effects could be expected from other economic factors such as distance from shore or pipelines. It is important, therefore, that these effects are investigated and that areas are selected for assessment in such a way that the effects are minimized.

ASSESSMENT OF UNDISCOVERED PETROLEUM RESOURCES OF THE GIPPSLAND BASIN

Forman and Hinde (1985) suggest an improved statistical method for predicting the number and size of future discoveries and the exploration effort required to make them. The method is based on an extrapolation of the linear relationship between the logarithms of oil or gas field sizes and discovery number.

The demonstration that there is also a linear relationship between the logarithms of the predrill areas of closure (log A_p) and new-field wildcat number suggests the possibility of using this effect as the basis for another method of assessing

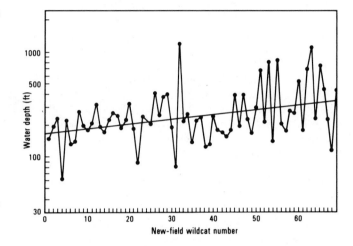

Figure 6. Logarithm of water depth in which wells have been drilled plotted against new-field wildcat number. The straight line has been fitted by the least-squares method.

undiscovered petroleum resources. Use of such a method requires determining the relationship between the predrill area of closure (A_p) and postdrill estimates of identified resources (V). The success rate for future drilling must also be estimated. The advantage of the second method is that it could be used either to assess areas where there are few or no discoveries or to assess individual trap or play types with few or no discoveries.

The potential for further oil and gas discoveries in the Central Deep of the Gippsland basin is first assessed by extrapolation of oil and gas field sizes, as outlined by Forman and Hinde (1985), except that oil and gas are combined to form estimates of the demonstrated resources of each field in BOE. The results obtained by this assessment are then used as a standard against which the results of assessment by the second method, using projections of the predrill area of closure, can be tested.

Assessment by Extrapolation of Field Size

The assessment of undiscovered petroleum resources is carried out using a computer program called LAMDA (Hinde, 1984a). The program extrapolates straight lines fitted to the plot of log field size versus discovery number. The extrapolations are made for particular values of λ taking into account a dependent relationship among λ, the average slope ($\bar{B}$) and intercept ($\bar{A}$) of the fitted lines, and the average standard deviation ($\bar{\sigma}_{res}$) of the residuals.

It has been found empirically that the dependent relationship is given approximately by the equation

$$\bar{B} = B_{MAX}(1 - e^{-s\lambda}), \quad \lambda > 0 \quad (5)$$

where s is the standard deviation of the logarithms of sizes of the entire population of fields and is in turn dependent on the number of fields in the population, $\bar{B}$ is the average slope for a given value of λ, and B_{MAX} is the slope of the fitted line when $\lambda = \infty$. The distributions of dependent values of slope (B), intercept (A), and standard deviation (σ_{res}) of the residuals corresponding to particular values of λ are unknown, but Eq. (5), in conjunction with the linear least-squares regression equations, may be used to calculate their approximate average values.

The method of extrapolation used in program LAMDA depends on using the maximum likelihood theory to determine the asymptotically normal distribution of the λ values that would account for the order in which the fields were discovered. The distribution is sampled at random to determine a value of λ and the corresponding values for $\bar{B}$, $\bar{A}$, and $\bar{\sigma}_{res}$ to be used in the extrapolation. Inclusion of a distribution of predicted success rates and the future number of new-field wildcat wells allows the number of new-field discoveries for each extrapolation to be determined probabilistically.

Forman and Hinde (1985) presented simulations to show, as approximations, that the distribution of log field size for each discovery number is normal and that each distribution has a common standard deviation. Therefore, a value for the field size can be estimated for each discovery number by random sampling during each extrapolation. After many extrapolations the results are expressed as a cumulative probability distribution of resources along with a distribution of the number of fields and a tabulation of the average size and standard deviation of the undiscovered fields.

Minimum and maximum success rates of 20 and 33%, respectively, were measured from selected parts of Figure 3b for future new-field wildcat drilling in the Central Deep of the Gippsland basin. We selected the average success rate for the last 56 new-field wildcat wells (28%) as most likely. The assessment for the next 50 wells is summarized in Figure 7, which shows a cumulative probability distribution of resources, a distribution of number of new fields, and a seriation of the average size and standard deviation of the undiscovered fields. The assessment indicates that the next 50 new-field wildcat wells drilled should discover, on average, a total of 145 million BOE in 13 or 14 fields.

Assessment by Extrapolation of Area of Closure

The method of extrapolation for the log A_p versus new-field wildcat number plot is identical to that used in the LAMDA program. The extrapolation is used to estimate the likely predrill area of future new-field discoveries. Simultaneous extrapolations of the log V/A_p versus discovery number plot are used to provide values of V/A_p so that the

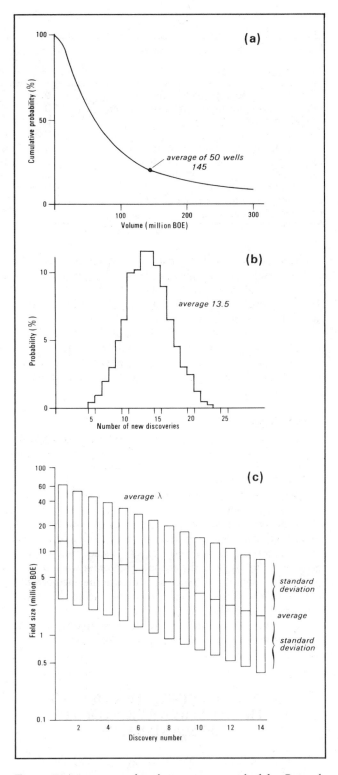

Figure 7. Assessment of exploration potential of the Central Deep of the Gippsland basin by projection of field size data: (a) cumulative probability distribution showing the total volume of hydrocarbons likely to be discovered by the next 50 new-field wildcat wells; (b) histogram of the number of new discoveries from the next 50 new-field wildcat wells; and (c) seriation showing average field size and standard deviation for next 14 discoveries based on the maximum likelihood estimate of λ.

possible size of future discoveries can be determined. The computer program that carries out these tasks is a variation of program LAMDA called VALAM (Hinde, 1984b). VALAM carries out the simultaneous extrapolations assuming no correlation between the maximum likelihood estimates of λ for each projection.

The assessment of the Central Deep of the Gippsland basin was carried out omitting stratigraphic prospects from the data, but using the same success rates as in the previous assessment. The assessment indicates that the next 50 new-field wildcat wells should discover, on average, a total of about 130 million BOE in 13 fields. The average value is lower than the 145 million BOE result obtained by projection of field sizes (V). Comparison of the two results was done assuming that the inconsistent treatment of stratigraphic prospects will make little difference. A t-test performed on the means of the two estimates ($t = 1.3$) indicates that there is no significant difference at the 95% level of confidence. Similar comparisons were made of the assessment values at the 20 and 80% lines of the cumulative curves. The main divergence between the two assessments appears at the 5% lines of the cumulative curve, where the second method gives higher values, indicating occasional selection of larger fields.

As with all methods, one must take care to check the field sizes generated by the computations to make sure that none greatly exceed the maximum size anticipated in the region. If necessary to avoid an overestimate, the standard deviation of the projection of V/A_p could be reduced or a maximum field size introduced into the computer program.

Davis and Harbaugh (1983) and others have shown that in some areas there is a strong correlation between the logarithm of the predrill area of closure (log A_p) of a prospect and the logarithm of the postdrill estimates of identified resources (log V). There also appears to be a statistically valid correlation ($z = 2.0$) in the Gippsland basin data. However, Eq. (4) demonstrates that the relationship between log V and log A_p is linear only where there is no correlation between log V/A_p and discovery number. Hence, use of the linear regression between log V and log A_p in this assessment would be inconsistent with our analysis of the data. For instance, an assessment carried out using the linear regression instead of the log V/A_p versus discovery number plot indicates that the next 50 new-field wildcat wells should discover, on average, a total of about 500 million BOE in 13 fields. These results are unlikely to be correct, because they suggest that the next 13 discoveries should contain more petroleum than the last 13 dicoveries.

Conclusions

Assessment using program VALAM yields results comparable to the results obtained using program LAMDA. This indicates that it is possible to develop a variation of the creaming method for use in areas where few or no discoveries have been made. The new method can be widely applied, because predrill maps are available to the government for all the new-field wildcat wells that have been drilled in Australia's prospective offshore areas. The method can use projections of the log A_p versus new-field wildcat number plot for the region in conjunction with projections of a log V/A_p versus discovery number plot determined from

relationships among fields in a geologically similar region. Future success rates and existence risk could be determined by geological risk analysis (Baker et al., 1984). The method can also be used to assess the hydrocarbon potential of individual trap types, even if few or no discoveries have been made in them.

RELIABILITY OF THE GIPPSLAND BASIN ASSESSMENT

We have shown that creaming methods can be used to project exploration trends for individual trap types as a basis for assessment of undiscovered potential. In the case of the Gippsland basin, where a number of fields have been identified in a variety of trap types, the combined field size trend can theoretically be used to indicate the sizes of future new-field discoveries, even including those in trap types where discoveries have not yet been made.

There are a number of reasons, however, why the Central Deep of the Gippsland basin could be much more prospective than the assessment indicates. For instance, the methods only provide estimates of potential resources that lie inside the area of study. The northern and western margins of the area, however, are poorly defined (Figure 1) and could be in error. Also, the eastern margin is the continental slope where comparatively large prospects could occur but remain undrilled at this time because of the expense of drilling in deep water. Furthermore, the methods do not provide estimates of how much oil and gas could be contained in parts of the area where data or interpretation may be inadequate.

There is also good reason to believe that the creaming method will underestimate future trends. Both of the creaming methods described use a field discovery model as a basis for projections that indicate the potential for additional new-field discoveries. Experience suggests, however, that the reserves of the fields will be revised upward from time to time and that these revisions will be greatest in the early years following a discovery. Therefore, projections based on field size data that do not take the possibility of future growth in reserves into account will most likely produce conservative estimates of further potential.

Other factors, not necessarily present in the Central Deep, that could lead to an underestimate of undiscovered potential include the following: (1) insufficient data—early trends based on data from one or two trap types could change markedly as exploration proceeds; (2) variation in water depth and distance from shore, pipelines, or platforms—significant variation will increase the creaming factors for field size or predrill area of prospect; and (3) future changes in economics, data quality, or geological interpretation—in many cases these changes will lead to an improvement of the . trends.

We conclude that this assessment of the petroleum potential of the Central Deep of the Gippsland basin is conservative. It should be carefully interpreted in the light of current geological knowledge to ensure that areas of significant potential have not been overlooked. Where possible, further assessment should be carried out by other methods to determine the full potential. Many of the other methods, however, suffer from the same limitations as those described in this paper.

ACKNOWLEDGMENTS

The writers thank John Houghton of the U.S. Geological Survey and an anonymous reviewer for critically reading the manuscript. The writers also thank BMR's computer-assisted cartography group for preparing the base map, and D. Lawry, N. Kozin, and C. Knight for drafting the figures. The paper is published with the permission of the Director, Bureau of Mineral Resources, Australia.

REFERENCES CITED

Baker, R. A., H. M. Gehman, W. R. James, and D. A. White, 1984, Geologic field number and size assessments of oil and gas plays: AAPG Bulletin, v. 68, p. 426–437.

Davies, O. L., 1961, Statistical Methods in Research and Production: London, Oliver and Boyd, 396 p.

Davis, J. C., and J. W. Harbaugh, 1983, Statistical appraisal of seismic prospects in Louisiana, Texas outer continental shelf: AAPG Bulletin, v. 67, p. 349–358.

Forman, D. J., and A. L. Hinde, 1985, Improved statistical method for assessment of undiscovered petroleum resources: AAPG Bulletin, v. 69, p. 106–118.

Hinde, A. L., 1984a, LAMDA: a computer program for estimating hypothetical petroleum resources using the "creaming method": Bureau of Mineral Resources, Australia, Record 1984/4, 23 p.

————, 1984b, VALAM and ARLAM: two computer programs for estimating hypothetical petroleum resources using prospect areas and field sizes: Bureau of Mineral Resources, Australia, Record 1984/29, 12 p.

Hocking, J. B., 1976, Gippsland basin in J. G. Douglas, and J. A. Ferguson, eds., Geology of Victoria: Geological Society of Australia special publication n. 5, p. 248–273.

Kendall, M. G., and A. Stuart, 1967, The Advanced Theory of Statistics, 2nd Edition, v. 2: London, Charles Griffin and Company, 690 p.

Meisner, J., and F. Demirmen, 1981, The creaming method: a Bayesian procedure to forecast future oil and gas discoveries in mature exploration provinces: Journal Royal Statistical Society A, v. 144, p. 1–31.

Schuenemeyer, J. H., and L. J. Drew, 1983, A procedure to estimate the parent population of the size of oil and gas fields as revealed by a study of economic truncation: Mathematical Geology, v. 15, p. 145–161.

Oil and Gas Resources of the U.S. Arctic

E. R. Schroeder[†]
Resource Assessment Task Group
National Petroleum Council
Washington, D.C.

The text presented is the final draft of the Report of the Resource Assessment Task Group to the Committee on Arctic Resources of the National Petroleum Council. The draft was submitted in July 1981. It provided a basis for further studies and analyses of real and potential problems by six other task groups during the remainder of that year. In December 1981, the National Petroleum Council submitted its report entitled "U.S. Arctic Oil and Gas" to the Secretary of Energy. The reader is referred to that publication for a complete synopsis of the work of all seven task groups.

This petroleum resource assessment is included in this volume to illustrate the organizational and data-handling problems encountered by the Resource Assessment Task Group and how these problems were overcome, as well as to present the results of our assessment of Arctic regions in which the petroleum industry will be actively exploring into the next millennium.

INTRODUCTION

Objectives

The fundamental objective of the task group was defined in a letter from the Secretary of Energy to Mr. C. H. Murphy, Jr., Chairman, National Petroleum Council dated April 9, 1980, in which the Secretary states (paragraph 3), "I request that the National Petroleum Council undertake a comprehensive study of the Arctic area oil and gas development. Specifically, the study should include: resource assessment information."

The scope of the task group's responsibility was further defined by the National Petroleum Council (NPC) Arctic Oil and Gas Resources Committee as follows:

The Resource Assessment Task Group is responsible for developing a comprehensive picture of the oil and gas resources in the defined Arctic area. In performing this function, they shall review and reconcile all pertinent information that is in the public domain and provide an analysis that will attempt to resolve conflicting information and define high potential areas. Where possible, a measure of the reliability of the estimates should be provided. Information shall be presented in a format that will relate to federal and state leasing schedules. Recommendation should be made for appropriate measures that should be taken to provide an early expansion of knowledge of reserves.

[†]E. R. Schroeder passed away in 1985, while in the People's Republic of China.

See Figure 1 for the organization of the committees established for this study.

Definition of Regions Studied

The U.S. Arctic region to be considered was defined by the NPC Arctic Oil and Gas Resources Committee as follows:

For the purpose of this study, the Arctic area is defined as territory under U.S. jurisdiction north of the Aleutians offshore and north of the Brooks Range onshore.

This sentence has been interpreted to include both the Beaufort and Chukchi seas portions of the Arctic Ocean north of Alaska (Figure 2).

Early in the study, the Coordinating Subcommittee of the Arctic Oil and Gas Resources Committee established the seaward outer limit for our consideration at 8,200 ft (2,500 m) water depth in the Bering Sea and at 8,200 ft (2,500 m), or lat. 74°N in the Arctic Ocean (Figure 2). The water depth limit is approximately in accord with a pending agreement among nation participants in the Law of the Sea Convention. I comment below on the matter of territorial boundary of the United States with Canada in the Beaufort Sea on the east, and with the Soviet Union in the Chukchi and Bering seas on the west.

Jurisdiction

A separate Jurisdictional Issues Task Group has considered political jurisdiction. The work of the Resource Assessment

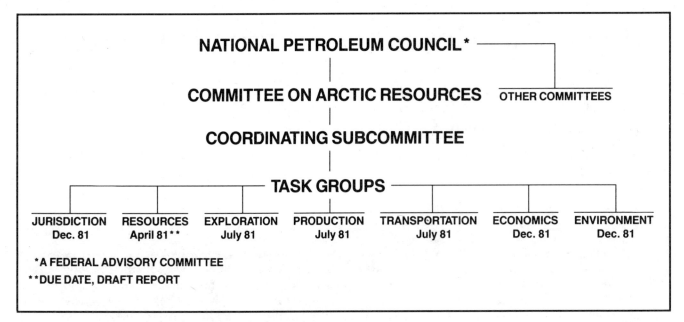

Figure 1. Organization of the Coordinating Subcommittee of the Committee on Arctic Resources, National Petroleum Council.

Task Group is of particular significance to both international and intranational entities. Onshore in Region I (Figure 2) there are three geographic areas (Figure 3): (1) the National Petroleum Reserve of Alaska (NPRA), (2) the Arctic National Wildlife Range (ANWR), and (3) that remaining area of the North Slope onshore, portions of which are under federal, state, and Arctic Slope regional corporation jurisdiction. The NPRA and the ANWR are both under federal jurisdiction, but they have different federal land designations. Consequently, they are under different rules and regulations as to their availability for oil and gas exploration and development.

The boundaries shown in Figures 2 and 3 are those applied by the U.S. Geological Survey (USGS) in its assessment work. USGS boundaries were used to ensure the opportunity for direct comparison of the task group's assessments with those of the USGS.

It is significant that some geologic basins that extend across international boundaries have, in the view of the task group, relatively high potentials for new petroleum resources. For example, the Beaufort shelf basin, which is bisected by the U.S.–Canadian border, is assessed as the most prospective of all individual basins for the discovery of new resources. In the area of the U.S.–Soviet Union, the North Chukchi and the Central Chukchi shelf basins in the Chukchi Sea and the Navarin basin shelf in the Bering Sea all have a high potential resource base. However, on the Beaufort slope and on both the Chukchi shelf and slope, attainable portions of the resource base are severely limited by the extremely hostile operating environment. This same observation applies, but to a lesser extent, in the case of the Navarin shelf on the west.

Sources of Technical Data

The charge to the Resource Assessment Task Group limited the group to data in the public domain. In accord with the data handling procedures of the National Petroleum Council, any proprietary information a member opted to provide the task group became, de facto, a part of the public domain. Because of the great amount of information on methods for execution of the study and the geologic and geophysical data already in the public domain on which to base estimates, this limitation was not an insurmountable obstacle to the task group in the execution of its work. Undoubtedly, many members also had access to and applied proprietary data in making their respective assessments on behalf of their companies. I comment further on limitations under the heading *Resource Assessment Methodology.*

Published references concerning resource assessment methodology are found in trade journals, professional publications, and special publications of government and professional societies. The great majority of information in the public domain concerning petroleum geology and geophysics in Alaska is in the memoirs, bulletins, circulars, and open-file reports of the USGS. Other useful information can be found in publications of the Alaska Geological Society, the state of Alaska, and AAPG as well as trade journals and other professional publications.

The bulk of the discovered resources in that part of Alaska considered by the task group occurs in the Prudhoe Bay oil and gas field on the North Slope, north of the Brooks Range. There are numerous published estimates of the reserves of oil and gas on the North Slope in Alaska in trade journals and government agency publications. The Resource Assessment Task Group elected to use reserve data as of August 1980 from the study prepared by W. D. VanDyke on behalf of the state of Alaska entitled "Proven and Probable Oil and Gas Reserves, North Slope, Alaska," dated September 25, 1980.

An important strength of the task group, which became evident in the course of the study, is the experience of the membership with Alaskan resources, as well as their broad experience as professional geologists, geophysicists, and

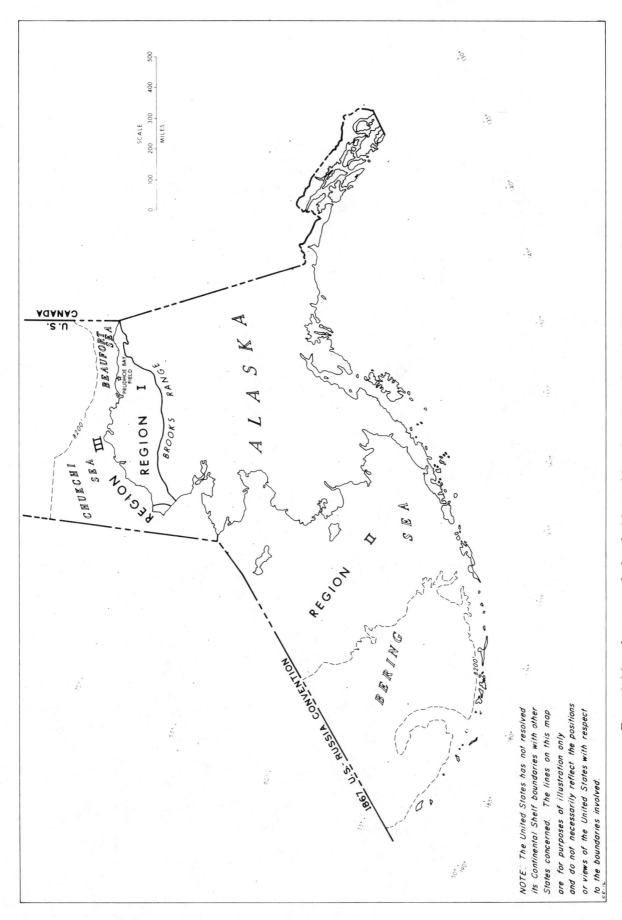

NOTE: The United States has not resolved
its Continental Shelf boundaries with other
States concerned. The lines on this map
are for purposes of illustration only
and do not necessarily reflect the positions
or views of the United States with respect
to the boundaries involved.

Figure 2. Map of regions studied in the National Petroleum Council's study of U.S. Arctic oil and gas.

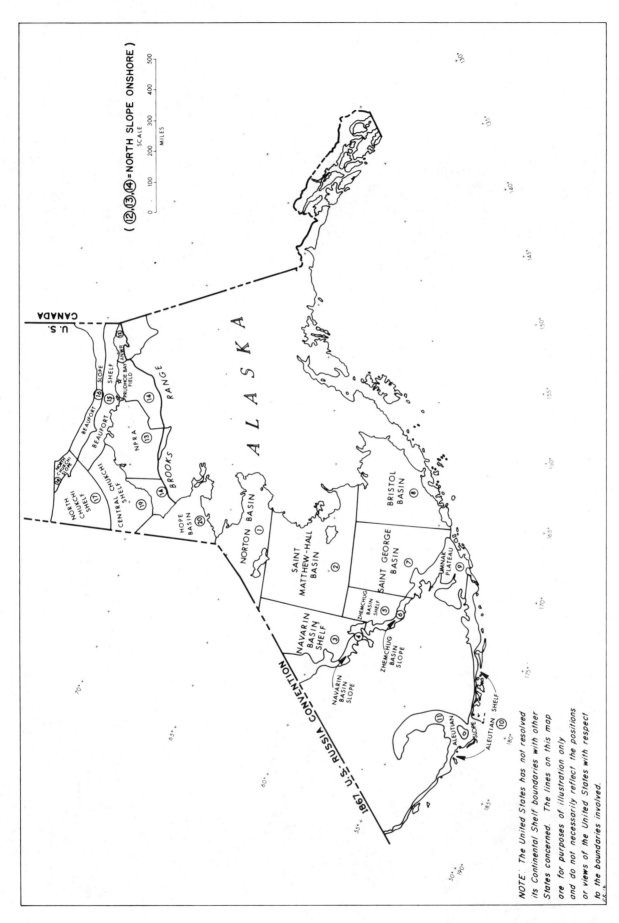

Figure 3. Map of basins and jurisdictional areas in the U.S. Arctic.

engineers active in the petroleum industry. The participants represent a total of 20 operating companies, consulting companies, and Federal and Alaska state government agencies active in industry.

ORGANIZATION

General

The primary task of the Resource Assessment Task Group, which is assessment of the oil and gas resources in the geologic provinces under consideration, required participation of the entire membership. The possibility of subdividing this main task into local areas and assigning teams to review these local areas was abandoned early as unworkable. Considering the experience of the membership, it was reasonably assumed that all were accustomed to using such concepts as petroleum source rock maturation, seismology, and reservoir rock development and would apply these in their assessments of Arctic Alaskan resources. A few members were called on to play a dual role in the Resource Assessment Task Group in addition to their basin assessment responsibility.

Definitions Work Group

A review of published resource assessments by other groups brought to light early in our work the need for terms and definitions of resources and categories of resources that are meaningful to nontechnical persons. To the extent possible, our definitions and text were to be as free as possible from industry jargon and words requiring constant recourse to specialized dictionaries. For this purpose the Definitions Work Group was established.

Methodology Work Group

The procedure (methodology) whereby assessment of resources in partially or unexplored geologic provinces is accomplished has become a specialized field within petroleum geology during the past decade. Application of computer-based statistical manipulation has allowed resource assessment to become progressively more sophisticated. Because of this evolution, a variety of concepts can now be applied to the task of resource assessment, and for this reason the Methodology Work Group was organized. The Resource Assessment Task Group was fortunate to have within its membership several individuals with extensive experience in resource assessment methodology. Given the nature of the study, much credit is due to these individuals for devising a statistically sound methodology within which the task group could function.

The Role of the U.S. Geological Survey

From the inception of the Resource Assessment Task Group, it was apparent that close communication would need to be established with the USGS. The USGS is unique in its variety of experience in Alaska, its volume of publications, and its publicly oriented professionalism; thus, it was our prime source of public information. Furthermore, the USGS has been occasionally responsible for assessments of Alaskan resources and has published evaluations of the methodology of resource assessment. Close communications between the Resource Assessment Task Group and the USGS was established via a liaison officer. It can be said without qualification that the task group enjoyed a most productive working relationship with the Survey.

The most important meeting of the Resource Assessment Task Group, in conjunction with the USGS, convened in Menlo Park, California, on January 28, 29, and 30, 1981. At this meeting, USGS staff members described to the task group all of the latest public information known to the USGS concerning the areas for analysis by the task group.

UNUSUAL HYDROCARBON OCCURRENCES

The oil and gas volumes that qualify as resources for consideration in this study are those which, according to data in the public domain, can be produced via conventional methods known and utilized in 1980. The term *conventional* as defined here means production of oil or gas to the surface from a well as a consequence of (1) natural pressure within the subsurface reservoir; (2) mechanical pumping of oil from the reservoir to the surface, where economically applicable; and (3) maintenance of reservoir pressure by means of water or gas injection, where economically applicable. (Definition of pressure maintenance as a conventional method is not uniformly accepted within the industry but is applied frequently, especially in frontier areas.)

The Resource Assessment Task Group briefly considered the possibility that both viscous oil and gas hydrates (both of which occur on the North Slope) could be considered resources for the purposes of this study. The conclusion in the case of viscous oil ($< 15°$ gravity API) is that while significant volumes have probably been encountered in the subsurface and perhaps will be produced economically, no information has been published that confirms this assumption. The volume of this unconfirmed resource cannot be quantified using published information.

Gas hydrates occur in uncalculated but reportedly large quantities in the soil and shallow subsurface of Alaska both onshore and offshore. This "frozen" gas is found in relatively dispersed form in permafrost (permanently frozen soil and subsoil), but nowhere in published literature, nor to our knowledge in any studies, has the means of economic production been reasonably postulated. Specifically excluded from the assessment are potentials from enhanced thermal or chemical oil recovery processes, as well as possible gas in low permeability formations, coal seams, or brines.

DEFINITIONS OF TERMS

Introduction

Early in the work of the Resource Assessment Task Group a request was made by a nontechnical member of the Coordinating Subcommittee of the NPC's Committee on Arctic Oil and Gas Resources that special effort be made to present the results of the work of all task groups in the most meaningful language possible for the benefit of those not accustomed to petroleum industry jargon nor to the technical vocabulary of the industry. Particular emphasis was placed on the need for lucid definition of the categories of resources that the Resource Assessment Task Group would identify. Thus, the following definitions were established. It is essential to keep in mind that these definitions are presented in the context of an assessment of the oil and gas resources both discovered and undiscovered in the Arctic areas under U.S. jurisdiction. Also, when the category of undiscovered oil or gas is elevated to the category of resource, certain assumptions are made concerning future economic conditions and technologic advances anticipated in the course of the discovery and development of these resources in Alaska. Special note should be made that all categories of reserves and resources reported in this study are *potentially recoverable*. The substantially larger amounts of oil and gas in place in reservoirs, of which recoverable quantities are a part, are not reported either in the text or tables.

Definitions

Resource: A known accumulation of or a new source of supply of naturally occurring oil or gas that is now or could be conventionally recoverable.

Proved reserves: The estimated quantities of naturally occurring oil or gas that geologic and engineering studies demonstrate with reasonable certainty are recoverable from known reservoirs under existing economic and operating conditions.

Probable reserves: The estimated quantities of naturally occurring oil or gas, additional to proved reserves, that geologic and engineering studies indicate are likely to be recoverable from partially defined reservoirs under existing economic and operating conditions.

Discovered resources: Proved reserves plus probable reserves.

Undiscovered resource base (undiscovered resources): A tentatively estimated, undrilled quantity of naturally occurring oil or gas that is presumed to exist in accord with regional geologic analyses and that would be conventionally recoverable from fields larger than 1 million barrels of oil equivalent (BOE) regardless of present accessibility or economics. Confirmation of the existence and quantity of this base is dependent on the result of future drilling.

Total oil and gas resources: The sum of discovered resources and undiscovered resources.

Economically attainable potential: the portion of the undiscovered resource base that is tentatively estimated to be economically recoverable by the year 2000, presuming no political restraints.

Oil equivalent gas: Gas volume that is expressed in its energy equivalent in barrels of oil and that is reconvertible to gas volume at 5,600 cu ft of gas per BOE.

Major field: An oil field that contains at least 50 million recoverable barrels of oil and oil equivalent gas.

Small field: An oil field that contains 1 to 50 million barrels of oil and oil equivalent gas.

Adequacy chance: The probability that at least one major field exists. (Corresponds in principle to the "marginal probability" of the USGS.)

Natural gas (categories): *Associated*—overlies and is in contact with crude oil in the reservoir. *Non-associated*—in reservoirs that do not contain significant crude oil.

Dissolved—in solution with crude oil in the reservoir.

Natural gas liquids: Portions of recoverable natural gas that are liquified at the surface in separators or in process plants.

RESOURCE ASSESSMENT METHODOLOGY

General

Among the activities of the Arctic Oil and Gas Resources Committee study, the work of the Resource Assessment Task group was unique in that it required response by each member individually to a series of geologically based problems. When aggregated these responses would provide the assessment that was the purpose of the task group. Prior to securing the input of the membership, however, a statistically sound and professionally acceptable methodology was devised which conformed with the requirements of the membership.

Methods for estimating undiscovered oil and gas resources prior to drilling are many and varied. Different approaches can be used for different levels of geologic knowledge and for different purposes of analysis. The recent trend is toward the use of sophisticated computer models incorporating a more realistic accounting of the geologic chance of success or failure and toward reporting the results as ranges of possible volumes in a probability curve format.

Assembly of the Resource Assessment Task Group under the sponsorship of the NPC offered an unusual opportunity to appraise oil and gas resources in specified areas in Alaska. Published assessments of resources in both partially explored and unexplored basins in the nation have been a responsibility of the USGS. Occasionally various companies companies and other entities have also published, what are of necessity, very general estimates of the resources in the state of Alaska. The main value of the various estimates was that they indicate persisting interest in Alaskan resources. It was made plain by the Department of Energy, however, that the fundamental purpose of our work was to provide policy and decision makers with some basis for future decisions and policies.

The Resource Assessment Task Group, incorporating the work of the USGS and representing the state of Alaska, was uniquely qualified to perform this task on the basis of geologic experience and knowledge of the fundamentals of industry economics. The Resource Assessment Task Group, however, labored under certain inherent limitations imposed on it by the nature of the NPC. The Council's study procedures require that all council work be performed under public scrutiny and that all data used be or become part of the public domain. It is unrealistic to expect that private industrial companies could or should present the results of proprietary exploration and engineering efforts to the public and to their competition.

It thus became the responsibility of the Resource Assessment Task Group, and particularly the Methodology Work Group within the task group to devise a method that would take advantage of the expertise offered by the Resource Assessment Task Group and at the same time observe stringent limitations imposed on that group by the Council's study procedures.

To ensure the broadest possible participation in the resource assessment, the NPC offered all members of the Council the opportunity for representation at the meetings in California in January 1981 and the opportunity thereafter to participate in the assessment. Several members accepted, bringing the total of task group members and new participants to 20.

Methodology Used

The analysis procedure devised and applied by the Resource Assessment Task Group can best be summarized by describing the sequence of steps pursued.

All members were

1. provided the opportunity (in meetings) to comment on the nature of the proposed project and express opinions as to how it could best be accomplished;
2. appraised of the limitations mentioned above and the significance of these limitations to the members and to operating companies;
3. provided bibliographies of the published technical data available concerning the regions to be assessed;
4. provided opportunity to concur in the methodology to be applied by the task group;
5. informed of most recent unpublished USGS geologic and geophysical information in the public domain; and
6. informed of the most recent USGS assessment of the basin to be assessed by the membership. (The assessment was subsequently released by the USGS in Open-File Report 81-192, February, 1981.)

Thereafter, each assessor privately

7. assessed the oil and gas resources of each basin for which he or she felt qualified to make an assessment and to respond (the assessments were expressed in a probability format); and
8. mailed his or her individual response directly to the accounting firm Price Waterhouse & Co., which was retained the NPC and was obligated to maintain the anonymity of each member's response.

Price Waterhouse & Co. then

9. aggregated the individual anonymous responses to arrive at a single set of probability curves, which expressed the composite estimate of the assessors for each basin; and
10. monitored the processing of these numerical data to arrive at a Monte Carlo–based distribution of possible total resources for basin groupings.

The methodology applied by the Resource Assessment Task Group is described as a modified Delphi technique. As implied in the name, the technique consists of averaging subjective expert opinion of potential resources. In our case, the assessors served as the experts, and aggregation of the results by Price Waterhouse & Co. provided the composite of these opinions. The departure from the standard Delphi

application was that the participants were not provided the opportunity to discuss among themselves the significance of the various geologic and geophysical factors that bear on the merit of the individual basins. This was precluded by the nature of the NPC study procedures. Considering the relatively large number of participants in the exercise, however, and the depth of their experience and expertise, we do not consider that this was a critical departure.

The main advantages of the Delphi technique are its ease of application and its full probability format. The principal disadvantages are that it contains no built-in scaling factor or documentation of the direct method, in other words, "one must know how expert are the experts in order to assess the assessment."

The results of the resource assessment by the task group can be generally compared with assessments published by the USGS. Certain differences in methodology preclude exact comparison. For example, the USGS uses a different method of averaging, which if applied to the task group's data, would yield a different composite probability curve. The task group used a consistent minimum field size in appraising geologic risk, whereas the USGS has at various times used different minimum field sizes. There also is an important difference in presentation format. The USGS estimate of high side resource was reported at the 5% probability level, whereas the high side estimate by the task group was reported at the 1% level. This inevitably results in different apparent high side potential estimates from the two groups.

ARCTIC RESOURCE ASSESSMENT

Basic Results

Table 1 provides the basic findings derived from the resource assessment. The results are expressed in potentially recoverable billions of barrels of undiscovered oil and oil equivalent gas (BOE). The 20 areas studied (Figure 3) are identified in Table 1 and three sets of data are tabulated. The USGS numbers are the latest risked mean assessment by the USGS of undiscovered conventionally recoverable oil and gas resources for each of the areas analyzed by the task group. The first set of NPC data is the task group's estimate of undiscovered resource base, which is the total recoverable volume of oil or gas in fields greater in size than 1 million BOE. The second set of NPC numbers illustrates the amount of the undiscovered resource base that, in the opinion of the task group, will be economically attainable.

There are three columns in each of the NPC sets of numbers in Table 1. The "adequacy chance" column illustrates the estimate of the assessors of the likelihood of there being at least one oil and gas field greater than 50 million BOE. The "risked mean" column illustrates the average of many possible values for the volume of undiscovered resource base that may exist in that basin. Realistically, these possibilities range from zero to the indicated high side potential. Thus, the risked mean is not the only answer; it is merely a statistical reference point in a wide range of uncertainty. Only the risked mean columns are for comparison with USGS assessments. The "risked high side" column is that volume for which there is, in the

Table 1. Undiscovered potentially recoverable hydrocarbons in the U.S. Arctic (in billion BOE, gas conversion 5.6 tcf/billion BOE).

Original Areas	USGS 1981 Risked Mean[c]	NPC Undiscovered Resource Base[a]			NPC Economically Attainable Potential[b]		
		Adequacy Chance Major Field[d] (%)	Risked Mean[e]	Risked[f] High Side	Adequacy Chance Major Field[d] (%)	Risked Mean[e]	Risked[f] High Side
1. Norton	0.38	43	0.87	7.6	29	0.47	6.1
2. St. Matthew	0.00	26	0.06	1.3	4	0.01	0.3
3. Navarin Shelf	1.68	41	4.00	44.0	30	2.22	36.0
4. Navarin Slope	0.15	31	0.23	2.8	6	0.04	1.2
5. Zhemchug Shelf	0.07	27	0.17	2.3	9	0.05	1.4
6. Zhemchug Slope	0.00	17	0.02	0.6	2	0.00	0.1
7. St. George	0.83	47	2.18	23.0	38	1.31	19.6
8. Bristol	0.39	47	1.32	10.8	38	0.75	8.9
9. Umnak	0.00	21	0.04	0.8	2	0.00	0.1
10. Aleutian Shelf	0.00	20	0.03	0.9	2	0.01	0.2
11. Aleutian Slope	0.00	22	0.07	1.5	2	0.01	0.3
12. Wildlife	3.29	70	3.71	21.7	67	2.89	20.2
13. NPRA	3.60	79	4.69	24.0	59	2.95	21.6
14. Rest of North Slope	4.77[g]	79	4.37	23.3	69	3.19	20.8
15. Beaufort Shelf	13.21	88	12.88	59.0	60	6.97	50.0
16. Beaufort Slope	1.54	57	2.46	20.4	29	0.73	12.0
17. North Chukchi Shelf	1.45	50	2.14	17.0	28	0.72	12.0
18. North Chukchi Slope	0.40	34	0.65	6.2	5	0.05	1.8
19. Central Chukchi	1.17	62	3.26	20.5	39	1.47	15.7
20. Hope	0.08	40	0.43	4.6	32	0.30	4.4
	33.01	100	43.58	99.0[d]	100	24.14	74.0[d]

[a]Includes NGL. Minimum field size included is 1 million bbl everywhere.
[b]Includes NGL. Minimum economic field size variable but generally >50 million bbl or more.
[c]Excludes NGL. Minimum field size included ranges from <1 to 400 million bbl.
[d]Percentage chance of finding at least one major field.
[e]Risked means are averages of a wide range of possibilities and are reported to two decimals so as to avoid rounding errors.
[f]Risked high side potentials are at 1% probability. *They cannot be added directly.*
[g]Equals total USGS assessment of North Slope onshore minus the special DOI assessments of NPRA and ANWR.

opinion of the task group, only a 1% chance of being exceeded. Note particularly that because Monte Carlo summation techniques were used to determine all high side estimates, it is statistically incorrect to add these estimates.

Note that the total risked mean of the USGS falls between the total risked undiscovered resource base and the total risked mean economically attainable potential of the Resource Assessment Task Group. It is also significant that on a per basin basis, there is general conformity between the assessment of the Resource Assessment Task Group and that of the USGS. The most notable difference is that the risked mean undiscovered resource base of the task group in the Bering Sea is more than twice that of the USGS, although there is good agreement on which basins in that region are most prospective.

To be included in the complete resource assessment, all assessors were required to submit their estimates of four basic parameters essential to the calculation of the undiscovered resource base. Several other optional estimates, were also solicited of the assessors. These optional estimates plus the first four, provided the basis for calculation of economically attainable potential and additional assessments (see Tables 2–5). The significance of the economically attainable potential, so derived, as a guidepost in planning resource exploration and development has been superceded by the work of the Economic Study Group. The reader is referred to the report of the Economic Study Group for detailed application of economic factors to the risked mean undiscovered resource base on Tables 1 and 2.

Summary of Assessments by Hydrocarbon Type

Table 2A presents a breakdown of the assessed undiscovered resource base by area according to hydrocarbon type. A summation of the total 20 areas is also presented on this table along with the risked mean and high side estimates for billion BOE, oil, total gas, natural gas liquids (NGL), associated and dissolved (A & D) gas, and nonassociated (NA) gas.

Table 2B presents the same breakdown applied to the economically attainable potential volumes. Tables 2A and B should be compared when considering preliminary priorities for further exploration. The estimated impact of technical problems and high operating costs is particularly apparent in the case of the attainable potential in the Beaufort Slope and North Chukchi shelf areas. Refer to the report of the Economics Study Group for detailed economic analyses.

Regional Subtotals

Regional subtotals in billion BOE are presented in Table 3. In the course of our work, we divided the area under

Table 2A. Summary of U.S. Arctic area assessments—undiscovered resource base.

Original Areas	Billion BOE			Oil (billion bbl)			Total gas (tcf)			NGL [a] (billion bbl)			A & D Gas (tcf) [b]			NA Gas (tcf) [c]		
	%[d]	Risked Mean	High Side	%	Risked Mean	High Side	%	Risked Mean	High Side	%	Risked Mean	High Side	%	Risked Mean	High Side	%	Risked Mean	High Side
1. Norton	43	0.87	7.6	43	0.32	2.8	43	2.74	24.0	43	0.06	0.6	43	0.44	3.8	43	2.30	19.9
2. St. Matthew	26	0.06	1.3	5	0.01	0.2	26	0.26	5.4	26	0.00	0.1	5	0.01	0.2	26	0.25	5.2
3. Navarin Shelf	41	4.00	44.0	41	2.31	25.0	41	8.29	91.0	41	0.22	2.4	41	3.38	37.0	41	4.91	54.0
4. Navarin Slope	31	0.23	2.8	24	0.10	1.2	31	0.64	8.0	31	0.02	0.2	24	0.15	1.9	23	0.49	6.1
5. Zhemchug Shelf	27	0.17	2.3	27	0.10	1.4	21	0.32	4.6	21	0.01	0.1	27	0.14	2.0	17	0.18	2.6
6. Zhemchug Slop	17	0.02	0.6	3	0.01	0.2	5	0.08	2.0	5	0.00	0.1	3	0.01	0.2	5	0.07	1.8
7. St. George	47	2.18	23.0	47	1.16	12.2	47	5.10	53.0	47	0.11	1.2	47	1.59	16.6	47	3.51	37.0
8. Bristol	47	1.32	10.8	47	0.62	5.0	47	3.49	28.0	47	0.08	0.7	47	0.85	6.9	47	2.64	21.0
9. Umnak	21	0.04	0.8	6	0.02	0.3	7	0.12	2.4	7	0.00	0.1	6	0.02	0.4	7	0.10	2.0
10. Aleutian Shelf	20	0.03	0.9	4	0.01	0.2	7	0.12	3.3	7	0.00	0.1	4	0.01	0.3	7	0.11	3.0
11. Aleutian Slope	22	0.07	1.5	6	0.02	0.4	8	0.25	5.2	8	0.01	0.1	6	0.03	0.6	8	0.22	4.6
12. Wildlife	70	3.71	21.7	70	2.34	13.7	70	6.55	38.0	70	0.20	1.2	70	3.30	19.3	70	3.25	19.0
13. NPRA	79	4.69	24.0	79	2.07	10.6	79	12.81	65.0	79	0.33	1.7	79	2.85	47.6	79	9.96	51.0
14. Rest of North Slope	79	4.37	23.3	79	2.14	11.4	79	10.96	58.0	79	0.28	1.5	79	3.10	16.5	79	7.86	42.0
15. Beaufort Shelf	88	12.88	59.0	88	8.18	37.4	88	22.38	102.0	88	0.70	3.2	88	11.70	54.0	88	10.68	49.0
16. Beaufort Slope	57	2.46	20.4	57	1.33	11.1	57	5.51	46.0	57	0.15	1.3	57	1.80	15.1	57	3.71	31.0
17. N. Chukchi Shelf	50	2.14	17.0	50	1.16	9.1	50	4.82	38.0	50	0.12	1.0	50	1.65	13.0	50	3.17	25.0
18. N. Chukchi Slope	34	0.65	6.2	34	0.33	3.1	34	1.54	14.7	34	0.05	0.4	34	0.47	4.5	34	1.07	10.3
19. Central Chukchi	62	3.26	20.5	62	1.74	10.9	62	7.47	47.0	62	0.19	1.2	62	2.48	15.5	62	4.99	31.3
20. Hope	40	0.43	4.6	40	0.18	2.0	40	1.21	13.0	40	0.03	0.3	40	0.27	2.9	40	0.94	10.1
Areas 1–20 Summation	100	13.58	99.0	100	24.15	55.0	100	97.66	203.0	100	2.56	5.5	100	34.25	79.0	100	60.41	130.0

[a] NGL is natural gas liquids.
[b] A & D gas is associated and dissolved gas.
[c] NA gas is nonassociated gas.
[d] % is the change greater than 0.05 billion BOE (gas conversion 5.6 tcf/billion bbl). Risked highsides are at 1% probability. Oil chance applies to A & D gas; total gas chance applies to NGL.

Table 2B. Summary of U.S. Arctic area assessments—undiscovered economically attainable potential.[a]

Original Areas	Billion BOE %	Risked Mean	High Side	Oil (billion bbl) %	Risked Mean	High Side	Total gas (tcf) %	Risked Mean	High Side	NGL (billion bbl) %	Risked Mean	High Side	A & D Gas (tcf) %	Risked Mean	High Side	NA Gas (tcf) %	Risked Mean	High Side
1. Morton	29	0.47	6.1	29	0.19	2.2	25	1.40	18.1	25	0.03	0.4	29	0.22	2.9	25	1.18	15.2
2. St. Matthew	4	0.01	0.3	0	0.00	00.0	2	0.06	1.4	2	0.00	00.0	0	0.00	0.1	2	0.06	1.4
3. Navarin Shelf	30	2.22	36.0	26	1.38	21.0	21	4.08	70.0	21	0.11	1.8	26	1.67	28.0	21	2.41	41.0
4. Navarin Slope	6	0.04	1.2	6	0.02	0.7	3	0.07	2.4	3	0.00	0.1	6	0.02	0.6	3	0.05	1.8
5. Zhemchug Shelf	9	0.05	1.4	8	0.03	0.8	6	0.10	2.5	6	0.00	0.1	8	0.00	1.1	5	0.10	1.4
6. Zhemchug Slope	2	0.00	0.1	0	0.00	00.0	1	0.00	0.2	1	0.00	00.0	0	0.00	00.0	1	0.00	0.2
7. St. George	28	1.31	19.6	32	0.76	10.3	26	2.73	41.0	26	0.06	0.9	32	0.85	12.6	26	1.88	28.0
8. Bristol	38	0.75	8.9	33	0.39	4.2	27	1.77	21.0	27	0.04	0.5	33	0.43	5.1	27	1.34	15.9
9. Umnak	2	0.00	0.1	1	0.00	00.0	1	0.00	0.2	1	0.00	00.0	1	0.00	00.0	1	0.00	0.1
10. Aleutian Shelf	2	0.01	0.2	1	0.00	0.1	1	0.02	0.4	1	0.00	00.0	1	0.00	00.0	1	0.02	0.4
11. Aleutian Slope	2	0.01	0.3	2	0.00	0.2	1	0.03	0.2	1	0.00	00.0	2	0.00	00.0	1	0.03	0.2
12. Wildlife	67	2.89	52.2	62	1.84	13.1	57	5.01	36.0	57	0.15	1.1	62	2.52	18.3	56	2.49	18.0
13. NPRA	59	2.95	21.6	59	1.43	9.8	49	7.44	58.0	49	0.19	1.5	59	1.65	12.8	49	5.79	45.0
14. Rest of North Slope	69	3.19	20.8	65	1.65	10.9	58	7.53	54.0	58	0.1,	1.4	65	2.13	15.2	58	5.4	39.0
15. Beaufort Shelf	60	6.97	50.0	59	4.74	32.0	49	10.63	82.0	49	0.33	2.6	59	5.56	43.0	48	5.07	39.0
16. Beaufort Slope	29	0.73	12.0	23	0.53	7.9	10	0.96	21.0	10	0.03	0.6	23	0.32	7.0	10	0.64	14.0
17. N. Chukshi Shelf	28	0.72	12.0	22	0.46	6.5	15	1.30	22.0	15	0.03	0.6	22	0.45	7.6	15	0.85	14.5
18. N. Chukshi Slope	5	0.05	1.8	4	0.04	1.2	2	0.07	2.4	2	0.00	0.1	4	0.02	0.7	2	0.05	1.7
19. Central Chukchi	39	1.47	15.7	34	0.88	8.7	26	2.88	34.0	26	0.07	0.9	34	0.96	11.1	26	1.92	22.0
20. Hope	32	0.30	4.4	21	0.10	1.6	31	0.98	12.3	31	0.02	0.3	21	0.22	2.7	29	0.76	9.5
Areas 1–20 Summation	100	24.14	74.0	100	14.44	45.0	100	47.06	145.0	100	1.25	4.0	100	17.02	61.0	100	30.04	88.0

[a]See Table 2A footnotes for further details.

Table 3. Regional subtotals—undiscovered billion BOE.

Region	Undiscovered Resource Base			Economically Attainable Potential		
	Adequacy Chance Major Field (%)	Risked Mean	Risked High Side[a]	Adequacy Chance Major Field (%)	Risked Mean	Risked High Side[a]
Region II—Bering (Areas 1–11)	99	12.77	37.0	96	9.03	32.0
Region I—North Slope Onshore I (Areas 12–14)	99	8.99	52.0	87	4.87	41.0
Region III—Beaufort and Chukchi (Areas 15–20)	100	21.82	67.0	92	10.24	53.0
Region Totals	100	43.58	99.0	100%	24.14	74.0

[a]Risked high side potentials are at 1% probability. *They cannot be added directly.*

Table 4. U.S. Arctic grand totals—NPC estimates of potentially recoverable hydrocarbons.

Type	Undiscovered Resource Base			Economically Attainable Potential		
	Adequacy Chance Major Field (%)	Risked Mean	Risked High Side[a]	Adequacy Chance Major Field (%)	Risked Mean	Risked High Side[a]
Oil (billion BOE)[b]	100	43.58	99.0	100	24.14	74.0
Oil (billion bbl)	100	24.15	55.0	100	14.44	45.0
Total gas (tcf)	100	94.66	203.0	100	47.06	145.0
NGL (billion bbl)	100	2.56	5.5	100	1.25	4.0
A & D gas (tcf)	100	34.25	79.0	100	17.02	61.0
NA gas (tcf)	100	60.41	130.0	100	30.04	88.0

[a]Risked high side potentials are at 1% probability. *They cannot be added directly.*
[b]Gas conversion 5.6 tcf/billion bbl.

consideration into three regions (Figure 2). Region I includes three areas of separate jurisdiction onshore. The jurisdictional boundaries cross geologic province boundaries. Region II encompasses 11 geologic basins or provinces offshore in the Bering Sea, and Region III encompasses 6 provinces offshore in the Beaufort and Chukchi seas. It is interesting to note the contrast between Regions I and II. Region I has, in the opinion of the task group, a higher risked mean potential but a lower risked high side potential as estimated at the 1% probability level. This difference most likely reflects the relatively greater quantity of geologic data in Region I where oil and gas reserves in significant volume have been proved and where relatively extensive exploration drilling and geophysical analysis have been performed. Consequently, the assessors have a better grasp of the future potential of Region I compared with Region II. Although there are numerous geologic provinces involved in Region II, there are no drilling data in the public domain for this region. Thus, while the risked mean assessment is thought to be conservative, the 1% probability of very favorable results among one or more of the geologic provinces has been considered.

Again, it is statistically incorrect and meaningless to sum arithmetically the risked high side numbers presented on Table 3.

Hydrocarbon Type

Table 4 illustrates the hydrocarbon types that make up the total of the risked mean assessment, which totals about 44 billion BOE for the undiscovered resource base. The types are displayed as oil in billions of barrels, total gas in trillions

of cubic feet (tcf), and natural gas liquids in billions of barrels. The total gas is made up of associated and dissolved (A & D) gas and nonassociated gas (both in tcf). The adequacy chance is placed at 100%, because under the assumption of statistical independence for the individual basins, the probability that all 20 basins are barren of any major fields is less than 1 in 1 million.

Again, the economically attainable potential is less than the undiscovered resource base because it reflects the limiting effect of economic factors on oil and gas exploration and production.

Average Assessment Factors

To complete the resource assessment, certain estimates of average percentages and ratios were requested of the task group members. These useful estimates, presented on Table 5, were not derived on a per basin basis. All factors except the economically attainable oil and gas fractions are remarkably similar to the corresponding historical averages experienced in extensively explored areas in the conterminous United States.

Assessment Limitations

It is important that those who use resource assessments of the type generated by this task group be aware that the numbers presented here do not constitute quantitative predictions of volumes of oil and gas that will be discovered in any one of the geologic provinces assessed. Many areas with estimated potential will be nonproductive. We can hope that one or more will yield ultimate recoveries approaching the corresponding high side estimate. It is more

Table 5. Average assessment factors for total study area.

Oil fraction of total billion BOE	55%	
Postulated oil recovery efficiency	33%	
Postulated gas recovery efficiency	77%	Of total Arctic
NGL bbl/million cu ft gas	27	resource
A & D gas cu ft/bbl oil	1420	base
Small-field % of major field mean	14%	
Economically attainable fraction		(Chiefly in major
of oil resource base	60%	fields)
Economically attainable fraction		(Chiefly in major
of gas resource base	50%	fields)

likely that the risked mean total undiscovered resource base for all areas (44 billion BOE) will be closest to reality. The assessment, however, does provide the basis for establishing priorities in future petroleum planning and policy. It also provides a quantitative basis for comparison of the assessed basins with others similarly assessed, and it indicates to the public and the energy industry the basic merit of the areas reviewed from a petroleum exploration point of view.

Basin Priority for Future Exploration

The results of our assessment clearly indicate that significant additional volumes of oil and gas can be anticipated in the three regions considered. If it is the primary objective of policy makers to enhance petroleum supply, then exploration in all high potential areas of the U.S. Arctic should be expedited.

In marine frontier areas the greatest likelihood for significant new petroleum discoveries is beneath the Beaufort Shelf, immediately offshore from the North Slope in the Beaufort Sea. Operational difficulties and the resulting high costs, however, may preclude thorough short-term exploration of this geologic province in the deeper water remote from the coast.

In the Bering Sea, the Bristol and St. George basins and the Navarin basin shelf all offer significant opportunity for new oil and gas resources. The remote location of the Navarin basin imposes limitations on what can be anticipated there in the short term, but this basin does have the second highest high side potential of all basins assessed.

National Significance of This Study

Resource assessments of undiscovered recoverable resources in the total United States and in just the conterminous United States have been made by other groups, but again direct comparison with the task group results in the Arctic is not technically valid. However, the risked mean undiscovered resource base for the U.S. Arctic, as determined by the Resource Assessment Task Group at 44 billion BOE, may constitute a significant portion (perhaps as much as 25–40%) of the total undiscovered recoverable oil and gas resources remaining within the jurisdiction of the United States. An even greater proportion of the total undiscovered recoverable *oil* resource in the United States (perhaps as much as 30–50%) may occur in the U.S. Arctic. Considering these high proportions and given that the bulk of the Arctic region assessed by the Resource Assessment Task Group falls within the jurisdiction of the U.S. Federal

Government, it is apparent that conclusive assessment by drilling of a significant portion of the undiscovered recoverable resources within the total United States is dependant on initiatives by the federal government.

It is the view of the Resource Assessment Task Group that among the many considerations bearing on the selection of areas to be offered for leasing by any government agency, the apparent relative potential for nationally significant volumes of new recoverable oil and gas resources among the areas available should be an important, if not the primary, consideration. In the U.S. Arctic in particular, the severe climatological, marine technological and environmental obstacles, and the consequent high cost and long lag time between leasing and resource production imposed by these obstacles, practically require that the most promising work areas be given prime consideration.

Table 6 lists the ten most promising areas, of the twenty considered in this study in the order of their map code number (Figure 3). It is the opinion of the task group that these areas collectively should be given priority consideration when lease sales are scheduled.

Until all of these areas are adequately explored by drilling, the real oil and gas resource potential of the U.S. Arctic will remain only partially defined.

CONCLUSIONS

The Resource Assessment Task Group presents the following conclusions in response to the request of the Secretary of Energy and in fulfillment of the charge to the task group by the NPC Arctic Oil and Gas Resources Committee. Substantiation of these conclusions can be found in the text of the report. Only significant conclusions on matters of importance to policy and decision makers are presented below.

Total U.S. Arctic Resources

• Among the 20 geologic provinces or jurisdictional areas considered collectively, it is estimated that 44 billion recoverable barrels of undiscovered oil and oil-equivalent gas (BOE) can be reasonably expected to be present, primarily in fields containing in excess of 50 million bbl. This is in addition to the 16 billion bbl of discovered resources.

• Taking the optimistic view, at the 1% level of probability, in excess of 99 billion BOE may be present among the 20 geologic provinces.

• There is an estimated 99% probability that the total undiscovered recoverable oil and gas resources in the U.S. Arctic will exceed approximately 13 billion BOE.

• Of the 44 billion BOE, approximately 24 billion may prove to be economically producible. Improvement in the economic climate or in technology could permit a higher proportion of the undiscovered resource base to become attainable.

• At the optimistic 1% probability level, as much as 74 billion bbl of the 99 billion undiscovered recoverable barrels may be economically producible.

• Of the undiscovered resource base, approximately 55% may be oil.

Table 6. Ten most promising areas to be offered for leasing.

Map Code	Name	Region
3	Navarin Shelf	Bering Sea
7	St. George Basin	Bering Sea
8	Bristol Basin	Bering Sea
12	ANWR	North Slope
13	NPRA	North Slope
14	North Slope (other)	North Slope
15	Beaufort Shelf	Arctic Ocean
16	Beaufort Slope	Arctic Ocean
17	North Chukchi Shelf	Arctic Ocean
19	Central Chukchi Shelf	Arctic Ocean

• Of the undiscovered resource base, 60% of the oil and 55% of the gas is considered likely to be economically recoverable. The economically attainable resources will occur primarily in very large fields.

Concerning Resources by Region

• The discovered resources in the area studied consist of approximately 16.5 billion BOE of which 10.2 billion (62%) is oil. All of these reserves are in Region I under Alaska state jurisdiction and the bulk of this is in the onshore area of the Prudhoe Bay field. Other more recent discoveries lie partly beneath state waters immediately offshore.

• The minor accumulations of both oil and gas discovered in the National Petroleum Reserve in Alaska (NPRA) are considered by the task group to be uneconomical, and thus these quantities are not included here as discovered resources.

• It can be reasonably anticipated that approximately 13 billion BOE of undiscovered recoverable resources may be present in Region I (North Slope, onshore).

• In Region II (the Bering Sea) it can be reasonably anticipated that 9 billion BOE of undiscovered recoverable resources may be present.

• In Region III (the Beaufort and Chukchi seas), 22 billion BOE of undiscovered recoverable resources may be present in the subsurface.

• The impact of technological obstacles and high operating costs in the marine areas (Regions II and III) is apparent in the reduced quantities that the task group considers economically attainable. This impact cuts the risked mean resource base in Region III by approximately 50% to a figure only slightly larger than the economically attainable risked mean for Region I.

Other Conclusions

• There is remarkable concurrence between the most recent assessments of the risked mean resource base assessed by the USGS and that assessed by the Resource Assessment Task Group (Table 1). There is qualitative agreement on the recoverable resources that may be present; there is agreement on the balance among the three jurisdictional areas assessed in Region I (North Slope, onshore), and there is also general agreement on which of the other geologic provinces offer the greatest opportunity. The most significant geologically based difference is that the NPC mean resource base in the Bering Sea region is more than double the USGS assessment mean.

Other apparent quantitative differences are attributable to differences in resource assessment methodology and in estimation of the impact of technological and economic factors.

• Certain geologic provinces that extend to or across international borders (the North Chukchi slope, the Hope basin, and Havarin basin shelf to the west and the Beaufort shelf to the east) offer significant promise for the discovery of new oil and gas resources.

• Governing agencies can best ensure the delineation of significant new oil and gas resources in the Arctic regions under U.S. jurisdiction in the short term by expediting continued exploration of the North Slope onshore, the Beaufort shelf immediately adjacent to the north, and the most promising areas in the Bering Sea.

• No further resource assessment analyses of the nature completed by this task group are warranted for these same regions until such time as the most promising geologic provinces and jurisdictional areas have been tested by drilling and significant new scientific data are available.

SUMMARY

The most significant results of the investigations and analyses by the Resource Assessment Task Group can be summarized as follows.

The discovered oil and gas resources (proved plus probable reserves) in the United States Arctic as of August 1980 are reported by the state of Alaska to be approximately 16 billion bbl of oil and oil equivalent gas. The total potentially recoverable, undiscovered oil and gas resources in the United States Arctic are estimated by the task group to be approximately 44 billion bbl of oil and oil equivalent gas.

The task group estimates that there is 1% probability that the total undiscovered recoverable oil and gas resources in the United States Arctic could exceed 99 billion bbl of oil and oil equivalent gas. There is an estimated 99% probability that the total undiscovered recoverable oil and gas resources will exceed approximately 13 billion bbl of oil and oil equivalent gas in the United States Arctic.

Resource assessments of the type generated by the task group are essentially estimates of the range of recoverable volumes that are likely to be present in the subsurface. They do not necessarily constitute predictions of volume that will be discovered in any one of the areas assessed.

To expand new and accessible United States Arctic resources in the shortest term possible, exploration leasing and drilling should be expedited in onshore areas of the North Slope and in offshore areas of the Beaufort shelf basin, as well as in those high potential areas in the Bering Sea identified in this study.

There is remarkable qualitative and reasonable quantitative similarity between the results of this task group's study and the most recent published assessments of these same areas by the U.S. Geological Survey.

A Play Approach to Hydrocarbon Resource Assessment and Evaluation[1]

L. P. White
Office of Minerals Policy and Research Analysis
U.S. Department of the Interior[2]

A methodology for assessing the oil and gas potential and forecasting the associated exploration activity in a frontier or partially explored petroleum basin is presented here. The models discussed are an integral part of a microeconomic simulation of petroleum endowment, exploration, development, production, transportation, and distribution. The first two components of this simulation, the geology (endowment) model and the exploration model, are the main topics of this paper. The geology model simulates an inventory of prospects and generates an associated resource assessment. The exploration model simulates the economic evaluation of these prospects and the drilling decisions for each, generating a sequence of discoveries (and dry wells) over time that form an inventory of deposits to be evaluated for development. The learning process in exploration is explicitly incorporated in the simulation through Bayesian revision of the simulated explorationist's perception of the geology. The approach has proved useful in the evaluation of alternative land use policies and should also be useful in exploration planning at the play and basin levels.

INTRODUCTION

This paper presents a methodology developed by the U.S. Department of the Interior to simulate oil and gas exploration activity for petroleum provinces. The analytical construct is an integral part of a larger microeconomic simulation of petroleum exploration, development, production, transportation, and distribution that has been used to evaluate public policy alternatives for the National Petroleum Reserve in Alaska (NPRA).

The NPRA is an area of approximately 37,000 sq mi located on the northern coast of Alaska west of Prudhoe Bay. The Naval Petroleum Reserves Production Act of 1976 required the President to conduct a study to determine the best overall procedures to be used in the development, production, transportation, and distribution of any petroleum resources in the Reserve and to access the economic and environmental consequences of alternative procedures. The Office of Minerals Policy and Research Analysis was directed to develop an economic and policy analysis in support of the legislative requirement. It has designed a petroleum process and decision model that probabilistically simulates the major activities involved in oil and gas exploration, development, production,

transportation, and distribution. Public policy alternatives are tested by the model to evaluate their economic consequences over an extended time period (e.g., 50 years). The modeling philosophy is process and decision oriented to capture the important interdependencies among the various petroleum activities and the sensitivity of economic decision making to various public policies. (For a more detailed discussion of the relative merits of a process-oriented approach, see Kaufman et al., 1975, and Eckbo et al., 1978.)

The model allows partitioning of the surface area of large basins or provinces into activity areas (delineated surface areas of arbitrary size, shape, and number); these serve as the basic geographic frame of reference for analyzing petroleum or other land use activities in the province. The size, shape, number, land use classification, connecting transportation corridors, and availability sequence of activity areas are major public policy options for which alternative policies and procedures can be tested. For example, a province can be divided into two sets of activity areas, one set representing areas closed to petroleum activity and the other set representing areas that could be opened for petroleum activity according to a particular schedule. Figure 1 presents an example set of activity areas for the NPRA.

Among the activities that the integrated model explicitly simulates are two of the major economic decisions that occur repeatedly in the development of a petroleum province. The first decision involves the determination of whether a particular prospect merits testing with an exploratory well;

[1]Published previously in "The Economics of Exploration for Energy Resources," JAI Press, 1981.
[2]Present address: Exxon Production Research Company, Houston, Texas.

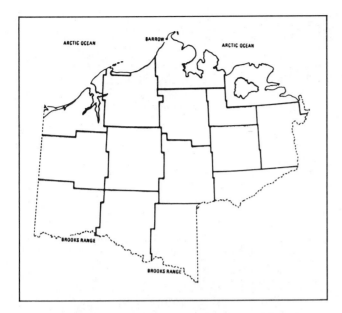

Figure 1. Example activity areas for the National Petroleum Reserve of Alaska (March 1, 1979).

the second, whether the resources contained in a discovered deposit or combination of deposits merit development, production, transportation, and distribution to market. This paper addresses the procedures developed to simulate the exploration decision. In the following sections an overview of the play approach is presented, followed by a discussion of alternative approaches. A geology model and an exploration model based on the play approach are then developed.

OVERVIEW OF THE PLAY APPROACH

The exploration process is simulated through the integration of two independent models: a geology model, which is based on a probabilistic assessment of the most important geologic parameters in a province, and an exploration model, which simulates the search for oil and gas in the province. The geologic model generates a list of prospects (potential drilling targets) and a resource appraisal of the oil and gas in place using subjective probability distributions developed by experts familiar with the geology of the area. The exploration model simulates both the economic evaluation of prospects by an explorer and the drilling decision, generating a sequence of discoveries that form an inventory of pools to be evaluated for development.

These two models are integrated in a Monte Carlo simulator of exploratory activity over an extended time period. Each Monte Carlo pass begins by the geologic model sampling from probability distributions for the important geologic parameters to simulate a possible state of geologic nature for the province. This state of geologic nature is composed of a particular number of prospects, some of which are simulated as actual deposits of oil and gas and some as dry. After simulating their expected size, these prospects are ranked according to expected volume to form a simulated target list for the discovery process.

The discovery process is then represented, on a year-by-year basis, as the sequential evaluation of prospects on the target list. The status of the prospects (as either having a deposit or being dry) is unknown to the simulated explorer. If the expected economic value of a particular prospect justifies drilling an exploratory well, the simulated decision is to test it and determine whether it contains hydrocarbons. This procedure continues each year in the Monte Carlo pass or until all prospects have been tested. The learning process in exploration is simulated by using the drilling results each year to update the simulated explorer's perceived state of geologic nature. The output of the exploration model each year is a list of dry wells and discovered deposits of oil and gas. The discovered deposits are added to an inventory of pools to be considered for development. A large number of Monte Carlo passes are made to generate frequency distributions for the important output variables, such as total oil and gas resources in place, discovered reserves, and production.

EXISTING APPROACHES TO RESOURCE APPRAISAL AND EXPLORATION MODELING

Existing approaches to resource appraisal and exploration modeling generally fall into one of three categories. A brief description and discussion of their usefulness to situations similar to the NPRA follows.

Historically, volumetric analysis has been the primary assessment procedure applied to a large province. First, an area that has received substantial exploration and development and that is geologically analogous to the area of interest is identified. Then, the results of the analog area per unit of sedimentary volume are used as a surrogate for the hydrocarbon potential per unit volume in the area of interest. A good example of the volumetric approach can be found in Jones (1975).

Two shortcomings exist with this approach when it is used to estimate the economic value of the potential resources of a frontier province such as the NPRA. First, since the degree to which the analog area matches the target area is difficult to assess, it is also difficult to quantitatively estimate the geologic uncertainty and, in turn, the economic uncertainty for the area of interest. Second, the level of geologic information is far too limited to be useful for economic analysis. Many of the geologic characteristics critical in determining economic value are lost (e.g., number and sizes of deposits).

Prospect analysis is a second approach to the assessment of an area, and it involves the identification and relatively detailed evaluation of all the potential targets for exploratory drilling that exist in an area. This approach is widely used and well developed (see, e.g., Wansbrough et al., 1976; Newendorp, 1975).

Again, there are two shortcomings in applying this technique to an area such as the NPRA. First, the levels of effort and data required in a prospect analysis are substantial and generally are not available for the initial economic evaluation of provinces or basins. Second, prospect analysis typically treats each prospect independently and ignores any regional correlation of geologic characteristics across prospects. The probability that a particular prospect is actually a deposit—the dry hole risk factor—is commonly used to independently risk each prospect. This implies that the presence of oil or gas in one prospect is totally independent of its presence in other deposits, even though the prospects exist in a relatively homogeneous geologic setting. Therefore, whereas prospect analysis is appropriate for certain applications, the large amount of information necessary for its application and that it ignores regional geologic correlation make it unattractive for the assessment of large areas such as the NPRA.

A more recent approach to modeling the exploration process uses the early results of exploration to estimate the returns to future drilling. In general, initial discoveries are used to estimate the parameters of a statistical model of the discovery process. Two examples of this approach can be found in papers by Kaufman et al. (1979) and Drew et al. (1979). A shortcoming of this approach, in its present state of development, is that it requires information that is not available until after some actual discoveries have been made in an area.

A GEOLOGIC MODEL BASED ON THE PLAY APPROACH

The geologic assessment procedure developed for application to the NPRA focuses on the play (a stratigraphic unit of relative homogeneous geology) as the basic unit of geologic analysis. A fundamental assumption is that the geologic characteristics within a play are significantly correlated but show substantially less correlation between plays. In particular, if all the regional geologic characteristics necessary for the occurrence of trapped hydrocarbons are present in the play area, it is likely that the play will contain deposits of oil or gas. However, if one or more of these regional characteristics are missing or unfavorable, it is likely that all the prospects within the play will be dry.

The play approach divides the geologic characteristics of a potential deposit into three categories: play specific, prospect specific, and reservoir specific. Play-specific attributes consist of geologic characteristics common to the play as a unit and include hydrocarbon source, timing, migration, reservoir rock, and the number of prospects. The occurrence of these attributes is a necessary, but not sufficient, condition for the existence of oil or gas deposits in the play.

Prospect-specific attributes are the geologic characteristics common to the individual prospects within the play and include the existence of a trapping mechanism, minimum effective porosity, and hydrocarbon accumulation. Conditional on the existence of the necessary play characteristics, the simultaneous occurrence of these three prospect attributes results in the presence of oil or gas in a prospect.

Reservoir-specific attributes are the reservoir characteristics of an individual deposit of oil or gas in the play and include the area of closure, reservoir thickness, effective porosity, trap fill, reservoir depth, water saturation, and hydrocarbon type (i.e., oil or dry gas). These reservoir attributes jointly determine the volume of oil or gas present in a deposit.

Probability judgments concerning each of these three sets of characteristics for the play are developed by experts familiar with the geology of the area of interest. The experts first identify the major plays within the basin or province, review all existing data relevant to the evaluation, and then make subjective probability judgments concerning the three sets of attributes for each identified play. For example, the USGS is responsible for the probability judgments for the NPRA.

The play-specific attributes are assessed first, in the following manner. A probability distribution is developed for the number of potentially drillable prospects that might exist in the play area. Then, a probability of existence or occurrence is assigned to each regional play characteristic. For example, the probability that a hydrocarbon source exists for the play may be assessed as 0.75, the probability that timing in the play area has been favorable at 0.8, the probability that hydrocarbons could have successfully migrated from the source to traps in the play at 0.9, and the probability that the play contains reservoir grade rock at 0.84. The product of these four probabilities is termed the *marginal play probability*. This is the joint probability that all of the regional geologic characteristics necessary for the accumulation of hydrocarbons in the play area are simultaneously favorable. For the above example, the marginal play probability would equal 0.454.

The second set of probability judgments required for the assessment concerns the likelihood that each of the three prospect-specific attributes is present in a prospect. These probability judgments are made on the condition that all of the play-specific attributes are favorable. The product of the three prospect-specific attribute probabilities is the joint probability that a prospect is a deposit; this is called the *conditional deposit probability*, that is, conditional on favorable play geology.

The third set of probability judgments required involves assessments of the range of values for the reservoir characteristics of an individual deposit within the play. Given that a prospect actually contains oil or gas, each characteristic is assessed as a probability distribution, with the exception of the hydrocarbon mix which is assessed as a point estimate of the likelihood that a deposit is oil rather than dry gas.

Thes three basic sets of probability judgments are made for each of the identified plays, and they compose the basic geologic input data to the geologic model. Figure 2 presents an example data form for recording these judgments for a particular play.

At the beginning of each Monte Carlo pass, the geology model uses the three sets of probabilities for each play to simulate one possible state of geologic nature. The geologic model proceeds in the following manner for each identified play in each pass.

Step 1. The probability distribution for the number of potentially drillable prospects is sampled to determine the number of prospects that will be simulated as existing in the play during the particular pass.

Step 2. Each of the reservoir volume distributions is sampled for each of the prospects to simulate the

EVALUATOR: _Red White_   PLAY NAME: _Alpha Sandstone_

DATE EVALUATED: _5/13/79_

	ATTRIBUTE		PROB. OF FAVORABLE OR PRESENT	COMMENTS
PLAY ATTRIBUTES	HYDROCARBON SOURCE		.75	
	TIMING		.80	
	MIGRATION		.90	
	POTENTIAL RESERVOIR FACIES		.84	
	MARGINAL PLAY PROBABILITY		.454	
PROSPECT ATTRIBUTES	TRAPPING MECHANISM		.90	
	EFFECTIVE POROSITY (≥3%)		.50	
	HYDROCARBON ACCUMULATION		.70	
	CONDITIONAL DEPOSIT PROBABILITY		.315	

HYDROCARBON VOLUME PARAMETERS	RESERVOIR LITHOLOGY	SAND		XX						
		CARBONATE								
	HYDROCARBON MIX	GAS		.75						
		OIL		.25						
	FRACTILES	PROB. OF EQUAL TO OR GREATER THAN								
	ATTRIBUTE	100	95	75	50	25	5	0		
	AREA OF CLOSURE (×10³ ACRES)	2	10	25	50	75	125	150		
	RESERVOIR THICKNESS/VERTICAL CLOSURE (FT)	5	25	50	100	150	180	200		
	EFFECTIVE POROSITY (%)	3	5	10	15	20	25	27		
	TRAP FILL (%)	5	10	20	30	40	55	75		
	RESERVOIR DEPTH (×10³ FT)	10	11	12	15	18	19	20		
NUMBER OF DRILLABLE PROSPECTS (Play Attribute)		30	32	35	45	55	58	60		
PROVED RESERVES (×10⁶ BBLS; TCF)		0.0								

Figure 2. Sample data form for recording probability judgments for a play.

amount of oil or gas present should the prospect be a simulated deposit during the pass.

Step 3. The marginal play probability is sampled to determine whether the play will be simulated as dry or as potentially productive during the pass.

Step 4. For each prospect in a productive play, the conditional deposit probability is sampled to determine whether it will be simulated during the pass as a deposit or as dry. All prospects in a dry play are simulated as dry.

The particular state of geologic nature determined in each pass in the geology model is not made known to the simulated explorer, who must proceed with an exploration drilling program to learn the status of the individual prospects and thus make discoveries.

An important input to the geology model is the percentage of prospects from each play that are likely to fall within each activity area. This can be directly assessed by the geologist or can be estimated as the percentage of the play covered by the activity area. The geology model uses this estimate to assign each specific prospect generated at the beginning of a Monte Carlo pass to a particular activity area. The resulting plays in

the three-dimensional geologic column are vertically stacked and projected onto a two-dimensional area at the surface; each activity area is assigned the number of prospects estimated to underlie it.

The list of prospects for each activity area is the primary output of the geologic model. A second major output, after completion of all Monte Carlo passes, is an appraisal of the oil and gas resources for the area. The resource appraisal is generated by accumulating the simulated deposits for each Monte Carlo pass over a large number of passes to develop a probability distribution for oil and gas resources. Figure 3 presents a flow chart of the geology model.

AN EXPLORATION SUBMODEL BASED ON THE PLAY APPROACH

The exploration model simulates the search for oil and gas deposits as a dynamic discovery process, which integrates the important elements of both the geology and the economics appropriate to the province. The primary function of the exploration model is to take the prospects generated by the geology model at the beginning of each Monte Carlo pass and simulate their evaluation and drilling by an explorer. Exploration is modeled as a decision process under uncertainty in that the simulated explorer does not know prior to drilling if a particular prospect has been simulated as a deposit or as dry by the geology model. In other words, perfect information about the simulated state of the geologic world is not available at the exploration decision point. Rather, the evaluation of a prospect and the decision whether or not to test it with an exploratory well are based on an imperfect perception of the geology derived from the original probability distributions. When the decision is made to test a particular prospect, the cost of an exploratory well is charged and the status of the prospect (a deposit or dry) is revealed to the explorer.

The discovery process continues each year until all prospects perceived to be economic have been tested or until the termination year for the particular Monte Carlo pass has been reached. The outcome, in terms of discoveries and dry prospects, represents the returns to exploratory drilling for one particular realization of geologic and economic conditions. A large number of such passes are made, each time using the geologic model to generate a possible state of geologic nature against which the exploration model simulates the discovery process. Outcomes of interest (e.g., total oil and gas in place or total oil and gas discovered) from each pass can be accumulated, and frequency distributions can be developed as outputs of the Monte Carlo simulation.

Once a particular geologic state of nature has been generated for each activity area in the geologic model, the year-by-year exploration process for the particular Monte Carlo pass is simulated in the exploration model. Associated with each activity area is the year that exploration is scheduled to commence. More than one activity area may be scheduled for initial exploration in the same year, whereas some activity areas may never be scheduled for exploration. This flexibility permits testing the economic consequences of alternative configurations for Federal lease sales in a province.

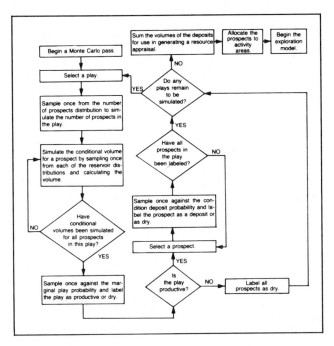

Figure 3. Flow chart of the geologic submodel based on the play approach.

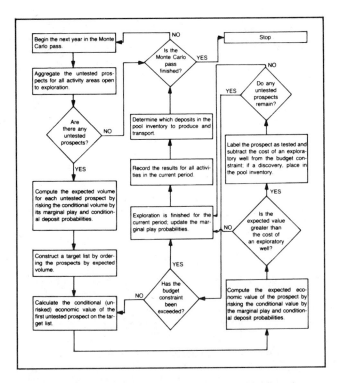

Figure 4. Flow chart of the exploration submodel based on the play approach.

The exploration model follows a four-step procedure simulating the discovery process over time in activity areas open to exploration (Figure 4 presents a more detailed flow chart).

Step 1. A target list containing all untested prospects in activity areas open to exploration is formed by ranking the prospects according to expected (risked) volume.

Step 2. Prospects on the target list are economically evaluated, taking into account the perceived geologic risks (the marginal play and conditional deposit probabilities).

Step 3. The decision to test a prospect, if any, with an exploratory well is made on the basis of expected net present value and an exploration budget constraint.

Step 4. The explorer's geologic perception is updated to reflect previous drilling results by revising the marginal play probabilities that were used in estimating the expected volume and economic value.

The four steps are repeated each year that exploration is allowed until either the termination year is reached or all of the prospects perceived to be economic have been tested.

Step 1 in the exploration submodel is the formation of a target list. All the untested prospects allocated to activity areas open to exploration are aggregated, and the expected volume of petroleum for each is computed by risking the conditional volume with the appropriate marginal play and conditional deposit probabilities. The target list is formed by ranking the prospects according to expected volume, and it represents the order in which the prospects will be evaluated

for exploratory drilling in Steps 2 and 3. Table 1 presents an example.

Step 2 involves the economic evaluation of a prospect on the target list. The expected net present value of the prospect, conditional on it actually being a deposit of oil or gas, is estimated by using a production model, a transportation model, and a distribution model. The production model estimates a simplified time profile of production conditional on the prospect becoming a developed discovery. The transportation and distribution models estimate the cost of moving oil or gas to market, conditional on any existing transportation network serving the area. The transportation model dynamically develops a pipeline network for the area as discoveries are made and placed in production. Field development, transportation, and distribution costs are estimated by the production, transportation, and distribution models to arrive at a time profile of the costs that must be incurred to market the oil or gas that the prospect might contain. The associated revenue stream is generated by matching the potential production profile with a time profile of market prices (an exogenous input to the model), and the present value of net revenue is computed using an appropriate discount rate.

It is necessary to incorporate geologic risk into the economic calculations, because up to this point the prospect has been economically evaluated on the condition that it is a deposit. The geologic risk consists of two components: the risk that the regional geology of a play is unfavorable and the risk that, even in cases where the overall play geology is favorable, the geology specific to the particular prospect is flawed. Therefore, it is necessary to risk the prospect by multiplying its conditional net present value by the marginal

Table 1. Example target list.

Prospect Rank	Play	Activity Area	Conditional Volume	MPP[a]	CDP[b]	Expected Volume	Status
1	Alpha	H	2,000	0.454	0.315	286	Dry
2	Beta	A	3,000	0.650	0.120	234	Dry
3	Alpha	B	1,500	0.454	0.315	215	Deposit
4	Alpha	B	1,000	0.454	0.315	143	Dry
5	Alpha	H	700	0.454	0.315	100	Dry
6	Beta	F	1,200	0.650	0.120	94	Dry
7	Alpha	A	600	0.454	0.315	86	Deposit
8	Beta	H	900	0.650	0.120	70	Dry
9	Beta	F	850	0.650	0.120	66	Dry
10	Alpha	A	400	0.454	0.315	57	Dry

[a]Marginal play probability.
[b]Conditional deposit probability.

play probability and the conditional deposit probability. The result is the expected economic value of the prospect.

Step 3 in the exploration model is the application of a decision rule to determine whether or not a prospect merits testing with an exploratory well. The decision rule is based on expected net present values and a budget constraint. The budget constraint sets an upper bound on the number of exploratory wells that can be drilled in any particular period. The expected economic value of the prospect is compared to the cost of an exploratory well, and only if the expected value equals or exceeds the cost of testing the prospect will an exploratory well be simulated as drilled.

During each time period, the process of comparing prospect expected values with exploratory well costs continues down the target list until one of the following three events occurs: (1) exploration in a particular year ceases when the sum of the costs for wildcat wells exceeds the budget constraint; (2) a prospect is reached for which the wildcat well costs exceed its expected net present value; or (3) no additional prospects remain on the target list. At any of these points the drilling decisions are assumed to be implemented, simulated exploration costs recorded, and the results of the drilling revealed. Any discoveries are placed in a pool inventory to await a development decision in following periods.

Step 4 involves updating the explorer's geologic perception to reflect the results of simulated exploratory activity in the current year. Results from all wells are assumed to become available simultaneously at the end of the period. Prospects that have been tested with an exploratory well during the period are removed from the target list, which reduces the number of available prospects for exploration consideration in following periods.

Furthermore, an attempt to model the learning process that takes place during the discovery process can be made by taking advantage of the way in which the geologic risk has been factored into two components. The marginal play probability for each play in which only dry prospects have been drilled is revised according to Bayes formula. Each time a dry prospect is drilled in a play, the marginal play probability is reduced as specified by the formula. (An example is presented in Figure 5.) The first time a prospect in the play is tested and revealed to be a simulated deposit of oil or gas, the marginal play probability for that play is set equal

to 1 to reflect that all of the regional play characteristics must be favorable for an oil or gas deposit to exist. These revisions in the perceived marginal play probability do not affect the simulated state of geologic nature, because it has been fixed at the beginning of the Monte Carlo pass. The revisions only influence expectations, that is, the perception of nature, and thus they simulate the learning process that will take place over time during exploration.

The revised marginal probabilities influence exploration decisions in future periods through their impact on the expected volumes and net present values computed in Steps 1 and 2. In the case of a play that is yielding an initial series of dry prospects, the marginal probability is continually revised downward. Thus, after a series of dry wells, the expected values of the remaining prospects in the play can become too small to justify further exploration in the play, although some of them may be simulated deposits and may even be commercial if discovered by the explorer. In the case of a play in which a prospect has been tested with an exploratory well and demonstrated to be a simulated deposit, the risking of remaining prospects in the play is based only on the conditional deposit probability because the marginal play probability is equal to 1. Remaining prospects will thus have higher expected values than would be the case if there had been no discoveries in the play.

The oil and gas discoveries generated by the exploration model form an inventory of pools that represent the input to the development decision. These pools are economically evaluated for possible production, transportation, and distribution, and those combinations of pools expected to be commercial are simulated as being produced, transported, and distributed to markets.

CONCLUSION

The procedures developed here have been designed to achieve an analytical melding of geology and economics; the objective has been to develop a resource assessment and economic evaluation methodology for a large area, under conditions of substantial uncertainty, and in a manner conducive to policy analysis.

A play approach to resource assessment and evaluation for large basins or provinces has been selected for several

Marginal play probability = p(favorable play) = 0.454

Conditional deposit probability = p(discovery|favorable play) = 0.315

p(dry well|favorable play) = 1.0 − p(discovery|favorable play) = 1.0 − 0.315 = 0.685

p(dry well|unfavorable play) = 1.0

p(unfavorable play) = 1.0 − p(favorable play) = 1.0 − 0.454 = 0.546

$$p\text{(favorable play|one dry well)} = \frac{[p\text{(dry well|favorable play)}][p\text{(favorable play)}]}{[p\text{(dry well|favorable play)}][p\text{(favorable play)}] + [p\text{(dry hole|unfavorable play)}][p\text{(unfavorable play)}]}$$

$$p\text{(favorable play|one dry well)} = \frac{(0.685)(0.454)}{(0.685)(0.454) + (1.0)(0.546)} = \frac{0.311}{0.311 + 0.546} = 0.363$$

$$p\text{(favorable play|two dry wells)} = \frac{(0.685)(0.363)}{(0.685)(0.363) + (1.0)(0.637)} = \frac{0.249}{0.249 + 0.637} = 0.281$$

$$p\text{(favorable play|three dry wells)} = \frac{(0.685)(0.281)}{(0.685)(0.281) + (1.0)(0.719)} = \frac{0.192}{0.192 + 0.719} = 0.211$$

p(favorable play|a discovery) = 1.0

Figure 5. An example of the Bayesian revision of the marginal play probability.

reasons. First, it provides a direct assessment of the geologic characteristics and uncertainty for the area of interest. While analogs are certainly of great use to a geologist in developing his or her judgments concerning an area, the final judgments are tailored explicitly to the information and perceptions of the target area. Second, the level of geologic detail provided by the play approach is rich enough to support meaningful economic analysis. Furthermore, the results of actual exploration in an area are easily incorporated into the play format. Third, although the play approach treats exploration as a process of prospect evaluation and decision, it does not require explicit identification and substantial detail for each prospect. Fourth, the play approach recognizes a regional component to the geology within a play that causes prospects to be geologically correlated. In essence, the play approach divides the traditional dry hole risk factor into two components. The first component is the risk that is common to all prospects in the play because they share a common potential for source material, migration, timing, and reservoir rock. The second component is the risk that an individual prospect may have a geologic flaw that is specific to it and independent of other prospects in the play. Finally, the approach does not require actual discoveries to have been made in a play; judgments can be based on whatever data exist and can explicitly reflect the uncertainty in the data.

Initial efforts to advance the analysis will focus on the simplifications that currently exist in the exploration model. For example, the target list is formed by ranking the prospects according to expected volume, and only a few prospects are fully evaluated to determine the expected economic value. The number of prospects fully evaluated is a function of the budget constraint. In principle, the target list should be based on economic value, but volume is used as a surrogate to save computation time. Alternative approaches that more closely reflect economic value (e.g., production rate) without adding substantially to the computation time are being investigated. The use of a budget constraint to limit the number of exploratory wells each period is a second simplification that can be improved—perhaps with a more explicit model of the investment process.

ACKNOWLEDGMENTS

The Office of Minerals Policy and Research Analysis has benefited from numerous discussions with Professor Gordon Kaufman of M.I.T. over the last several years.

Our selection of the play as the appropriate geologic unit was in large part the result of the work of and discussions with the Energy Subdivision, Institute of Sedimentary and Petroleum Geology of the Geological Survey of Canada. In particular, the NPRA Study Team is much indebted to Drs. Robert McCrossan and Richard Procter (see Roy et al., 1975).

The importance and implications for modeling of regional geologic correlation was brought to our attention by Gil Mull of the Alaska Department of Natural Resources, Division of Geological and Geophysical Surveys.

REFERENCES CITED

Drew, L. J., E. D. Attanasi, and D. H. Root, 1979, Importance of physical parameters in petroleum supply models: The Economics of Exploration for Energy Resources Conference, May 17–18, 1979, New York University, New York.

Eckbo, P., H. Jacoby, and J. Smith, 1978, Oil supply forecasting—a disaggregated approach: Bell Journal of Economics and Management Science 9, p. 218–235.

Jones, R. W., 1975, A quantitative geologic approach to prediction of petroleum resources: AAPG Studies in Geology, n. 1, p. 186–195.

Kaufman, G. M., Y. Balcer, and D. Kruvt, 1975, A probabilistic model of oil and gas discovery: AAPG Studies in Geology, n. 1, p. 113–142.

Kaufman, G. M., W. Runggaldier, and Z. Livne, 1979, Predicting the time rate of supply from a petroleum play: The Economics of Exploration for Energy Resources Conference, May 17–18, 1979, New York University, New York.

Newendorp, P. D., 1975, Decision Analysis for Petroleum Exploration: Tulsa, Oklahoma, Petroleum Publishing Company, 668 p.

Roy, K. J., R. M. Procter, and R. G. McCrossan, 1975, Hydrocarbon assessment using subjective probability: AAPG Research Symposium, Probability Methods of Oil Exploration, Stanford University, Aug 20–22, p. 56–60.

Wansbrough, R. S., E. R. Price, and J. L. Eppler, 1976, Evaluation of dry hole probability associated with exploration projects: 51st Annual Fall Technical Conference and Exhibition of the Society of Petroleum Engineers of AIME, New Orleans, Oct 3–6, SPE Paper n. 6081, 16 p.

A Comparison of the Play Analysis Technique as Applied in Hydrocarbon Resource Assessments of the National Petroleum Reserve in Alaska and the Arctic National Wildlife Refuge

Kenneth J. Bird
U.S. Geological Survey
Menlo Park, California

A play analysis method of petroleum resource assessment has been developed by the U.S. Department of the Interior and was used successfully in the evaluation of two northern Alaska frontier areas: the National Petroleum Reserve in Alaska (NPRA) and the coastal plain of the Arctic National Wildlife Refuge (ANWR). The assessment procedure entails the input of subjective probabilistic geologic judgments into a computer, which then quickly generates a set of probabilistic resource estimates. The two assessment areas are part of the same North Slope petroleum province and are generally similar except for their size and the absence of seismic and well data for the ANWR. The 17 plays in the NPRA and the 10 in the ANWR were defined stratigraphically, with the exception of one tectonically defined play in the NPRA.

Our results show that although the assessment area of the NPRA (about 37,000 sq mi) is approximately ten times larger than that of the ANWR coastal plain, the undiscovered in-place oil and gas resources are estimated to be nearly the same in both areas, although pool sizes are estimated to be larger in the ANWR.

These two assessments were the first ever undertaken by the USGS using the play assessment technique, and our experience suggests that several small modifications in the method would improve its efficiency and enhance its reliability. The advantages of this method over conventional procedures include its capability to furnish a record of probabilistic geologic judgments on large amounts of data and its ease of revision and updating as new information becomes available.

INTRODUCTION

The USGS and the Office of Minerals Policy and Research Analysis, U.S. Department of the Interior, have developed a play analysis method for petroleum resource assessment. The USGS utilized this method for assessing the undiscovered hydrocarbon resources of two frontier areas of Alaska: the National Petroleum Reserve in Alaska (NPRA) and the coastal plain of the Arctic National Wildlife Refuge (ANWR).

In this play analysis method, geologists provide (1) their professional judgments on the relative favorability of various geologic conditions necessary for petroleum accumulation within a given play area and (2) quantification of a set of geologic variables. An automated Monte Carlo technique generates a set of probabilistic resource estimates including a pool size distribution.

The play, which is the basic unit of assessment in this method, is defined as an area consisting of one or more prospects in a common or relatively homogeneous geologic setting, the prospects of which can be explored by conventional methods. Use of the play method yields probability distributions of undiscovered in-place hydrocarbon resources and pool sizes for each play, as well as a prospect list for each play that can be combined with other data to generate an economic analysis.

Our play analysis method is an adaptation of the method used by the Geological Survey of Canada to assess Canada's petroleum resources (Canada Department of Energy, Mines, and Resources, 1977). It was developed with a twofold purpose in mind: (1) to provide an assessment of the undiscovered hydrocarbon resources of frontier petroleum basins; and (2) to formulate the assessment results in the manner most compatible with a procedure for economic analysis.

This paper discusses two play analysis assessments that were conducted at the request of the U.S. Congress. The initial study was of the Petroleum Reserve, (Department of the Interior, 1979) and consisted of a resource estimate and an economic analysis. The second project, an assessment of the hydrocarbon resources of the ANWR (Mast et al., 1980), did not include an economic analysis.

The purpose of this report is to compare these two frontier area assessments in terms of their geology, available data, assessment procedures, and assessment results. This paper includes recommendations for procedural improvements, as well as suggestions for selected areas of research that might facilitate the desired refinements in procedure. Finally, validation of this method is presented in the form of a comparison of postassessment exploration results with those values predicted by the play method.

Other publications describing this play analysis method include Bird (1981), Mast et al. (1980), Miller (1981, 1982, and in press), and White (1981). Detailed information on the assessment of the coastal plain of the ANWR is presented by Mast et al. (1980) and on the assessment of the Petroleum Reserve by Bird (in press). The economic and policy analysis of the NPRA is described by Bugg et al. (in press) and the Department of the Interior (1979). A comparison of the several hydrocarbon resource estimates of the NPRA by means of various methods (including the play analysis method) is presented by Bird and Powers (in press). A concise review of hydrocarbon assessment methods is given by White and Gehman (1979).

GEOLOGIC COMPARISON

The NPRA and the petroleum prospective area of the ANWR (Figure 1) are each a part of the Alaskan North Slope petroleum province. They have grossly similar sedimentary rock sequences and a shared tectonic history. The petroleum geology of the NPRA is summarized by Bird (1981), Carter et al. (1977), and Gryc (in press) and that of the ANWR by Mast et al. (1980).

The petroleum prospective rocks in both areas consist of two distinct sedimentary assemblages: (1) the older, relatively thin, continental margin assemblage derived from the north known as the Ellesmerian sequence, and (2) the younger, thick orogenic assemblage derived from the south known as the Brookian sequence (Lerand, 1973). The economic basement in both areas consists of pre-Mississippian metamorphosed sedimentary rocks and minor amounts of igneous rocks.

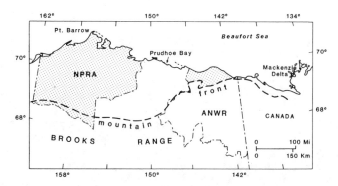

Figure 1. Regional index map showing location of the National Petroleum Reserve in Alaska (NPRA), the Arctic National Wildlife Refuge (ANWR), and the areas of resource assessments (stippled).

The Ellesmerian sequence, Mississippian to Early Cretaceous in age, is generally 6,000 ft (~1,800 m) thick and consists of fluvial clastic or shallow marine clastic and carbonate deposits in the northern third of the province, grading southward to relatively deep marine chert and shale. Limited mostly to the northern third of the province, reservoir rocks in this assemblage are compositionally mature and exhibit fair-to-excellent porosity and permeability; they include the productive Prudhoe Bay reservoirs.

The areal distribution of Ellesmerian rocks in the northern part of the province is irregular. In the NPRA the distribution is controlled by onlap of the source terrain and erosion prior to rifting; however, in the ANWR the distribution is controlled by erosion prior to rifting and possibly by the rifting itself.

A regional tectonic event in the north during Late Jurassic and Early Cretaceous time produced uplift, rifting, and subsidence, thus eliminating the northern sediment sourceland and terminating Ellesmerian deposition. In the south, by comparison, subsidence, continental subduction, and a rising orogenic landmass (the ancestral Brooks Range) initiated Brookian deposition.

The edge of the Barrow platform is interpreted as the rift margin, and the Barrow arch is interpreted as a hingeline along which rocks sagged or were faulted down to the north. The Colville trough, the thrust belt, and the adjacent fold belt are all products of the Brooks Range orogeny. These tectonic features are shown in Figure 2.

The Brookian sequence, latest Jurassic or earliest Cretaceous to Tertiary in age, consists of 20,000 ft (6,100 m) or more of fluvial to deep marine clastic sediments shed from the ancestral Brooks Range and deposited in the adjacent Colville trough (Figure 2). Filling of the trough was by progradation, which proceeded with a pronounced northeasterly trend. A relatively thin distal marine shale having a relatively high percentage of organic material was deposited over a wide area at the base of the Brookian sequence and ahead of the prograding trough fill. These relationships are illustrated in Figure 3.

From the foregoing discussion, and as illustrated by Figures 2 and 3, we can draw the following comparative conclusions. The NPRA consists of three east-trending, subparallel

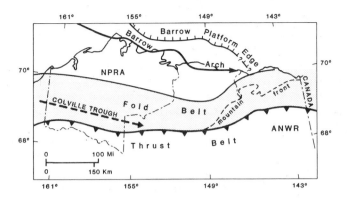

Figure 2. Map of northern Alaska showing the Mesozoic and Cenozoic structural features referred to in this paper.

tectonic elements: the gentle south flank of the Barrow arch, the fold belt, and the thrust belt. In contrast, the ANWR assessment area lies at the intersection of the Barrow platform margin, the Barrow arch, and the fold belt. Ellesmerian reservoir rocks in the NPRA onlap the Barrow arch and display deteriorating reservoir quality southward, at increasingly greater depths. Although the Ellesmerian reservoir rocks in the ANWR assessment area may be mostly missing by erosion or faulting, where they are present they exhibit facies ideal for good reservoir development. Brookian rocks in the NPRA are older and display generally poorer reservoir characteristics.

DATA COMPARISON

When we compare data from these two Alaskan frontier areas, one major difference that is readily seen is the existence of subsurface information (wells and reflection seismic) for the NPRA and the lack of such data for the ANWR. Data from surface geologic mapping, source rock geochemistry, porosity and permeability measurements, and aeromagnetic and gravity surveys are otherwise similar for both areas. These data types are summarized for each area in Table 1. The assessment of the ANWR was aided immeasurably by several wells along its western border and by offshore reconnaissance seismic data.

COMPARISON OF PLAYS

Plays in both areas were defined stratigraphically by their individual potential reservoir rock units, except in the disturbed belt of the Brooks Range and adjacent foothills in the NPRA where one play was defined as a tectonic assemblage. Where structural relationships or source rock to reservoir rock relationships were considered to be significantly different, more than one play was established within a single stratigraphic interval.

In the NPRA 17 plays were established and assessed; this compares with 10 plays in the ANWR. The relationship of plays to stratigraphy for each area is summarized by the

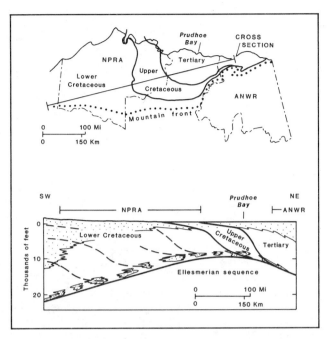

Figure 3. Map and cross section of northern Alaska showing the areal distribution and generalized style of deposition of the Brookian sequence sediments (southern source) that fill the Colville trough (shown in Figure 2).

Table 1. Comparison of data available for oil and gas resource assessments in the NPRA and the ANWR.

	NPRA	ANWR
Reflection seismic	6 × 12-mi grid	None
Wells	78	None
Surface Geology	Yes	Yes
Gravity	Yes	Yes
Aeromagnetics	Yes	Yes
Source rock geochemistry	Yes	Yes
Porosity and permeability measurements	Yes	Yes

time-stratigraphic diagrams shown in Figure 4. Because of geologic similarities, many of the plays in both areas are in the same stratigraphic intervals. In the ANWR, however, one play was established for pre-Mississippian intrabasement carbonate rocks (play number 10, Figure 4). No comparable plays are known to exist in the NPRA. Play descriptions and maps are presented in Mast et al. (1980) for the Refuge and in Bird (in press) for the Reserve.

ASSESSMENT PROCEDURES

The Department of the Interior play analysis method of resource appraisal divides the geologic characteristics of potential hydrocarbon accumulations into three categories: (1) play specific, (2) prospect specific, and (3) reservoir specific (see data form, Figure 5). When combined, categories 1 and 2 provide the risk factor. Category 3 (plus the number

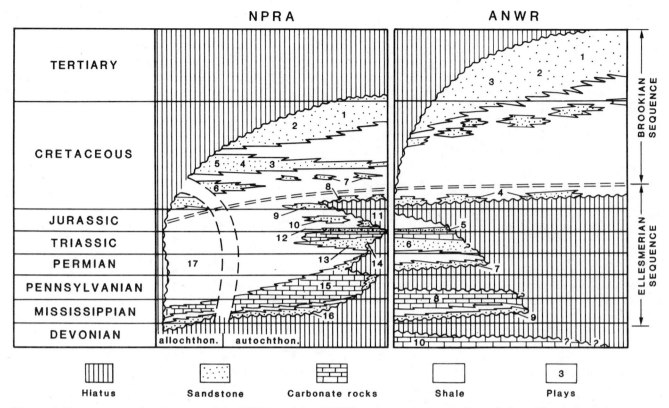

Figure 4. Time-stratigraphic diagrams of the NPRA and the ANWR comparing stratigraphy and plays. The stratigraphic relationships in the ANWR diagram are representative of its westernmost part. Numbers assigned to plays in each area are mutually exclusive.

of drillable prospects, which is an attribute of category 1) provides the data necessary for volumetric calculations of resource and pool size. Subjective probability judgments are made for each of these three categories by geologists familiar with the local geology, and these probabilities are then combined by a Monte Carlo method to yield probability distributions of pool size and in-place hydrocarbon volumes for each play. Further use of the Monte Carlo method provides an aggregation of hydrocarbon volumes for all plays, yielding a total resource estimate for the area. Descriptions of the assessment method are presented in Bird (1981 and in press), Mast et al. (1980), Miller (1981, 1982), and White (1981). Definitions of assessment terms on the data form (Figure 5) are presented by Bird (in press) and Mast et al. (1980). A flow chart showing the steps involved in a single Monte Carlo pass is presented in Miller (1981 and in press) and White (1981).

As part of the assessment procedure, two committees were organized to provide the input. Committee members included (in nearly equal numbers) experts on the petroleum geology of the area in question and experts in petroleum geology and resource appraisal. The first committee, after a thorough review of the geology and establishment of the plays, reviewed the definitions of assessment terms and probability concepts and then gave subjective probability judgments on each category for each play. The second committee reviewed the work of the first to ensure adherence to established procedures and definitions of terms. This

procedure was employed in the separate assessments of the NPRA and the ANWR using many of the same experts for both areas.

The assessments of both areas were supported by the expertise in statistics, probability theory, and computer science of the personnel from the Office of Minerals Policy and Research Analysis, Department of the Interior. These experts had previously developed the formal computer model for the method, and they were present to provide the on-site computer capability that made it possible for the resource estimates to be generated within minutes of completion of a play assessment data form.

The data forms were completed according to committee consensus after group discussion of each item; individual analysts' judgments were not recorded per se. Judgments were made by referring to appropriate maps, cross sections, well logs, or other data displays. Where data were sparse or nonexistent (such as for trap fill), analog comparisons were made with other North Slope fields or with Canadian data, or the data were supplemented by the experience of the committee members.

Because the play method assessments were in many respects "learning experiences" for the scientists involved, the computer-generated resource distributions were accepted with a proper skepticism. Because of the short time lapse (several minutes) from scientific judgment to computer output, the scrutiny of the computer-generated results was conducted with the details of the geologic discussions within

OIL AND GAS APPRAISAL DATA FORM

EVALUATOR: __U.S. Geological Survey__   PLAY NAME: _____

DATE EVALUATED: _____   REVISED: _____

ATTRIBUTE		PROB. OF FAVORABLE	COMMENTS
PLAY ATTRIBUTES	HYDROCARBON SOURCE (S)		
	TIMING (T)		
	MIGRATION (M)		
	POTENTIAL RESERVOIR FACIES (R)		
	MARGINAL PLAY PROBABILITY		
PROSPECT ATTRIBUTES	TRAP OCCURRENCE (TM)		
	EFFECTIVE POROSITY (≳3%) (P)		
	HYDROCARBON ACCUMULATION (C)		
	CONDITIONAL DEPOSIT PROBABILITY		

				PROB. OF EQUAL TO OR GREATER THAN							COMMENTS
HYDROCARBON VOLUME PARAMETERS	RESERVOIR LITHOLOGY	SAND									
		CARBONATE									
	HYDROCARBON MIX	GAS									
		OIL									
	FRACTILES / ATTRIBUTES		100	95	75	50	25	5	0		
	AREA OF CLOSURE (x10³ ACRES)										
	RESERVOIR THICKNESS (FT)										
	EFFECTIVE POROSITY (%)										
	TRAP FILL (%)										
	RESERVOIR DEPTH (x10³ FT)										
NUMBER OF DRILLABLE PROSPECTS											
PROVED RESERVES (x10⁶ BBLS; TCF)											

Figure 5. An example of a data form employed in recording judgments for the play analysis method of the Department of the Interior.

easy recall. An unacceptable computer result was one that was at odds with subjective professional judgments. Such a situation provoked intense discussion and review of the input judgments, and occasionally resulted in a revision of these judgments. This facet of the procedure improved both the final resource estimates as well as the geologists' understanding of the method.

COMPARISON OF ASSESSMENT RESULTS

Assessment results for the NPRA and the ANWR are compared in Figure 6, which shows estimated volumes of in-place undiscovered oil and gas and estimated pool sizes. The similarity in the estimated volumes of oil and gas is remarkable in view of the tenfold difference in size of the two assessment areas: 23.6 million acres for the NPRA and 2.2 million acres for the coastal plain of the ANWR. Pool size estimates are significantly larger for the ANWR than for the NPRA.

The similar hydrocarbon volume estimates can be explained by a consideration of regional geologic characteristics. The assessed portion of the ANWR is located mostly within a fold belt structural province that affects young (Tertiary) basin-filling deposits with fair-to-excellent reservoir characteristics overlying rich Cretaceous source rocks. However, Ellemerian rocks, which are productive at Prudhoe Bay, may be absent from part or nearly all of the area. In contrast, the much larger assessment area of the NPRA shows widespread occurrences of both the Ellesmerian and Brookian rocks. The Ellesmerian rocks here, are generally more deeply buried than at Prudhoe Bay, they thin to the north, and they occur in the generally structureless coastal plain province. Brookian rocks in the Reserve are older than Brookian rocks in the Refuge; they display poor-to-fair reservoir quality and fair source rock quality, and the structural traps that are present are often highly faulted.

It is evident from the estimates that the preponderance of hydrocarbon resources occurs in Brookian rocks both in the NPRA (Bird, in press) and in the ANWR (Mast et al., 1980). The Prudhoe Bay area, in apparent contrast, has proven commercial reserves only in Ellesmerian reservoirs. However, the recent discoveries in the Point Thomson area, 50 mi (80 km) east of Prudhoe Bay (van Dyke, 1980) and the announcement of 18–40 billion bbl in-place of low-gravity oil in Brookian reservoirs just west of Prudhoe Bay (Oil and Gas Journal, 1982) suggest that even in the Prudhoe Bay area the proportion of in-place hydrocarbon volumes in Brookian and Ellesmerian reservoirs may be similar to the proportions for our two assessment areas.

Pool size estimates in this play analysis method are dependent on: (1) area of closure, (2) reservoir thickness, (3) reservoir depth, (4) effective porosity, (5) trap fill and (6) connate water saturation. The larger estimated pool sizes for the ANWR as compared to the NPRA may be supported by consideration of the same regional geologic characteristics as discussed above. The number of drillable prospects and area of closure parameters for the ANWR were most difficult to estimate because of a sparsity of this type of information. We made estimates for these two parameters by counting and

measuring the few surface mapped structures within the ANWR and by counting and measuring any offshore seismically controlled structures that appeared to project onshore. On the basis of detailed examinations of offshore seismic records, the committee increased this relatively small number of onshore prospects by an order of magnitude, reasoning that when the area was fully explored with a close-spaced seismic grid, the few large structures that were perceived would be found to actually consist of numerous smaller structures.

SUGGESTIONS FOR PLAY METHOD IMPLEMENTATION

Suggested modifications to the play method are directed toward improving the understandability of the method, facilitating the required geologic judgments, and refining the reliability of the estimates.

1. A clearly written set of definitions of play method terms, including examples of probability judgments, is critical to the proper employment of the method by any geologist/assessor.

2. Along with a clear understanding of what is required in forming a judgment, the geologist/assessor needs a complete, well-illustrated summary of the essential petroleum geology data. Such a comprehensive data summary is time consuming to assemble, but experience suggests that its availability improves the credibility of the estimates and lessens the time required for the actual assessment process. Any given individual or committee can recall only a limited number of data items at any one time; thus, data summaries are mandatory.

3. One facet of the data summary and data review process should be a standardized, objective system for the evaluation of dry holes (failed prospects). Although such an evaluation was not part of our methodology, this evaluation could be accomplished by using that part of the data form (Figure 5) concerned with prospect attributes. For the target interval(s) in each dry hole, judgments could be made on the favorability of these attributes (trap occurrence, effective porosity, and hydrocarbon accumulation). This procedure would ensure that each dry hole is critically examined and a judgment rendered on why it is dry. A collection of such judgments for a play would also be helpful in making the difficult play and prospect attribute judgments required on the data form.

4. An assemblage of analogs would be most helpful for assessments of frontier areas for which data are often incomplete or nonexistent. At present, the lack of such data required for estimating the number of drillable prospects and the area of closure makes the task more difficult than if these data were available.

5. The possibility that resource estimates by the play analysis method are statistically biased is a concern of several committee members. Their concern is with the effect of dependent relationships among geologic input parameters, which in this method are assumed to be independent. Dependent relationships are those in which a change in one parameter implies a value (or limited range of values) in another parameter. The Monte Carlo simulation technique

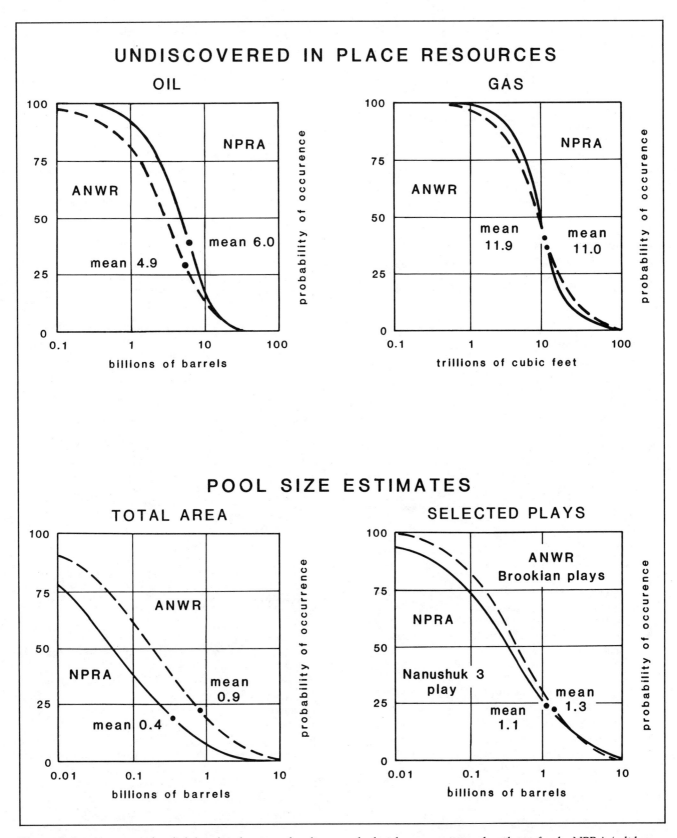

Figure 6. A comparison of probability distributions of undiscovered oil and gas resources and pool sizes for the NPRA (solid curve) and the ANWR (dashed curve). The larger ANWR pool sizes compared to the NPRA pool sizes (lower left diagram) are the result of very large pool sizes in ANWR Brookian sequence plays. These plays are geologically most similar to the NPRA Brookian Nanushuk 3 play (number 5 on Figure 4), and a comparison shows that they have similar pool sizes (lower right diagram).

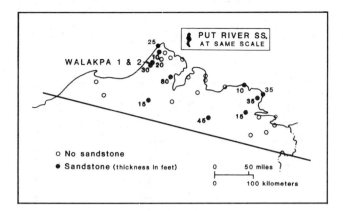

Figure 7. Map of the Pebble Shale play area (shaded) in northern NPRA showing the location of wells that penetrate this play interval and the thickness of sandstone encountered in this play interval. This play is number 8 in Figure 4.

employed in this play analysis method assumes most parameters to be independent and that any combination of values is equally likely. For example, effective porosity generally decreases as reservoir depth increases. The Monte Carlo technique causes any combination of values to be selected and thus may combine maximum reservoir depth with maximum porosity or minimum reservoir depth with minimum porosity. Both combinations may be geologically unrealistic for a given play. The committee's concern is that these two unlikely combinations, once selected, may not cancel each other and may therefore introduce bias into the assessment results. The effect of dependent parameters could be investigated by changing one dependent parameter while holding all others constant through multiple computer runs and observing the change in the resource estimates. Dependence of varying degrees exists among the following assessment parameters: (1) effective porosity and reservoir depth, (2) area of closure and trap fill, (3) area of closure and reservoir thickness, (4) trap fill and reservoir thickness, (5) effective porosity and trap fill, and (6) number of drillable prospects and area of closure. Documentation of dependent relationships could simplify the data form by the elimination of some (dependent) parameters.

POSTASSESSMENT EXPLORATION RESULTS

Only time and continued exploration can ultimately determine how accurate the resource estimates are. For 1 year after the final resource estimate was made, the exploration drilling program continued in the NPRA. The results of this exploration, combined with North Slope analog data, offer encouraging validation of our method.

Prior to the last assessment (May 1980), gas was discovered in Walakpa 1 (the discovery well of the Walakpa gas field) in a sandstone reservoir in the pebble shale play (Figure 7). Data indicated that this is a stratigraphically trapped accumulation of unknown size. In the drilling season following that last assessment, a successful follow-up well (Walakpa 2) was drilled, which encountered gas in the same reservoir 3 mi (5 km) to the south and 600 ft (180 m) deeper. To date the areal

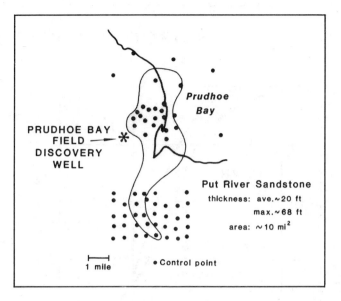

Figure 8. Map of an area adjacent to Prudhoe Bay showing the areal distribution of the Lower Cretaceous Put River Sandstone. Modified after McIntosh (1977.)

Table 2. Calculations of in-place gas volume for the Walakpa Gas Field.

Volume Equation	Put River Ss. Analog	Aeromagnetic Anomalies
Area (acres)	6400	42,000
×	×	×
Reservoir thickness (ft)[a]	20	20
×	×	×
Gas volume (million cu ft) per acre-ft of reservoir[b]	0.49	0.49
=	=	=
In-place volume (bcf)	63	412

[a]Estimated average thickness from Walakpa 1 and 2.
[b]Value calculated from porosity and pressure measurements from Walakpa 2 (Gruy and Associates, 1981).

extent of the reservoir and its average thickness (and thus the size of the accumulation) remain unknown.

Although present information is insufficient to calculate the total volume of the Walakpa gas field, useful estimates of this volume can be obtained through the use of an analog and a set of aeromagnetic anomalies. The analog is a similar sandstone in the Prudhoe area. The geologic details of the Walakpa gas-bearing reservoir compare favorably to the Lower Cretaceous Put River Sandstone (Ellesmerian) in the Prudhoe Bay field (Jamison et al., 1980). The areal distribution and thickness of the Put River Sandstone, summarized in Figure 8, are adapted from McIntosh (1977). These data, when combined with data from the Walakpa wells, result in a calculated in-place gas volume of 63 bcf (billion cubic feet) (Table 2). The gas accumulation may be even larger if the high wave number magnetic anomalies reported by Donovan et al. (in press) result from microseepages outlining the accumulation. The Walakpa wells lie near the center of a roughly circular area bordered

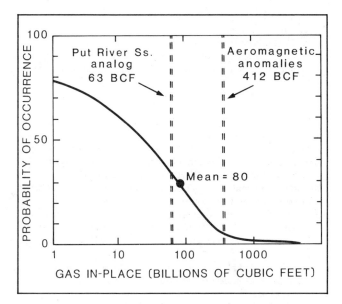

Figure 9. A comparison of two calculated in-place gas volumes (bcf = billions of cubic feet) for the Walakpa gas field (dashed vertical lines) with the May 1980 play analysis estimates of in-place gas for the entire pebble shale play (solid curve and mean value).

by a discontinuous ring of magnetic anomalies. If we assume that these anomalies mark the areal limits of the field and we combine this amount of area with data from the Walakpa wells, we obtain a calculated in-place gas volume of 412 bcf (Table 2). In terms of probability, the smaller volume (Put River Sandstone analogy) would have a greater chance of occurrence than the larger volume (aeromagnetic anomalies) because more smaller fields occur than larger fields. In Figure 9 these two calculated volumes are compared to the estimated volumes for the entire pebble shale play by the play analysis method. A reassessment of the pebble shale play at this time would probably result in increased hydrocarbon resource estimates.

SUMMARY AND CONCLUSIONS

The play analysis method of resource assessment as utilized by the USGS has been successfully applied to two frontier areas in northern Alaska. In this method the geologist/assessor formulates judgments on pertinent geologic characteristics of a given play area that are input into a computer, which generates the resources estimates. This procedure best utilizes the geologist's expertise in purely scientific speculation and the computer's recognized capability for sophisticated manipulation of data. Another advantage to this method is that it provides a record of geologic judgments on large amounts of data and includes provision for easy revision as new information becomes available.

In this method an absence of data produces an increase in the range of resource estimates. Substantial increases in the range of results occur when seismic data are not available for

determination of prospect number and size. Likewise, a lack of subsurface control for determination of reservoir characteristics also increases the range of the resource estimates. These data limitations influenced the ANWR assessment, but in this instance of a relatively small assessment area, they are believed to be compensated for by projection of data from nearby areas and by use of analog comparisons.

Most of the hydrocarbon resources for both assessment areas are estimated to occur in southern source (Brookian) reservoirs. Recent oil discoveries in Brookian reservoirs in the Prudhoe Bay sector of the North Slope suggest that a similar relationship may even exist in this area where all commercial oil presently comes from Ellesmerian reservoirs.

Post assessment exploration of the NPRA pebble shale play supplied new information, which combined with North Slope analog data demonstrated a hydrocarbon volume calculation in remarkably good agreement with the computer-generated resource estimates.

ACKNOWLEDGMENTS

I thank C. D. Masters for the invitation to participate in the 1982 Circum-Pacific Energy Conference workshop and symposium on resource appraisal methodology in Honolulu, Hawaii, where this paper was first presented. It has since benefitted from the critical review, comments, and suggestions of Paul Bugg, Gordon Dolton, C. M. Molenaar, and R. F. Mast.

REFERENCES

Bird, K. J., 1981, Petroleum exploration of the North Slope, Alaska, in J. F. Mason, ed., Petroleum Geology in China: Tulsa, Oklahoma, Pennwell Publishing Company, p. 233–248.
———, in press, The geologic basis for the appraisal of undiscovered hydrocarbon resources in the National Petroleum Reserve in Alaska, by the play-appraisal method, in G. Gryc, ed., The geology of the National Petroleum Reserve in Alaska: U.S. Geological Survey Professional Paper 1399.
———, and R. B. Powers, in press, Comparison of six hydrocarbon resource assessments of the National Petroleum Reserve in Alaska, in G. Gryc, ed., The geology of the National Petroleum Reserve in Alaska: U.S. Geological Survey Professional Paper 1399.
Bugg, P., S. Miller, and L. P. White, in press, Policy analysis of the National Petroleum Reserve in Alaska—Methodology and applications; in G. Gryc, ed., The geology of the National Petroleum Reserve in Alaska: U.S. Geological Survey Professional Paper 1399.
Canada Department of Energy, Mines, and Resources, 1977, Oil and Natural Gas Resources of Canada: Report EP77-1, 76 p.
Carter, R. D., C. G. Mull, K. J. Bird, and R. B. Powers, 1977, The Petroleum Geology and Hydrocarbon Potential of Naval Petroleum Reserve No. 4, North Slope, Alaska: U.S. Geological Survey Open-File Report 77-475, 61 p.
Department of the Interior, 1979, Final report of the 105(b) economic and policy analysis: Prepared by the U.S. Department of the Interior, Office of Minerals Policy and Research Analysis, U.S. Government Printing Office, Washington, D. C., 145 p.

Donovan, T. J., J. D. Hendricks, A. A. Roberts, and P. T. Eliason, in press, Low-level aeromagnetic surveying for petroleum in Arctic Alaska, *in* G. Gryc, ed., The geology of the National Petroleum Reserve in Alaska: U.S. Geological Survey Professional Paper 1399.

Gruy, H. J. and Associates, Inc., 1981, Well test analysis for Walakpa test well No. 2, National Petroleum Reserve No. 4, Alaska: Unpublished report prepared for the U.S. Geological Survey, March, 1981, Houston, Texas.

Gryc, G., ed., in press, The geology of the National Petroleum Reserve in Alaska: U.S. Geological Survey Professional Paper 1399.

Jamison, H. C., L. D. Brockett, and R. A. McIntosh, 1980, Prudhoe Bay—A 10-year perspective, *in* M. T. Halbouty, ed., Giant Oil and Gas Fields of the Decade 1968-1978: AAPG Memoir 30, p. 289–314.

Lerand, M., 1973, Beaufort Sea, *in* R. G. McCrossan, ed., The Future Petroleum Provinces of Canada—Their Geology and Potential: Canadian Society of Petroleum Geologists Memoir 1, p. 315–386.

Mast, R. F., R. H. McMullin, K, J. Bird, and W. P. Brosgé, 1980, Resource appraisal of undiscovered oil and gas resources in the William O. Douglas Arctic Wildlife Range: U.S. Geological Survey Open-File Report 80-916, 62 p.

McIntosh, R. A., 1977, Prudhoe Bay Unit Operating Plan, Reservoir Description: State of Alaska, Oil and Gas Conservation Commission, May 5th Conservation Hearing No. 145, Exhibit No. 8, Anchorage, Alaska.

Miller, B. M., 1981, Methods of estimating potential hydrocarbon resources by the U.S. Geological Survey—Case studies in resource assessment in the National Petroleum Reserve in Alaska and the William O. Douglas Arctic Wildlife Range: Proceedings of the Southwestern Legal Foundation Exploration and Economics of the Petroleum Industry, v. 19, p. 57–95.

——— , 1982, Application of exploration play analysis techniques to the assessment of conventional petroleum resources by the USGS: Journal of Petroleum Technology, v. 34, n. 1, p. 55–64.

——— , in press, Methodology for petroleum resource assessment for the National Petroleum Reserve in Alaska, *in* G. Gryc, ed., The geology of the National Petroleum Reserve in Alaska: U.S. Geological Survey Professional Paper 1399.

Oil and Gas Journal, 1982, Texas seen key snag for Alaskan heavy oil work: Oil and Gas Journal, v. 80, n. 12, p. 52.

van Dyke, W. D., 1980, Proven and probable oil and gas reserves, North Slope, Alaska: State of Alaska, Department of Natural Resources, Division of Minerals and Energy Management, 11 p.

White, D. A., and H. M. Gehman, 1979, Methods of estimating oil and gas resources: AAPG Bulletin, v. 63, n. 12, p. 2183–2192.

White, L. P., 1981, A play approach to hydrocarbon resource assessment and evaluation, *in* J. B. Ramsey, Economics of exploration for energy resources, *in* E. I. Altman, and W. Ingo, eds., Contemporary Studies in Economic and Financial Analysis, v. 26, (JAI Press, Greenwich, Connecticut), p. 51–67.

Estimation of Potential Gas Resources—Methodology of the Potential Gas Committee

Harry C. Kent
Director, Potential Gas Agency
Colorado School of Mines, Golden, Colorado

J. C. Herrington[1]
General Chairman, Potential Gas Committee
ARCO Exploration Company, Dallas, Texas

Since 1960, the Potential Gas Committee (PGC) has prepared periodic estimates of the undeveloped and undiscovered natural gas resource potential remaining in the United States. In preparing its estimates, the PGC uses available information regarding past discoveries and production of natural gas as well as geologic and engineering data about prospective geologic provinces. Three categories of estimates are made. First, probable supply is associated with discovered fields and represents the remaining gas that could be developed and produced with additional drilling and field development. The estimate of probable supply is based on extrapolation of characteristics of the previously developed parts of the field. Second, possible supply is related to projections of additional new fields that are anticipated to be amenable to discovery during the continued exploration within provinces and trends that are already established as productive. The estimate of possible supply is based on analysis of the geologic characteristics of the undeveloped part of the province or trend and application of the geologic and engineering characteristics that have been established in the fields discovered to date. Finally, speculative supply is any additional supply that is anticipated might be discovered and developed in formations and geologic provinces that are not presently productive; this estimate is based on the known geologic characteristics of the speculative province and analogy with known productive areas. The estimates of speculative potential resources are subject to much greater uncertainty than the estimates of either possible or probable potential.

INTRODUCTION

The Potential Gas Committee (PGC) comprises volunteer members from the natural gas industry, government agencies, and academic institutions who are concerned with natural gas resources. The objective of the PGC is to provide estimates, based on expert knowledge, of the potential supply of natural gas, which, in conjunction with estimates of proved reserves of natural gas, make possible an appraisal of long-range gas supply.

For many years estimates of the future supply of natural gas have been developed by numerous individuals and organizations. Since these estimates of future supply have varied widely and often have had entirely different definitions, objectives, and meanings, they have sometimes confused rather than clarified the long-range outlook for natural gas supply. With the view of achieving a more unified and industry-sponsored effort, many of the leaders of the natural gas industry, as well as various financial and governmental agencies, have expressed a strong desire for a continuing study of the supply of natural gas in the United States. As a result of this concern, the work of the PGC was initiated in 1960. It has continued to function since that time and has prepared regular reports presenting estimates of

---

[1]Currently Chairman, Board of Directors, and Past General Chairman, Potential Gas Committee. Mr. Herrington has retired from ARCO.

the recoverable natural gas believed to exist in addition to that amount included in proved reserves. Two of the most recent reports, for example, cover all information available for 1982 through 1984 (PGC, 1983, 1985). Since 1967 the work of the PGC has been guided and assisted by the Potential Gas Agency of the Colorado School of Mines. The Potential Gas Agency is supported by the American Gas Association.

Estimates of the PGC are "base-line" estimates in that they attempt to provide a reasonable appraisal of the total natural gas resource potential of the United States and are not limited by assumptions of the time of development of the natural gas resource, the life span of the natural gas industry, or the specific price to be paid for the produced gas. There is no consideration of whether or not this resource *will* be developed; rather, the estimates are of the resource that *could* be developed if the need and economic incentive exist.

To achieve uniformity in the work done by a large number of individuals, the PGC has adopted standardized procedures to be followed by its members in preparing the resource estimates that it publishes. The members are asked to follow these procedures as much as possible, but sufficient flexibility is permitted to allow for large variations in the available data for resource evaluation in diverse areas of the country. This paper summarizes the procedures adopted by the PGC. A more complete coverage is contained in the publication, "Definitions and Procedures for Estimation of Potential Gas Resources" (PGC, 1984).

NATURAL GAS RESOURCE CLASSIFICATION

At any given time, a finite volume of natural gas-in-place exists within the rocks of the shallow portion of the earth's crust. This finite volume of gas-in-place is the total natural gas resource. The amount of this resource that is recoverable is a function of technology and economics. Today, technology has advanced to the degree that a significant part of the natural gas resource base can be recovered. A major part of this resource, however, is dispersed throughout the earth's crust in such minute accumulations or under such conditions that it cannot be extracted with existing or foreseeable technology. Also, the cost of recovery of such gas would be too high compared to the value of the gas. Even with likely advances in technology, production of some deposits would not be feasible because recovery of such accumulations would require more energy than that provided by the gas produced.

Estimates of the PGC include only that part of the natural gas resource that is susceptible to discovery and production during the life of the industry using current or foreseeable technology and under the condition that the price:cost ratio would be favorable. This total quantity is referred to as the ultimately recoverable resource.

The above limitations of size of accumulation and economics are applied as follows. For situations in which a distribution of sizes of individual accumulations is envisioned (particularly in estimation of probable and possible resources, as described below), a minimum size of recoverable accumulation is determined on the basis of the estimator's current judgment of the relationship between the value of

the resource and the cost of drilling and production. This relationship is based on extrapolation of the prevailing price:cost ratio at that time for the given geologic province and depth. This often takes the form of an estimate of the minimum recovery per well, which is necessary in order for an operator to determine that it is economically feasible to complete and produce the well in a given geographic area and depth. Such a well may not, in total, be an economically viable venture, but given that the well has already been drilled to the target, the available resource will be produced and will probably recover the costs of completion and production, even if the total costs of exploration and drilling are not recovered. At the other end of the size and economic spectrum, a maximum size of accumulation is estimated on the basis of the estimator's evaluation of the geologic factors controlling the size and effectiveness of traps and the reservoir conditions.

The size and economic limitations are evaluated in a much more general way in situations in which it is not possible to estimate the number and sizes of individual accumulations. This is particularly the case in the estimation of speculative resources (as discussed below). In these situations, analogies are made with provinces or formations in which natural gas production has already been established and which have similar geologic characteristics to the unexplored province or formation. It is hoped that, if these analogies are drawn carefully and accurately, the analogous conditions of size and economics will also apply.

The ultimately recoverable resource is composed of both the discovered and undiscovered natural gas resources. Discovered gas consists of the quantity produced in past years plus that proved by drilling and engineering tests and included in present proved reserves. Discovered gas also includes the gas remaining in known fields that will be recovered through extension and complete development of known pools and reservoirs. The undiscovered resource consists of the potential gas supply that could become productive with further exploration and development. The undiscovered resource includes both the gas remaining in undiscovered pools and reservoirs within known fields and the gas that may be discovered in new fields and reservoirs within geologic provinces that are presently productive and those that are as yet unproductive.

The subdivision of the ultimately recoverable resource into categories reflects the variation in the amount of geologic and engineering information on which the estimate is based. Proceeding from left to right in Figure 1, the level of geologic and engineering knowledge decreases.

The potential supply includes all of the undiscovered resource plus the portion of the discovered gas resource that is not included in proved reserves. Three categories of potential supply are recognized and reported by the PGC: probable, possible, and speculative. These three categories are differentiated on the basis of variation in available geologic, geophysical, and engineering information. Estimates of both nonassociated and associated or dissolved gas are included in all categories.

Limitations to the estimates have been established by the PGC on the basis of current technological capabilities and an estimate of the expansion of those capabilities in the foreseeable future. No estimate is made for onshore areas of natural gas that might exist at drilling depths in excess of 30,000 ft (~9150 m). Estimates for onshore areas are also made for two depth intervals, 0–15,000 ft (~0–4575 m) and 15,000–30,000 ft (~4575–9150 m), in recognition of the greater costs and technical requirements for wells drilled below 15,000 ft (~4575 m). No estimate is made for offshore areas of any natural gas that might exist beneath water depths > 1,000 m (> 3280 ft). Also, separate estimates are made for offshore areas of resources below water depths of 0–200 m (0–640 ft) and depths of 200–1,000 m (640–3280 ft), respectively, because of the higher costs associated with drilling platforms, production facilities, and pipelines in deeper water.

Probable potential supply is associated with existing fields. It is the most assured of potential supplies because it is associated with known accumulations. A relatively large amount of geologic and engineering information is available to aid in the estimation of the resource existing in this category. Probable supply bridges the boundary between discovered and undiscovered resources. The discovered portion of it includes the supply from future extensions of existing pools in known productive reservoirs. The pools containing this gas have been discovered, but the extent of the pool has not been completely delineated by development drilling. Therefore, the existence and quantity of gas in the undrilled portion of the pool are as yet unconfirmed. The undiscovered portion of probable supply is expected to come from future new pool discoveries within existing fields either in reservoirs productive in the field, or in shallower or deeper formations known to be productive elsewhere in the same geologic province or subprovince.

Possible potential supply is a less assured supply because it is postulated to exist outside of known fields, but it is associated with a productive formation in a productive province. Its occurrence is indicated by projection of plays or trends of a producing formation into a less well explored area of the same geologic province or subprovince. New plays or trends developed within a producing formation are also included. Supply is from new field discoveries postulated to occur within these trends or plays under both similar and different geologic conditions (i.e., the types of traps and/or structural settings (or both) may be the same or may differ in some aspect).

Speculative potential supply is the most nebulous of the three categories of potential supply. This supply comes from formations or provinces that have not yet proven to be productive. Geologic analog are developed to ensure reasonable evaluation of these unknown quantities. Supply is from new pool or new field discoveries in formations not previously productive within a producing geologic province or subprovince, or from new field discoveries within a geologic province not previously productive.

TECHNIQUES FOR ESTIMATION OF POTENTIAL GAS SUPPLY

Geologic Provinces

A total of 122 standard geologic provinces have been defined by the PGC. The boundaries of these provinces in

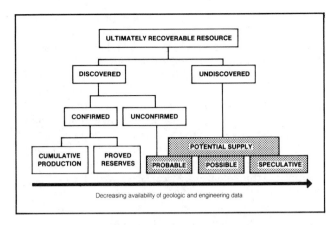

Figure 1. Classification of natural gas resources used by the Potential Gas Committee (PGC).

the conterminous United States correspond closely to the geologic province boundaries established by the Committee on Statistics of Drilling (CSD) of the American Association of Petroleum Geologists. In some instances, adjacent CSD provinces have been combined by the PGC where thes provinces have only limited potential for natural gas resources. Consideration was given to making PGC province boundaries correspond to those used by the Resource Appraisal Group (RAG) of the U.S. Geological Survey. In the conterminous United States, however, the CSD boundaries conform more closely to the geologic boundaries recognized by the working geologists of the PGC and have the advantage of more accessible drilling, discovery, and geologic information for the province. Minor changes to province boundaries have also been made where, in the judgment of PGC members, geologic conditions required modification. Provinces for Alaska have been established independently by the PGC on the basis of the current understanding of geology and resource potential and do not correspond to provinces established by the CSD or the RAG. Resource estimates are made for each geologic province by members of the PGC.

For purposes of organization, the PGC is subdivided into seven Area Work Committees that correspond to geographic areas of the country. Each area contains a number of geologic provinces, and each committee is responsible for preparation of the resource estimates for the geologic provinces that fall within that area's limits. Membership for the Area Work Committees is recruited from among the geologists and engineers who are actively engaged in exploration and development of natural gas in the various geologic provinces. Since the judgment of the estimator is the most significant single factor in the preparation of resource estimates, every effort is made to recruit the most knowledgeable experts for membership.

Estimation Procedures

The basic technique for estimating potential gas supply is to compare the factors that control known occurrences of gas with factors present in prospective areas. This technique is applied to each of the categories adopted by the PGC. In each case, what is known about the prospective area is evaluated relative to what is known about natural gas accumulations and the factors known to control their occurrence. Natural gas occurrences are related to conditions that favor their formation and accumulation, such as the existence of source rocks, sufficient maturation of organic material, and presence of reservoir rocks and traps. Studies of producing areas have provided information on the productive capacity of particular formations and the average size of accumulations. In the most simplified form, the estimate of the potential supply is derived by: (1) estimating the volume of potential gas-bearing reservoir rock, (2) multiplying this volume by a yield factor (yield factors represent the the amount of gas expected to be produced from a given unit volume of rock), and (3) discounting to allow for the probability that the traps and/or the accumulations exist. Modification of the basic technique is necessary to accommodate variations in the amount of information available for prospective areas.

Each estimator considers three separate situations in preparing estimates:

1. The quantity of gas that might exist and be recoverable is present under the most favorable conditions possible. The probability that such conditions might prevail is assumed to be approximately 10% or less, and thus there is only a 10% or less probability that at least this much gas resource is present. This assumes the maximum number of potential traps with favorable source bed and reservoir conditions, the maximum reasonable yield factor, and that each trap is filled with recoverable accumulation. Such assumptions would lead to a reasonable estimate of the maximum resource quantity.

2. It is assumed that the minimum number of traps exist, the most marginal of source bed and reservoir conditions are present, and the minimum reasonable yield factor should be applied; also, the possibility is assumed that many of the traps that might exist do not contain recoverable accumulations of natural gas. In this case it is assumed that there is approximately a 90% or greater probability that at least this much gas resource is present. Such assumptions would lead to a reasonable estimate of the minimum resource quantity.

3. The most reasonable estimate is made of the existence of traps and accumulations and the most reasonable assessment is made of the source bed, yield factor, and reservoir conditions. These conditions would lead to the most likely estimate of the resource.

By considering each of the three situations outlined above, the estimator is best able to express both the limiting conditions in the geologic region as well as the most likely estimate of the resource potential. Estimates prepared by the individual estimators are reviewed at meetings of Area Work Committees and by the entire PGC at the national level to maintain a reasonable degree of consistency and uniformity.

SPECIFIC ESTIMATION TECHNIQUES FOR RESOURCE CATEGORIES

Probable Potential Resources

Probable potential resources bridge the division between discovered and undiscovered resources. The discovered portion includes extensions of pools estimated to be partially developed. The undiscovered portion is composed of new pools that may be discovered in the future within an existing field.

In the case of discovered, but unconfirmed, probable potential resources, the volume of potential reservoir rock is calculated to the estimated limits of future pool development on the basis of geologic, geophysical and engineering data. The estimated volumes of undeveloped reservoir rocks for all fields that are not yet completely developed, within the province and depth interval under consideration, are multiplied by appropriate yield factors. These factors are determined from the areas of the fields which are presently producing from the same formations as the extensions being evaluated; they are adjusted for possible variations as changes in lithology, porosity, permeability, hydrodynamic conditions, or relationships among gas, oil, and water in the reservoir. The product of the reservoir volumes and the yield factors for the entire province and depth interval under consideration is then multiplied by a probability factor which expresses the probability that the accumulation actually exists. This probability of the existence of the accumulation is determined by the estimator on the basis of previous experience in development of similar fields in the province. This probability is the only one assigned because the existence of traps has already been proved. Figure 2 illustrates this procedure diagrammatically.

The factors on which the resource estimate is based are adjusted by the estimator within reasonable limits to allow the determination of the maximum, minimum, and most likely estimates. The uncertainty of the resource estimates in this category is relatively small because many of the controlling factors have already been determined by actual drilling, testing, and production. Therefore, the range between maximum and minimum estimates is usually smaller for this category than for other resource categories.

Estimating undiscovered probable potential supply involves the application of geologic and geophysical interpretations to delineate all possible new pools within the limits of discovered fields (Figure 3). The volume of potential gas-bearing reservoir rock is calculated for known but untested traps (e.g., suspected multizone development of an anticlinal trap) and for potential but unknown traps (e.g., postulated pools separated from current production by faults or pools separated by facies changes, such as a series of overlapping barrier bars). The calculation of this volume includes consideration of possible changes in lithology, such as sand pinchouts or facies changes, and variations in thickness and lateral extent of the potential reservoir rocks. Proper evaluation requires a familiarity with the nature of the prospective reservoir formation within the geologic province as well as the local facies relationships and the structural setting. The estimated volume for the entire province and depth interval under consideration is multiplied by an adjusted yield factor obtained by analogy with a producing area in the same formation as the potential new pools. When adjusting yield factors to prospective areas, one should consider differences in porosity and permeability that might be due either to differences in original depositional environment or to diagenetic changes after deposition. Presence and quality of source rocks and their maturation levels must also be considered.

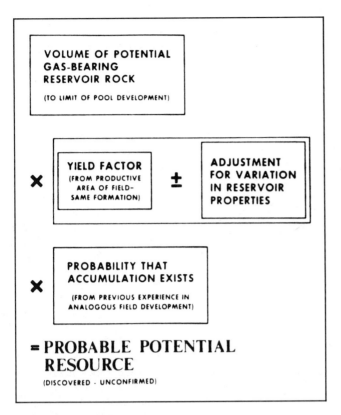

Figure 2. Schematic diagram of the procedure used by the PGC for estimation of probable (discovered but unconfirmed) potential resources.

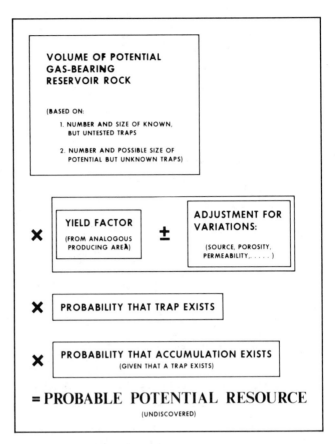

Figure 3. Schematic diagram of the procedure used by the PGC for estimation of probable (undiscovered) potential resources.

Estimators include two discount factors in their calculation of the estimates of this potential supply. The first is a measure of the certainty of the existence of the traps in the province and depth interval under consideration. In the case of potential multizone anticlines, there may be absolute certainty that the traps exist, and thus, there would be no discount (100% probability). In the case of potential pools separated structurally or stratigraphically from existing production, the actual existence of the prospective traps may be in question. In this instance the estimator gives an evaluation of the probability of trap existence based on the quantity and quality of data available and his or her experience in the area. The probability of trap existence, for example, may only be 50% or less if there are problems with seismic interpretation or if little or no geophysical work has been done. In contrast, if seismic interpretation has been shown to be highly reliable and if extensive geophysical work has been done, the probability of trap existence may be as high as 80–90%.

The second discount factor covers the possibility that although the trapping conditions necessary for gas accumulation may exist, the trap may not actually contain any gas. This situation might result if the timing of gas generation was not consistent with the time of trap formation and if the trap did not exist at the time of gas migration. Alternately, although geometric and timing conditions relating to trapping may be favorable, there may not have been an adequate seal and gas that had previously accumulated has subsequently escaped from the potential trap.

As with estimates of all categories of potential gas resources, evaluations of the minimum, maximum, and most likely quantities must be included in the consideration of undiscovered probable resources. The estimator considers each of the resource factors separately (i.e., volume of prospective reservoir rock, yield factor, and discount factor) and adjusts them within reasonable limits. Use of the most pessimistic values for each of the factors will yield the minimum estimate; conversely the maximum estimate is derived from the most optimistic values. The most likely estimate represents the estimator's best judgment concerning the value of each of the factors.

In a number of instances, members of the PGC will not have access to the detailed field data necessary to follow the procedures outlined above for the estimation of probable potential resource. An estimate of the probable resource can be made in these cases by applying historical experience regarding the growth of the size of natural gas fields in the region. In most instances, except in the case of very small fields, the ultimate size of the field (or pool) is not determined by the proved reserves measured by the initial discovery well or by the other wells drilled in the field (or pool) during the year of discovery. Historical data are available for various formations and regions which relate the quantity of proved reserves measured during the year of discovery to the ultimate size of fully developed fields. These relationships can be utilized to approximate the ultimate

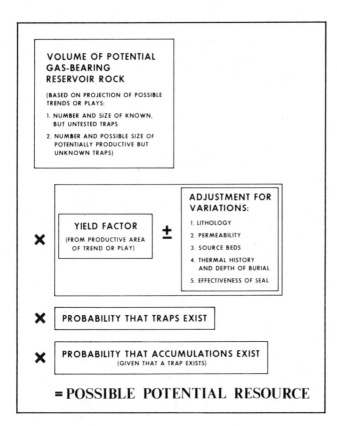

Figure 4. Schematic diagram of the procedure used by the PGC for estimation of possible potential resources.

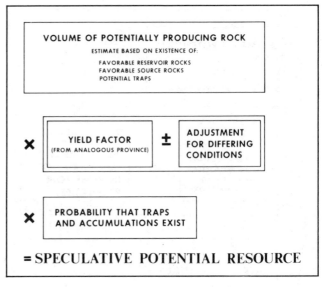

Figure 5. Schematic diagram of the procedure used by the PGC for estimation of speculative potential resources.

resources of a newly discovered field through comparison with other fields at a similar stage of development. The difference between the estimated ultimate resources of the field and the presently confirmed resource (sum of production and current proved reserves) is the probable potential resource.

Because probable resources span the gap between proved reserves and undiscovered resources they may be one of the reasons for the differences among various published natural gas resource estimates. These differences have occurred both in definition and in estimating technique. Some estimates have not included any resources comparable to the probable category and have been limited solely to undiscovered resources (e.g., Moody and Geiger, 1975, for Mobil). The USGS RAG has a category called inferred reserves that is similar to the PGC's probable category, and they have used the method of extrapolation of historical growth rates exclusively (Dolton et al., 1981). Other estimators have referred to this category as "growth of known fields" and have variously used extrapolation of historical growth rates either alone or in conjunction with geologic estimates, or have not stated their methodology (e.g., Langston, 1976, for Exxon).

Possible Potential Resources

The technique for estimating possible potential resource is essentially the same as for the undiscovered probable resource category (Figure 4). Both situations involve untested traps in formations that are already productive within the

geologic province under consideration, but they differ with respect to proved production.

Possible potential resources are located in a productive geologic province but are associated with anticipated new field discoveries. They are normally related to the extension of a producing trend or play. By considering the extension of a trend or play from a productive area into a relatively unexplored area, a geologist can make a reasonable interpretation the number and types of traps to be found in the region. Available geophysical data can also help to indicate the location and possible size of potential traps.

On the basis of these determinations, an estimate is made of the total potential reservoir volume for the province and depth interval under consideration. This total potential volume is multiplied by a yield factor determined from presently developed portions of the trend or play or by analogy with similar trends. The probabilities of the existence of traps and trap fill are also estimated on the basis of the estimator's experience in the developed portion of a trend or play or by analogy with similar trends. Yield factors and probabilities must be applied between analogs and prospective areas on the same basis so that similar volumes and conditions are compared. Careful evaluation of the trend or play is necessary to ensure proper reservoir volume and yield adjustments and reasonable estimates for trap and accumulation discounts. The probabilities of the existence of traps and accumulations are dependent on how strongly the trend or play is established and the quantity and quality of available exploration data.

Speculative Potential Resources

The estimation of speculative potential resources (Figure 5) involves the size of accumulations that may be found in untested formations located in either producing or nonproducing provinces. Estimates are based on whether or not conditions are considered favorable for gas accumulations in (1) formations which are potential reservoir rocks, but that have not yet been demonstrated to be productive in geologic provinces where production has been established in other formations, or (2) formations that are potential reservoir rocks in geologic provinces where no previous natural gas production has been established.

The criteria considered in making estimates of speculative potential include the type of sedimentary rocks, structural and stratigraphic relationships, tectonic history, and thermal maturation. Yield factors are estimated using known production statistics from other geologic provinces that are believed to be reasonable analogs of the prospective formation or province under consideration. In these other provinces, gas may be produced from formations different from those in the speculative province, but characteristics of the formations in both provinces are believed to be reasonably similar. Adjustments are made to allow for dissimilarities between the two provinces. Yield factors may be used comparatively either for the anticipated volumes of productive reservoir rocks or for the gross volume of the basin depending on the quantity and quality of basinal data available. In any case, when used to compare the analog and the prospective province they must be applied consistently.

The probabilities that the traps and the accumulations exist are generally grouped together as one discount factor in the estimation of speculative resources. It is usually not feasible to discriminate between the probability of trap existence and the probability that accumulations exist in the traps in regions where little exploration has been conducted and for which there is little geologic or geophysical data.

The large spread between the minimum and maximum estimates for speculative resources reflects the great amount of uncertainty regarding the magnitude of the potential resource and even its very existence. It must be recognized that in many unexplored or poorly explored regions, there is a high probability that no recoverable resource exists.

CONCLUDING REMARKS

The judgment of the estimator is the most significant factor in making estimates of potential supply. Estimators must make the necessary decisions concerning the extent and number of traps, reservoir conditions, probability of existence of traps, and probability of accumulations. They must also choose appropriate analogs where necessary. These judgments must necessarily be based on the experience of the estimators and the data available to them. No other considerations can supplant the need to involve the most knowledgeable and experienced professionals in the preparation of credible resource estimates.

ACKNOWLEDGMENTS

The authors wish to acknowledge the contributions of the many members of the PGC who have been involved in the development of the concepts presented in this paper over the 20 years of its existence. Additionally, we are grateful for the repeated and continuing discussions held with many people outside of the committee, especially Harry L. Thomsen and Betty M. Miller.

REFERENCES CITED

Dolton, G. L., et. al., 1981, Estimates of undiscovered recoverable conventional resources of oil and gas in the United States: U.S. Geological Survey Circular 860, 87 p.

Langston, J. D., 1976, A new look at the U.S. oil and gas potential, in Proceedings of the 14th Institute on Exploration and Economics of the Petroleum Industry, Southwestern Legal Foundation, March 10–11, 1976, Dallas, Texas: New York, Matthew Bender & Co., Inc., p. 33–50.

Moody, J. D., and R. E. Geiger, 1975, Petroleum resources: how much oil and where?: Technology Review, March/April, p. 39–45.

Potential Gas Committee, 1983, Potential supply of natural gas in the United States (as of December 31, 1982): Potential Gas Agency, Colorado School of Mines, Golden, 74 p.

Potential Gas Committee, 1984, Definitions and procedures for estimation of potential gas resources: Potential Gas Agency, Gas Resources Studies, n. 2, 16 p.

Potential Gas Committee, 1985, Potential supply of natural gas in the United States (December 31, 1984): Potential Gas Agency, Colorado School of Mines, Golden, 161 p.

Conventional U.S. Oil and Gas Remaining To Be Discovered: Estimates and Methodology Used by Shell Oil Company

R. A. Rozendal
Shell Oil Company
Houston, Texas

As part of its long-term planning effort, Shell Oil Company and its subsidiaries make an annual forecast of conventional U.S. oil and gas resources remaining to be discovered. The forecast is based on a summation of forecasts for 234 separate provinces in the conterminous United States and Alaska. The most recent estimate (mean value) is 55 billion bbl of oil and condensate and 310 trillion cubic feet (tcf) of gas remaining to be found. The methodology and assumptions used in preparing this forecast are reviewed and the results are compared with other published estimates.

INTRODUCTION

To make the necessary long-range decisions that will affect an energy company's future, there is a need to make forecasts that attempt to relate economic and social factors to patterns of future energy consumption and supply. To do this, Shell Oil Company and its subsidiaries (hereafter referred to as Shell) have developed a planning system that provides an organized method for projecting exploration and production investment alternatives and results. This system includes a coordinated set of premises and forecasts covering the principal factors that might impact Shell's future. One important part of this coordinated set is a forecast of conventional oil and gas resources remaining to be discovered in the United States.

DOMESTIC CONVENTIONAL OIL AND GAS RESOURCES

Shell's assessment of total recoverable U.S. oil and gas resources is 271 billion bbl of crude oil and condensate and 1215 trillion cubic feet (tcf) of gas (Figure 1; Table 1). Of the estimated 271 billion bbl of ultimately recoverable crude and condensate, 151 billion bbl (or 56%) has already been produced and an additional 65 billion bbl (or 24%) already exists as discovered reserves, leaving 55 billion bbl (or 20%) remaining to be discovered. For natural gas, the estimate of total ultimate recovery is 1215 tcf, with 633 tcf (52%) already produced and 272 tcf more (22%) as reserves already

discovered, leaving 310 tcf (26%) remaining to be discovered.

The forecast is based on certain definitions and internal assumptions as stipulated by Shell. These include (1) definitions of terminology used; (2) boundaries of geographic areas, especially in offshore boundaries; (3) technological assumptions; and (4) economic assumptions.

Most of the definitions used in the Shell forecast are the same as those used by the Department of Energy, Energy Information Administration (DOE/EIA), in the report "U.S. Crude Oil, Natural Gas, and Natural Gas Liquids Reserves— 1982 Annual Report" (DOE/EIA, 1983, p. 29–34).

Cumulative Production

Cumulative production (as used in this paper) includes all hydrocarbons produced through December 31, 1982. Crude oil and condensate includes production of crude oil and natural gas liquids plus lease condensate and plant liquids. Natural gas production is dry natural gas excluding impurities such as CO_2.

The cumulative production figures used by Shell come from published data. Through 1979, the American Petroleum Institute (API) and the American Gas Association (AGA) published reports on reserves of crude oil, natural gas liquids, and natural gas in the United States and Canada (API/AGA/CPA, 1967–1980). This provided a historical record of production and proved reserves according to the year of discovery and the state. For production figures through December 31, 1979, Shell agrees with the figures derived by the U.S. Geological Survey from the

Table 1. Shell's assessment of total conventional recoverable oil and gas resources in the United States.

Resource Type	Crude Oil and Condensate			Natural Gas		
	billion bbl		%	tcf		%
Cumulative production[a]	151		56	633		52
Discovered reserves	65[b]		24	272[c]		22
Proved		35			202	
Unproved		30			70	
Future discoveries	55		20	310		26
Total	271		100	1,215		100

[a]As of Dec. 31, 1982.
[b]Includes additional resources that may be recovered by tertiary enhanced recovery methods.
[c]Excludes underground storage of 5 tcf. Does not include a Shell estimate of 52 tcf that might be obtained from "old gas" decontrol (Matthews, 1983).

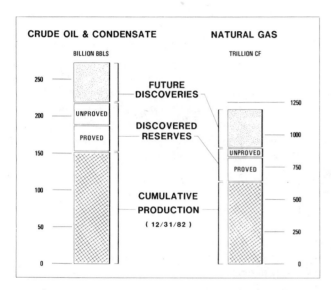

Figure 1. Shell's assessment of total conventional recoverable oil and gas resources in the United States.

API/AGA/CPA data and published in U.S. Geological Survey Circular 860 (Dolton et al., 1981, Table 1). Production figures for 1980–1982 are from the DOE/EIA annual reports on U.S. crude oil, natural gas, and natural gas liquids reserves (DOE/EIA, 1981–1983).

Discovered, Proved Reserves
Discovered, proved reserve figures included are those defined by the DOE and published in the 1982 annual reserve report (DOE/EIA, 1983).

Discovered, Unproved Reserves
Discovered but unproved reserves as defined by Shell are essentially the same as the indicated reserves plus the inferred reserves as defined by the U.S. Geological Survey (Dolton et al., 1981). This category includes reserves that will be added to known fields through extensions, revisions, and new pay zones. The discovered but unproved reserves of crude oil also include supplemental reserves that will be recovered by secondary and tertiary methods. The discovered but unproved reserves for natural gas do not include

underground storage of 5 tcf of gas. The estimate of discovered, unproved reserves is highly dependent on internal assumptions of Shell regarding technology and economics.

Assumptions of Technology
The crude oil reserves that may result from secondary and tertiary recovery include those from thermal and chemical/miscible techniques such as the use of CO_2 or caustic agents. Included are additions only from currently available technology or new technologies that are already under development and appear to have a good chance of success.

Assumptions of Economics
The timing of future price increases is critical to the amount of additional oil or gas that might come from existing fields, particularly those approaching depletion and abandonment. Any additions that may result from improved economics as seen by Shell are included in the discovered but unproved volumes. The discovered but unproved reserves for natural gas, however, do not include a Shell estimate of 52 tcf that might be obtainable from "old gas" decontrol (Matthews, 1983).

Future Discoveries
For the 55 billion bbl of crude and condensate and the 310 tcf of gas (mean values) that Shell estimates is yet to be discovered, a longer list of assumptions and conditions needs to be stated.

Conventional Oil and Gas
Included in Shell's forecast is the estimate of oil and gas that can be recovered using conventional oil- and gas-producing technology that is now available or that is under development and is known to have a reasonable chance of being successful. Unconventional sources of oil and gas are not included in the forecast. Hydrocarbons from tight sands that cannot be produced using conventional oil and gas completions are not included. The forecast volume also does not include (1) tar sands, (2) gas from geopressured aquifer zones, (3) reserves from coal bed methane or coal liquification, or (4) hydrocarbons from Devonian shales or oil shales.

Table 2. Distribution of future conventional recoverable oil and gas discovery volumes.

Location	Crude Oil and Condensate		Natural Gas	
	billion bbl	%	tcf	%
Onshore conterminous U.S.	15	27	160	52
Offshore conterminous U.S.	10	18	75	24
Onshore Alaska	5	9	32	10
Offshore Alaska	25	46	43	14
Total	55	100	310	100

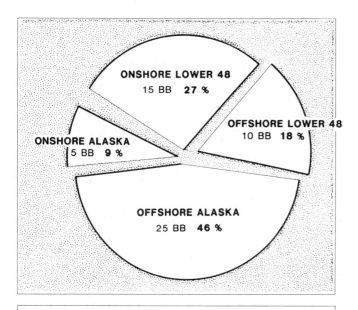

CRUDE OIL & CONDENSATE

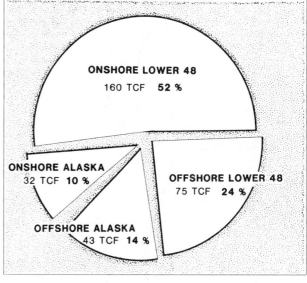

NATURAL GAS

Figure 2. Distribution of future conventional recoverable oil and gas resources in the United States (BB = billions of barrels).

Economics

In the long term, economics play an important part in the Shell forecast of ultimate discoveries of recoverable oil and gas. In the forecast, the factor that determines the ultimate volume of conventional oil and gas that will be produced is the cost of developing alternate fuels, such as tar sands, oil shales, nuclear energy, or solar energy, to replace them. Future oil and gas prices are based on a long-term forecast made by Shell that takes into account future energy supply and demand. The Shell forecast of ultimate recoverable hydrocarbon volume is more dependent on the forecast of future oil and gas prices and on the forecast of the economics of alternate fuels than it is on the limits of a finite resource base. For instance, if Shell is seriously in error on the low side of the cost of developing alternate fuels, more of the resource base may be economically recoverable than is now foreseen.

Geographic Boundaries

It is assumed that most areas that are currently legally inaccessible in the United States, such as wilderness areas, offshore military withdrawals, and the Great Lakes, will become available in the future and volumes from these areas have been included in the forecast.

The offshore conterminous United States future discovery volume estimate was divided in two parts: (1) areas in < 1200 ft (366 m) water depth, and (2) areas in > 1200 m (366 m) water depth. The 1200 ft break was chosen as the depth at which a change in the producing system was assumed to be necessary (existing technology allows for bottom-founded producing platforms to be used in water depths < 1200 ft). Volumes from water depths > 1200 ft (366 m) were included only if they were within a certain distance from the depth line that would allow some of the producing facilities to be placed in shallower water. This was done for economic reasons, because on the basis of forecast economics and present technology, deep-water, open-ocean completions far from shelf areas may not be able to compete economically with alternate fuel sources. This is one area, however, in which technology is rapidly advancing, and major breakthroughs may increase future discovery volumes produced in deep water.

For offshore Alaska, no volumes in water depths > 1200 ft (366 m) have been included.

Another way to look at future discovery volumes is by geographic distribution. In the Shell forecast, over 60% of crude oil and condensate discoveries are expected to be in offshore areas (Figure 2; Table 2), and more than two-thirds of

Figure 3. Geologic regions and subregions in the United States used in making the Shell forecast.

the offshore oil is expected to be found in Alaska. Alaska is expected to provide 55% of future discoveries and the conterminous United States the remaining 45%. In contrast, the remaining potential for natural gas is mostly in the conterminous United States (76%), with only 24% coming from Alaska. Of the 310 tcf of natural gas remaining to be discovered, over half (52%) is expected to be found in onshore conterminous United States.

METHODOLOGY

Procedure

To prepare the forecast, forecasters at Shell divided the United States, including Alaska, into 11 broad geologic regions and 28 subregions for use as the summary areas in the forecast (Figure 3). The forecasts, however, were made by geologic province, which were then summed to obtain the

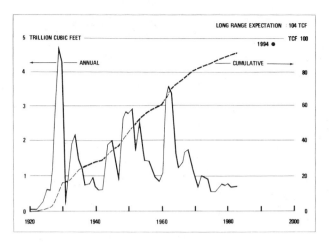

Figure 4. Industry annual and cumulative associated and nonassociated natural gas discovery volumes in the Permian basin.

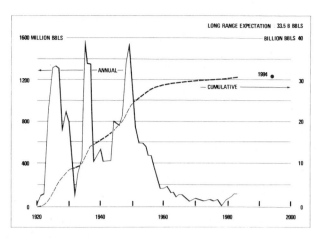

Figure 5. Industry annual and cumulative oil and condensate discovery volumes in the Permian basin.

subregion totals. In onshore conterminous United States, a total of 147 provinces were evaluated. In offshore conterminous United States, a total of 52 geologic trends or provinces were used. In onshore Alaska 20 provinces were evaluated, and in offshore Alaska 15 areas were used. The province breakdowns in offshore Alaska, in the offshore Atlantic, and in the offshore Pacific are closely tied to the outer continental shelf (OCS) planning areas as delineated by the Department of Interior for past and future OCS sales.

The method of evaluation used in each geologic province varies depending on the maturity of exploration in the province, the level of Shell activity in the area, and the amount of data available to make the evaluation. In a number of the provinces in onshore conterminous United States where exploration is in a mature stage and where Shell has done little or no recent work and has little or no in-house expertise, future discovery volumes were estimated by extrapolating trends of present discovery rates as a function of time.

Figure 4 is an example of a gas discovery graph for the Permian basin subregion of onshore conterminous United States. Annual discoveries are shown by the solid line (left-hand scale), and cumulative discovered volumes are shown by the dashed line (right-hand scale).

Annual discovery volumes for years prior to 1980 are based on AGA data as published in the API/AGA/CPA (1967–1980) reserve reports. These reports contain historical records of produced and proved reserves by year of discovery and by state; by using appropriate growth factors, these figures can be used to generate an estimate of the ultimate recovery of discovered reserves. The growth factors applied have been generated from API/AGA/CPA reserve data from fields discovered since January 1, 1950. After 1979, the DOE/EIA reserve reports were used. The DOE/EIA data are more difficult to work with because reserves are no longer listed by date of discovery.

The graph in Figure 4 portrays a typical shape for a curve of cumulative discovered volumes: it increases rapidly before 1970 when gas was easy to find, but the growth rate diminishes with time after 1970 when it became harder to

find the remaining gas. Large discoveries, such as the 5 tcf discovery that occurred in 1929, cause the cumulative discovered volume curve to shift. Since this type of discovery is difficult to forecast, no future "big surprises" in the mature producing provinces have been included in the forecast.

Figure 5 is the oil discovery graph for the Permian basin. Since this is a mature basin and there is only a finite resource available, the curve has flattened. The expected long range discovery volumes have been forecast to be only 6% higher than the cumulative amount discovered to date.

The same procedure of extrapolating historical discovery trends was also used for the geologic provinces of offshore Texas and offshore Louisiana (Gulf of Mexico), which are in mature stages of exploration. Figure 6 shows the portion of the forecast (shaded) that has been supported by statistical analysis of discovery history (see Table 3).

Onshore Conterminous United States

Included in the forecast for onshore conterminous United States is an additional 4 billion bbl of crude oil and condensate and 44 tcf of natural gas that are not supported statistically. These volumes were estimated by geologic and geophysical methods from areas that are lightly explored (frontier areas) or from provinces in which Shell is actively exploring and has a good geologic and geophysical data base. In these areas, a range of uncertainty of the discovery volume magnitude was estimated, assuming the presence of hydrocarbons, from which an expectation given success was derived. Then a probability of failure (no developable hydrocarbons in the province) was determined, taking into account such parameters as source rocks, temperature history, the presence of reservoir rock, and the availability of traps. From this, a risk-discounted, expected discovery volume was determined. Overall, this method is very similar to that described by the U.S. Geological Survey in Circular 860 (Dolton et al., 1981).

Offshore Conterminous United States

For offshore conterminous United States, Shell used historical trends of discovery volumes in only the more

Table 3. Future discovery volumes that are supported by statistical analysis of discovery history.

Resource and Location	Total	Supported By Discovery History	Other
Crude oil and condensate (billion bbl)			
Onshore conterminous U.S.	15	11	4
Offshore conterminous U.S.	10	1	9
Alaska	30	0	30
Total	55	12	43
Natural gas (tcf)			
Onshore conterminous U.S.	160	116	44
Offshore conterminous U.S.	75	21	54
Alaska	75	0	75
Total	310	137	173

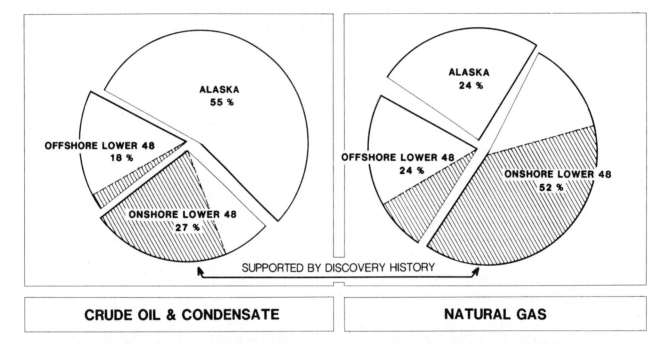

Figure 6. Future discovery volumes that are supported by statistical analysis of discovery history (shaded areas).

mature provinces in the Cenozoic in offshore Texas and Louisiana. In the partially evaluated and unevaluated areas of the Cenozoic, Shell's extensive grid of seismic data was used to generate prospect-oriented, potential discovery volumes. Mature trends having known results and extensive seismic data were used as analogs to evaluate other less explored trends. The results of these comparisons were then discounted. Richness factors (recoveries per acre) derived from producing analog areas were used in some trends to determine potential discovery volumes as a check against the prospect-derived numbers.

In the Mississippi–Alabama–Florida area of the Gulf of Mexico, in the offshore Atlantic, and in the offshore Pacific, Shell has good seismic data and has done detailed sale preparation work for most areas. The results of this work were used to generate discovery volumes (assuming success) on the basis of the prospects and leads present, which were then risk discounted.

Offshore Alaska

Although Shell does not have as extensive a seismic net in offshore Alaska as in offshore conterminous United States, the available data provide a good reconnaissance grid for most areas. Again, a prospect-oriented, risk-discounted, discovery volume was generated for each province.

Onshore Alaska

In areas other than the North Slope and Anchorage basin, there is little seismic or well information available for onshore Alaska. In the North Slope and the Anchorage basin areas, a modified prospect-oriented determination was used. In the other areas of onshore Alaska, a variety of geologic analyses and analogs were used.

Summary

A few points should be emphasized regarding the estimation procedures used by Shell in the various subregions

Table 4. Comparison of estimates of future conventional recoverable oil and gas discovery volumes made by Shell, U.S. Geological Survey, and Exxon Corporation (oil and condensate in billions of barrels and natural gas in trillions of cubic feet).

Location	Shell 1983 Oil[a]	Gas	USGS 1980[b] Oil[a]	Gas	Exxon 1976[c] Oil[a]	Gas
Onshore						
Conterminous U.S.	15	160	59	390	17[d]	160[d]
Alaska	5	32	8	37	7[e]	26[e]
Total	20	192	67	427	24	186
Offshore						
Conterminous U.S.	10	75	19	102	n/a	n/a
Alaska	25	43	14	65	n/a	n/a
Total	35	118	33	167	39	101
Total U.S.	55	310	100	594	63	287

[a]Includes condensates and natural gas liquids.
[b]From Dolton et al. (1981).
[c]From Langston (1976).
[d]From White et al. (1975).
[e]Determined by difference.

and provinces to develop the future forecast. In onshore conterminous United States, the analyses do not allow for a "big surprise." Most of the projections of discovery rates are influenced by the more recent data and have been projected conservatively. Also, in the less well evaluated areas and in the frontier areas, Shell has discounted the potential success volumes on a conservative basis. In offshore conterminous United States and in offshore Alaska, the forecasts are strongly oriented toward prospect and lead evaluation. This makes little allowance for hydrocarbons in stratigraphic traps and for totally new concepts or big surprises. In addition, limited area in depths of water beyond 1200 ft (366 m) are included in offshore conterminous United States, and no volumes beyond this depth are included for Alaska. Therefore, Shell's forecast of future discovery volumes probably tends to be conservative.

COMPARISON OF FUTURE DISCOVERY ESTIMATES

Table 4 is a comparison of Shell's current estimate of future discoveries subdivided by geographic areas with estimates published by the U.S. Geological Survey (Dolton et al., 1981) and by Exxon (Langston, 1976).

Compared to the U.S. Geological Survey, Shell's estimates are substantially lower for undiscovered oil and gas resources in onshore conterminous United States where ours are largely supported by discovery history. Shell's estimates are substantially higher for oil discoveries in offshore Alaska, where our estimates are supported by seismic data. Shell's estimates in the remaining areas are all slightly lower than those of the U.S. Geological Survey.

Shell and Exxon estimates are similar except that Exxon's 1976 estimate is slightly lower for gas and slightly higher for oil. However, an updated estimate by Exxon of 300 tcf was reported (Platt's Oilgram News, 1984). No updated figure for oil was indicated.

Table 5 is a comparison of the present Shell estimate of future discoveries versus the Shell forecast made 5 years ago (Blackburn, 1978). It must be remembered that according to the DOE/EIA reserve reports, during that 5-year period almost 1.5 billion bbl of oil and condensate and 16 tcf of gas have been discovered in new fields. These are probably minimum figures because growth in reserves with time is expected. Viewed in this light, the Shell forecast of future discoveries for oil has been slightly reduced, but for gas it has increased.

USE OF THE FORECAST

As a part of long-term planning, Shell makes a forecast of annual industry discovery volumes for the future by geographic subregions. From this industry forecast an estimate is made of the magnitude of Shell discoveries and reserves that can be expected from each subregion. This estimate is based on the expected level of Shell exploration effort and the historical share of industry discoveries made by Shell in each subregion. This portion of the forecast contributed by Shell is then used to determine Shell's future development program for these reserves and the hydrocarbon production rate to be expected from them. The resulting necessary expenditures and forecast revenues are then used to predict future cash flows, income, and profitability. The forecast made each year of total future industry discovery volumes answers three primary questions involved in making yearly future projections:

1. Is there a significant recoverable resource yet to be found to justify a long-term exploration and production effort?
2. Are the forecasted annual industry discovery volumes reasonable compared to the total recoverable resources remaining to be discovered?
3. Where are the future discovery volumes most likely to be found?

Table 5. Comparison of estimates made by Shell in 1978 and 1983 of future conventional recoverable oil and gas discovery volumes (oil and condensate in billions of barrels and natural gas in trillions of cubic feet).

Location	Shell 1978[a]		Shell 1983	
	Oil[b]	Gas	Oil[b]	Gas
Onshore				
Conterminous U.S.	15	150	15	160
Alaska	10	50	5	32
Total	25	200	20	192
Offshore				
Conterminous U.S.	10	70	10	75
Alaska	25	45	25	43
Total	35	115	35	118
Total U.S.	60	315	55	310

[a]From Blackburn (1978).
[b]Includes condensate and natural gas liquids.

In Shell's analysis, the answers to these questions are as follows:

1. The recoverable resource yet to be found does justify a long-term exploration and production effort.
2. The annual industry discovery volumes, as projected by Shell, are thought to be reasonable and to justify forecasting a sustained and continuing industry exploration and production effort for at least the next 21 years, which is the time frame for Shell's strategic planning.
3. As previously noted, Alaska becomes very important as the location of a significant part of future discoveries.

On the basis of this information, our exploration department can set priorities and plan allocation of exploration resources and manpower. The production department, by analyzing the estimate of the size and location of future reserves, can plan research needed to solve the technological problems associated with these areas.

In addition, the discovery volume projections are translated into a forecast of annual conventional oil and gas production for the U.S. petroleum industry. Comparing these projections with our analysis of energy demand gives insight into future supply and demand balances and interfuel competition, which is then used as a framework to develop long-term strategies regarding alternate fuels. These strategies and forecasts are the assumptions that are used as a foundation for Shell's long-term investment plan.

CONCLUSION

The forecast of future discovery volumes made by Shell is part of a planning system that gives Shell an organized method for projecting its exploration and production programs. The objective of this paper has been to provide insight into how Shell makes the forecast of future discovery volumes, what is included in this forecast, and how the results are used. This should allow the reader to place Shell's forecast in a better context with similar forecasts made by other groups.

ACKNOWLEDGMENTS

I wish to thank Shell Oil Company for permission to attend the 1984 AAPG Convention at San Antonio, Texas, and present a talk on the subject of this paper and for permission to publish this paper. I also wish to thank R. M. Sprague and W. C. Hauber for reviewing the manuscript and offering comments that greatly improved the final version.

REFERENCES CITED

American Petroleum Institute, American Gas Association, and Canadian Petroleum Association, 1967–1980, Reserves of Crude Oil, Natural Gas Liquids, and Natural Gas in the United States and Canada. Annual volumes 1967–1980.

Blackburn, C. L., 1978, Long range potential of domestic oil and gas: paper presented at National Association of Petroleum Analysts/Petroleum Investor Relations Assocation Fall Conference, Boca Raton, Florida, October 19, 1978.

Department of Energy, Energy Information Administration, 1981–1983, U.S. Crude Oil, Natural Gas, and Natural Gas Liquids Reserves—1980–1982 Annual reports: Washington, D.C.

Dolton, G. L., et al., 1981, Estimates of Undiscovered Recoverable Conventional Resources of Oil and Gas in the United States: U.S. Geological Survey Circular 860, Washington, D.C.

Langston, J. D., 1976, A new look at the U.S. oil and gas potential: presented to the 16th Annual Institute on Petroleum Exploration and Economics, Dallas, Texas, March 10, 1976.

Matthews, C. S., 1983, Increase in United States "old gas" reserves due to deregulation: testimony presented to Senate and House Energy Committees, 1983.

Platt's Oilgram News, 1984, 300 tcf of undiscovered U.S. Gas: Exxon, v. 62, n. 82, p. 20.

White, D. A., et al., 1975, Three methods assess regional oil, gas potential: Oil and Gas Journal, v. 73, n. 35, p. 143.

Oil Discovery Index Rates and Projected Discoveries of the Free World

L. F. Ivanhoe
Novum Corporation
Santa Barbara, California

This paper discusses many of the problems involved in making economically realistic appraisals of future U.S. and foreign oil discoveries. The old "discovery index" (DI) method (i.e., annual barrels new recoverable reserves added per foot exploratory drilling) can be modified to evaluate any country's DI trends and to project when future oil discoveries may occur. The free world's DI rate has been declining since 1969 at an average rate of 7% per year. In 30 years, new oil found may be at only 10% of today's rate. Only the super oil-rich Persian Gulf region can count on having significant reserves of conventional oil thereafter. The end of our globe's oil age will be in sight by 1999 and will effectively end during the first half of the twenty-first century.

INTRODUCTION

Reserves versus Resources

Our present oil supply comes from reserves discovered many years ago, most of which will be produced and consumed by 1999. New reserves must be continually found to provide a future supply of conventional crude oil for the world. The essence of petroleum exploration is to convert unknown resources to recoverable reserves.

Misuse of two technical terms, *reserves* and *resources*, has caused much confusion in the general public; they must be carefully distinguished. *Reserves* of petroleum are the amount that petroleum engineers know they can produce at a profit from known fields, using known techniques, in a known time. *Resources*, in contrast, are theoretical estimates of all of the oil and gas that may exist in any given area, most of which are unlikely ever to be converted into reserves.

Resource prediction is not a precise science, because results are not repeatable by other appraisers within an acceptable margin of error. Government planning agencies estimate resources, whereas industrial companies normally appraise only reserves. The total theoretical resources of any nation or region are largely irrelevant to operating oil companies, which are critically concerned with the next drilling prospect's potential reserves. Busy oil men publish little of what they know or plan to do.

The public appears to be interested mainly in the immediate economic availability of oil and gas rather than in geologically postulated petroleum resources. The key issue to the public is not how much oil may exist theoretically, but when it can be produced and what it will cost. Water and many metals can be repeatedly recycled, but once petroleum is burned, it is gone forever. The majority of the public does not appear to be educated on this difference. For all practical purposes the world will have to operate on its known (proven) reserves of oil and gas during the next decade, because of the long lead times required for exploration and development of discoveries. Under such conditions, discussion of potential resources badly misleads the public and politicians, many of whom assume that reserves and resources are synonymous. Well-intentioned but careless scientists who continue to discuss resources instead of reserves may be one of the causes of our government's lack of realistic energy policies (Figure 1) (McKelvey, 1973).

Short-Term versus Long-Term View

Optimism is always encouraged in petroleum exploration, but it is misleading to imply to any government that oil discoveries are probable where geologic prospects are poor. It is preferable in such cases to hope for the best but plan for the worst. Overoptimism as a policy may be disasterous. Also, the important difference between technical and economic feasibility is often ignored. For example, it is technically feasible to send a person to the moon, but it is not economically feasible to do so.

A long-term view is academic when we are faced with critical short-term deadlines. Time is of the essence during energy shortages when fuel and electricity uses are necessarily curtailed. In general, Americans tend to be short-term oriented, and because of our political system, any time beyond the next election seems to be long-term for many of our politicians. The table below shows the only

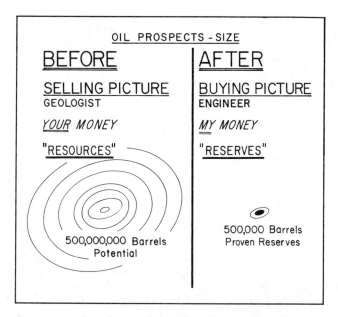

OIL PROSPECTS - SIZE

BEFORE	AFTER
SELLING PICTURE	BUYING PICTURE
GEOLOGIST	ENGINEER
YOUR MONEY	*MY MONEY*
"RESOURCES"	"RESERVES"
500,000,000 Barrels Potential	500,000 Barrels Proven Reserves

Figure 1. Illustration of the differences between resources and reserves: selling versus buying and the view point of the geologist versus the engineer, both of which are needed. (From Ivanhoe, 1980e.)

sources of oil that will be available for various time periods in the future:

Term	Time (years)	Oil Sources
Mini	0–0.3	Tanks, tankers, pipelines
Short	0–5	Reserves
Medium	5–15	Reserves (plus resources)
Long	+15	Resources (plus reserves)

Regional Variations in Oil Richness

Oil prospects are not equal in all countries, nor do they stop at national borders. It is commonly overlooked by the public and many economists that some parts of the world are much richer in oil than others. Only a few areas have the high "A-factor" that puts conventional petroleum into subsurface strata. It is statistically possible, but geologically unreasonable, to apply an "average" amount of oil recovery per square mile to an untested basin. The huge resources of the statisticians result when they multiply large alluvium- or water-covered areas by an estimated uniform barrels per square mile factor. Unfortunately, this method is not geologically valid, since some of the small oil-rich areas are far richer than the extensive poor ones. (For example, tiny Kuwait has far more reserves than all of South America and Australia combined.) Oil and gas accumulate under very special circumstances in relatively small traps, leaving the remainder of the earth barren. The world's producing areas amount to <1% of the total surface of the continents and offshore shelves.

The sizes of oil fields in a given basin have roughly log normal distributions, meaning that there are only a few large fields but many tiny ones (Halbouty, 1970; Klemme, 1971b; Ivanhoe; 1976a). The five biggest fields usually contain most

of the oil in any basin (Klemme, 1983). Also, the largest fields are the easiest to find because they are the biggest targets and are usually discovered early in the exploration if all of the basin is "politically accessible," (open for leasing). The major producing fields that comprise "big oil" are much more critical to any nation or international company than the very small fields having <5 million bbl recoverable. But every bit helps, particularly to a less developed country (LDC) that imports oil.

Exploration in the super oil-rich countries is much more of a sure thing (by several orders of magnitude) than long-shot exploration in the still unproductive nations. Large countries, such as the United States, the Soviet Union, Canada, Mexico and China, have a much better chance of finding commercial production simply because they have many basins, some of which may be rich in oil, among their extensive barren regions. This is in sharp contrast to the numerous small nations that have only one tiny basin in which to find oil. The few apparent exceptions to this size generalization are the small countries that share part of a rich petroleum province with large neighbors (e.g., Kuwait). Small isolated islands do not usually have any petroleum unless they are located within a producing basin (e.g., Trinidad) (Ivanhoe, 1979).

U.S. versus Foreign Oil Business

Two things are required to develop new oil reserves: geologic prospects and exploratory drilling. This study shows that from 1958 to 1980, the oil discovered per unit of effort in foreign, noncommunist, oil-producing nations averaged 30 times higher than in the 48 conterminous states of the United States. But in 1982 the exploration footage drilled in the conterminous United States totaled ~95 million ft versus 28 million ft for the rest of the noncommunist world combined. This indicates that there are strong political and economic reasons for the international oil industry's preference to drill in the conterminous United States, although the nation is long past its prime oil discovery years (which occurred during the 1930s) (Nehring and Van Driest, 1981). This business preference is due to the legal peculiarity that the conterminous United States is the only place in the world where farmers usually own the onshore mineral rights and where the politically independent courts are impartial in law suits in resolving the contractural obligations between oil land lessors and lessees. This results in many oil deals and high drilling rates by small independent entrepreneurs who find it difficult to operate elsewhere. Comparable drilling rates and the number of discoveries of tiny fields will never be approached anywhere else in the world (Odell and Rosing, 1975; Grossling, 1976).

When free companies explore with their own money there is always a compromise between the gross local business terms (including politics, laws, royalties, taxes, and infrastructure) and the geological merit of an area. The overall foreign investment restraints must be tougher than in the conterminous United States or else the companies would be looking for oil overseas instead.

Countries lacking petroleum delude themselves if they hope to entice small U.S. independent oil companies to drill in their basins. Foreign nations must realize that they are in competition with all other countries, including the United States for the international oil companies' exploration funds.

Non-oil-producing LDCs often set their asking terms to oil companies at the high levels used in oil-rich countries, which are economically unrealistic for the nation's lacking in oil. These less fortunate countries may have to offer extremely favorable terms to stimulate interest by independents in their unattractive prospects.

Oil-producing nations have similar problems for any untested basins. Many public servants and politicians ignore or forget that oil and gas companies operating on their own money are not designed to be scientific institutions or semipublic services. Independent commercial oil companies are profit-oriented groups that are in business to earn money for their shareholders. The goal is money, not oil. They will risk their funds in exploration only in places where they can hope to find oil and gas that they can sell at a profit: no profit, no exploration. They can and will go elsewhere or will instead, diversify into any other business that has a better profit potential.

Less developed countries are not necessarily non-oil-producers. All of the 13 members of the Organization of Petroleum Exporting Countries (OPEC) and Mexico are oil-rich LDCs, whereas many industrial nations are lacking in oil (e.g., Japan, Sweden, Belgium, and South Africa). Mexico provides a good example of effective exploration by a big government oil company (Pemex) in an oil-rich LDC. It is the presence of major petroleum reserves, however, that matters, not that the country has a national petroleum company. Many LDCs do not acknowledge that merely setting up a government oil company and getting World Bank loans will not ensure them of any discoveries if their country has no oil (Meyerhoff, 1976).

Modern Petroleum Exploration Technology

Modern petroleum exploration is a very efficient process, and political and leasing delays often consume more time than the necessary geophysical work and wildcat drilling. Maturity of a basin's oil prospects comes quicker than it did in the early history of oil production. Digital seismic surveys have decreased the average finding time required to discover 100% of the reserves in a new basin's five largest fields after the geology is understood and the first large field found from 14 years in the 1940s to 6 years in the 1970s (if all of the basin is politically accessible) (Klemme, 1983). Modern digital seismic surveys give a much more accurate delineation of structure and stratigraphy for the estimated ultimate recovery (EUR) than did older methods, particularly offshore. Large stratigraphic trap fields now seem to be possibilities rather than probabilities (Halbouty, 1982). If large commercial-sized stratigraphic oil fields (like East Texas), however, are common, geologists would have discovered more of them in the highly drilled conterminous United States or in other areas via digital seismic "bright-spots" during the last 25 years.

Since the 1960s, digital seismic techniques have reached such a stage of development that a virgin area covered with water (e.g., North Sea) or with permafrost (e.g., northern Alaska) can be easily surveyed and its poor areas eliminated, its best land leased, and its discovery wells accurately located. Modern geophysical and drilling techniques are very effective in finding oil in such areas and thus are just as valid in eliminating poor areas in other regions. There are practically no virgin areas left, and there is no place where an unknown basin the size of the Persian Gulf could go undetected. The recently discovered giant offshore Mexican oil fields all lie within the previously known Gulf of Mexico basin (Ivanhoe, 1979, 1980a,b). The world has now reached a semimature (i.e., "middle aged") stage of petroleum exploration.

Methods of Estimating Undiscovered Oil Resources

Estimates of undiscovered petroleum resources are usually made by one of three basic methods or combinations of them, which we can term Economic, Geologic, or Engineering. Each of these approaches has its problems and must be carefully reviewed to determine its degree of reliability. There is no sure way to predict the future.

Semantics are critical. It is often overlooked that there are few provable facts that are 100% certain. A geologic "fact," for example, may be 80% certain and 20% assumption (common sense). Common sense, however, is not a true fact in the scientific sense, but a product of our education and experience. If we change the assumptions, then we will correspondingly alter the final "facts." Thus, we must always separate the provable facts from the assumptions (estimates) to see whether changing the latter influences any conclusions.

Economic Estimates

The inherent assumption of economic estimates is that the petroleum supply is infinite and that all basic problems are those of distribution, which can be controlled by money. The need to actually discover oil and gas before it can be produced is not considered in the estimations. In the past, petroleum geologists found so much oil and gas that much of the public now takes discoveries for granted. Most laymen and many geologists are unaware of how the world's petroleum prospects have decreased since 1970 (Ivanhoe, 1984a,b,c).

Geologic Estimates

Geologic estimates of petroleum resources attempt to predict where the oil and gas are and how much may still exist. A critical shortcoming of such predictions is that the timing (i.e., between now and eternity) of discoveries is not indicated. Results are usually displayed on a cumulative probability graph. The amount of petroleum (zero to infinity) is graphed against the cumulative probability of its occurrence (0–100%) (Figure 2). A 75% cumulative probability on such graphs means that there is a 75% chance that there is more and a 25% chance that there is less than the indicated volume. Low, median, and high estimates (95%, 50%, and 5% probabilities) are often published. Some companies refine their predictions by discounting for the likelihood (e.g., 30%) of any commercial oil being found.

Geologic estimates can take several basic forms, including volumetric appraisals, geologic consensus, and exploration plays.

Volumetric appraisals are scientific estimates of a region's potential resources (not reserves). The basic procedure is to calculate the basin's area or volume and then multiply it by some assumed geologic factor to get the total oil and gas that might be in the region. Many assumptions are involved in the process (Mast et al., 1980).

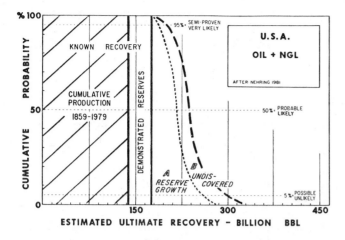

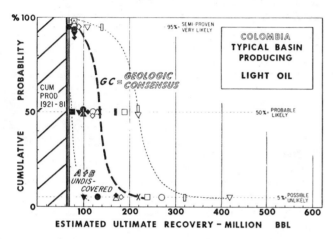

Figure 2. Graphs showing cumulative probability versus estimated ultimate recovery of (a) U.S. oil (after Nehring, 1981), and (b) a typical producing basin in Colombia. The latter is a typical curve of a geological consensus. Individual opinions are distinguished by different symbols to show the range of optimism. X indicates the computed average. (From Ivanhoe, 1984f.)

Geologic consensus appraisals are simply the averages of the opinions of various experts about the undiscovered resources of an area (see other paper by Ivanhoe, this volume). The quality of a consensus can vary tremendously and critically depends on the experts' geologic knowledge about the area being studied. A minimum of six independent opinions, from both qualified petroleum geologists and engineers, are probably required to produce a valid geologic consensus. Each expert's opinions should be shown and identified on the final graphs. Individual appraisals for any basin are always suspect because there is no way to establish an expert's degree of optimism until his appraisal is compared with those of a group of his peers. Petroleum engineers are usually less optimistic than geologists, but both are much more conservative in areas where they have been personally humbled by the actual local problems of finding oil than in foreign areas where they lack first-hand knowledge (Ivanhoe, 1984f) (Figure 2).

A current scientific consensus is that the world will ultimately recover about 2,000 billion bbl of conventional crude oil (Hubbert, 1979; Wood, 1979; Nehring, 1980; Grenon, 1982). This figure, however, is probably slightly high because it was estimated by committees of scientists who had no way of being familiar with all of the world's onshore and offshore petroleum potential. The degree of optimism tends to be in inverse relationship to local knowledge.

The third form of geologic estimates, exploration plays (sets of drilling prospects), can only be realistically appraised if considerable amounts of seismic and geologic expertise have been built up in a given basin to provide a substantial amount of hindsight. An exploration staff develops several plays for a basin (e.g., the Santa Barbara Channel) for which such geologic parameters as structure, stratigraphy, depth, and size are reasonably well known. An estimate of the possible size of the postulated prospects is then made. This gross volumetric estimate may then be reduced by the assumed probability (e.g., 30%) of any commercial oil being found. (Operating oil company staffs are much more conservative in predicting probable commerciality than are government committees who are rarely humbled by dry holes.) An entire region (e.g., the North Sea) can be approached in the same manner (Roadifer, 1975; Miller, 1982; Baker et al., 1984). Credibility dwindles, however, as hindsight decreases. At times it seems that all of the geologic effort involved in calculating and discounting the size of postulated plays in virgin areas may be no more reliable than simply scaling the qualitative reserve potential from graphs of maximum field expected versus minimum economic field (Ivanhoe, 1976a, 1979; Klemme, 1983; Baker et al., 1984).

Engineering Estimates

Engineering estimates of future oil and gas discoveries are much more realistic and conservative than economic or geologic appraisals. They attempt to tell us where, how much, and approximately when petroleum may be found and produced. Such appraisals are repeatable within acceptable variations and are usually made by petroleum engineers with extensive industrial experience. Some of the exploration play appraisals mentioned above approach being engineering estimates if they forecast when the oil can be produced.

M. K. Hubbert is the dean of engineering appraisers. While working at Shell Oil Company 30 years ago, he predicted by projecting finding rates that the United States would reach its production peak in about 1970 and discoveries would decrease thereafter (Hubbert, 1950, 1956). This prophecy was largely rejected by the oil industry because it ran against the conventional optimistic outlook of the time. Hubbert was proven correct, however, when the oil production peaked exactly as he predicted. There is no longer any scientific question of whether, but only when and how, our globe's oil age will effectively end during the first half of the twenty-first century (Hubbert, 1982; Ivanhoe, 1984e).

Hubbert's ideas were adapted to provide the basis for a series of engineering appraisals of most oil-producing countries by the U.S. Department of Energy/Energy Information Administrations/Foreign Energy Supply Assessment Program (FESAP). The small group of practical FESAP petroleum engineers and geologists patiently assembles and synthesizes the known data on reserves and oil and gas production of all major petroleum countries so that all this information will be available to the U.S. government when the next energy crisis strikes. Their practical engineering reports, with an indefinite shelflife, will be invaluable at that time (Dietzman et al., 1979, 1982, 1983a,b,c,d, 1984a,b). (The FESAP reports are gradually superseding earlier preliminary assessments by the USGS (see Masters et al., 1981–1984).

Another important but simple engineering method is the oil discovery index (DI) rate method, which is discussed in the following section.

FOREIGN OIL DISCOVERY INDEX RATE METHOD

The Hubbert and FESAP engineering estimate methods are both quite mathematical and require knowledgeable technicians to make, update, and interpret the computations. Consequently, a simpler method was needed for routine planning purposes. I modified the old discovery index (DI) rate method (Lahee, 1946; Moody, 1978; Haun, 1981) for Occidental Exploration and Production Company (Oxy), to quantify the effectiveness of the past petroleum exploration of different nations. Projections of future oil and gas production in noncommunist countries were not the original objectives, but evolved as the study progressed (Ivanhoe, 1983, 1984, a,b,c).

Discovery Index Rationale

The rationale of the DI method is very simple. The technical effectiveness of oil exploration (new recoverable oil per unit of exploratory effort) and general oil richness of any country can be measured by the annual oil DI rate which was defined for foreign countries (Ivanhoe, 1983) as the annual number of barrels of new recoverable reserves added per foot of exploration drilling. The new recoverable oil added is equivalent to the annual production plus the recoverable reserves at the end of the year, minus the reserves at the first of the year.

Natural gas was not included in the DI because foreign gas is not the economic equivalent of easily transported oil, and foreign gas statistics are usually incomplete and untrustworthy. Communist nations were not covered because of the lack of reliable drilling statistics.

A country's DI rate is basically dependent on geology, not drilling. It is the discovery rate that matters. If the geologic conditions for finding oil are so poor that the DI rate is zero, then no amount of drilling will discover any oil. If twice as many exploratory wells are drilled in a producing basin in a given year, these wells will probably discover about twice as much oil, which would result in approximately the same DI rate for that year. Inevitably, the quality of prospects decreases as time goes by because the biggest fields (largest targets) are discovered first, which reduces the DI rate in future years.

The DI graphs summarize the oil industry's worldwide discovery record over many years, which is the true state-of-the-art of petroleum exploration. Figure 3 shows DIs for the conterminous United States on a nonlogarithmic

graph. Semilogarithmic graphs were used for all other countries and regions to cover the range of DI values in Figures 4, 5, and 6. Extrapolation of the historical DI record will tell us roughly when and how much future oil may be discovered in any region. This is in contrast to the annual AAPG statistics on North American drilling activities that since 1946 have reported the number of new-field wildcat wells that have discovered "significant" reserves, which they define as new fields containing > 1 million bbl of oil. The AAPG's qualitative field sizes (A, B, C, D, E, and F) do not actually tell us specifically how much oil was found in a given year and neglect the giant (> 500 million bbl) and supergiant (> 5,000 million bbl) fields that are most important (Ivanhoe 1980a; Nehring, 1981). The AAPG method was designed for the United States, and cannot be used in many foreign countries where detailed statistics about new-field wildcat wells are not available. My foreign DI includes primary and secondary reserves added by pool extensions, whereas the AAPG method does not. No reliable foreign statistics before 1945 exist.

The old DI method probably had not been used more recently by others because the foreign data were only ~ 80% complete or accurate; the other 20%, which were biased by political, economic, and technological factors, required critical analysis and reconciliation by one objective person to produce consistent and usable statistics. The final results on my semilogarithmic DI graphs are not significantly affected by errors in the foreign DI computation procedure. The DI results are substantially repeatable by others. Consequently, the simple DI method will most likely provide realistic oil discovery rate projections for countries and regions during the next 0–15 years, with an estimated accuracy of about ± 20%. This is better than for most other forecasting methods used in short- and long-term planning purposes (Grenon, 1982).

DI Methodology

The oil potential of different nations and regions can be compared from their DIs. Discovery indexes for foreign countries are more difficult to calculate than for the United States, because information on past drilling activities and reserve estimates are incomplete and scattered. Nevertheless, enough data exist in commonly available publications (which can be tabulated by year and country) to provide the basis for calculation of subquantitative (logarithmic) DIs.

Statistics on annual production and drilling are actual facts, whereas those for reserves are only estimates that must be critically audited. My reserve audit procedure is similar to balancing old checking accounts; that is, one starts at the last balance and works backward to find and correct obvious discrepancies. The latest (1981) reserves were presumed to be reliable (because they have had the most time to self-correct), and they provide an up-to-date base for future DI studies.

There is always a delay between the discovery year and the date that a new oil field's reserves are finally declared (booked). This delay, however, is not critical when comparing long-term trends between countries as long as the same computation method is used for all nations in a DI study. Several significant economic factors, such as financing, infrastructure, and delineation drillings, are incorporated when delayed reserves are booked that are not

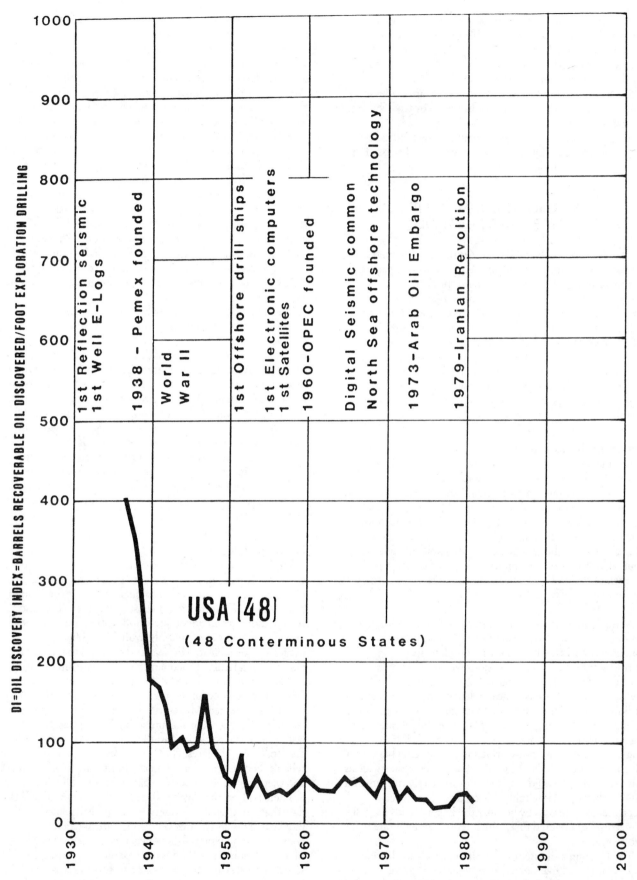

Figure 3. A nonlogarithmic plot of oil discovery indexes (DI) versus years for the conterminous United States, showing significant political, economic, and technical developments. (From Ivanhoe, 1983.)

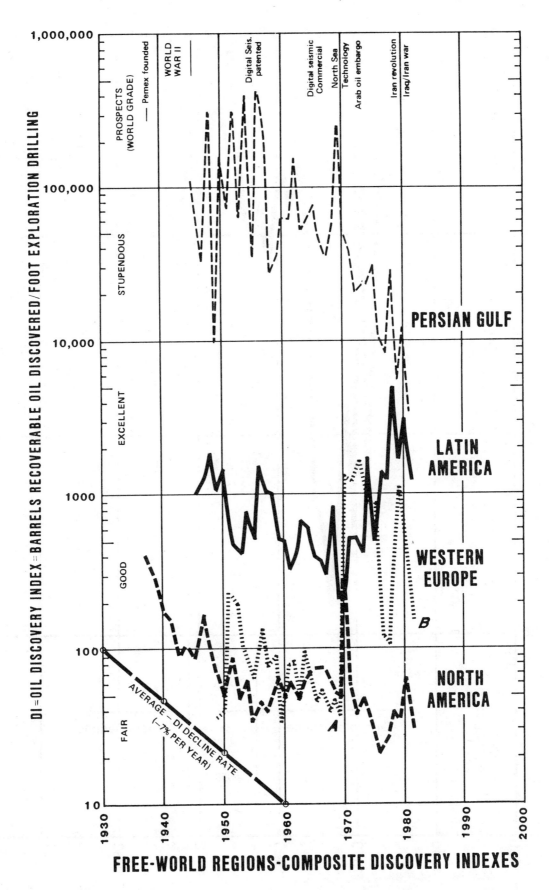

Figure 4. A semilogarithmic plot of regional composite oil discovery indexes (DI) versus years for North America, western Europe, Latin America, and Perisan Gulf. The empirical average DI decline rate is −7% per year. (From Ivanhoe, 1983.)

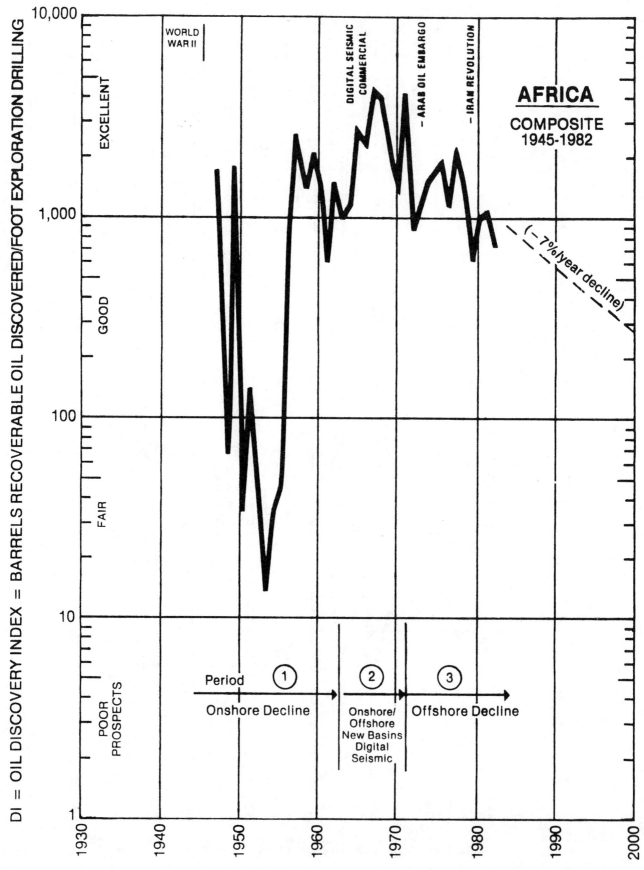

Figure 5. Semilogarithmic plot of composite oil discovery indexes (DI) versus years for Africa showing three DI decline periods. (From Ivanhoe, 1984b.)

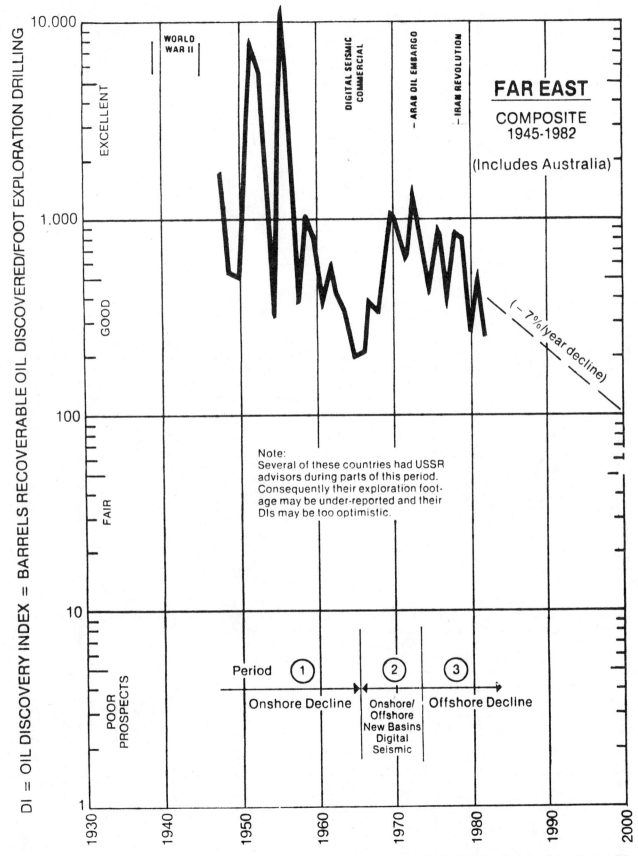

Figure 6. Semilogarithmic plot of composite oil discovery indexes (DI) versus years for the Far East showing three distinct DI decline periods. Several nations had Soviet advisors during the period 1945–1981, consequently their DI statistics may be too optimistic. (From Ivanhoe, 1984b.)

Table 1. Example of oil reserve adjustments for discovery index computations for Greece.

Year (end)	Crude Oil Production (million bbl)	D&M Reserves (million bbl)	Remarks	Adjusted Reserves (million bbl)	Adjusted New Oil (million bbl)	AAPG Exploration Footage (million ft)	DI Rate (bb/ft)
1970	—	—	—	—	—	23	0
1971	—	—	—	—	—	7	0
1972	—	—	1st Discovery, offshore	—	—	16[a]	0
1973	—	—	—	—	—	21	0
1974	—	600	Geopolitical estimate	52	52	34	1530
1975	—	137	100% oil in place	52	0	No report	0
1976	—	137	100% oil in place	52	0	8[a]	0
1977	—	137	100% oil in place	52	0	40[a]	0
1978	—	137	100% oil in place	52	0	24[a]	0
1979	—	70	50% oil recovery	52	0	32	0
1980	—	55	40% oil recovery	52	0	40	0
1981	3	49	38% oil recovery	52	0	40[a]	0

[a]Estimated at 8000 ft per well.

available the year fields are first discovered. Such delays provide a much better idea of the EUR than does a single discovery well (Ivanhoe, 1984d).

It is an axiom that recoverable reserves plus total production can never decrease with time, so the true increase in reserves can never be a negative number in any year. Thus, proceeding back in time in 1-year increments from the 1981 tabulations, the reported reserves are adjusted each year as required to balance the accounting. Major corrections are made at erratic plus or minus DI "spike" years, which are easy to identify on a long-term tabulation. A common problem occurs when fields finally go into production: at these times the more conservative petroleum engineers take over for the optimistic geologists and politicians and begin to report "recoverable" reserves rather than "oil-in-place." The application of a realistic 10–40% oil recovery factor abruptly reduces the reserves and the DIs for that year. If the DI is ever negative for a given year, it is invalid and must be corrected by adjusting the reported reserves at the last major positive DI spike. Table 1 gives an example of the oil reserve adjustments for one country (Greece). Fallow years without reserve additions between discoveries can be handled for a few years by averaging the new reserves per total exploratory footage over all of the barren period. Such averaging can also be used for any rare cases in which reserves have been backdated to the year of discovery or in which drilling statistics are estimated. This smoothing can be recognized on the DI graphs as a horizontal line for several years.

The final tabulation and computation procedure can be outlined by these steps:

Foreign gas without an established market is not the commercial equivalent of easily transported oil and belongs in the economically inaccessible "inactive reserve" (resource) category (e.g., Australia northwest shelf, offshore Trinidad, and Arctic Alaska). It costs from three to four times more to transport 1 Btu of natural gas than it does 1 Btu of crude oil in the conterminous United States (Megill, 1984). Lack of long-term trust from one government to the next is a critical obstacle for the international financing and use of natural gas. Consequently, gas is not included in the DI computations. The often-used factor of 6 million cu ft per barrel of oil equivalent (BOE) is the calorific, not the economic (usability) conversion of their values. Natural gas liquids (NGL) that are gas by-products should be assigned to the "inactive" gas reserves (resources) until they actually come on production. Gas hydrates in deep ocean basins are not even in the inactive reserve or resource categories.

A higher DI occurs for any year in which local exploratory drilling was under-reported. This is common for those nations (e.g., Cuba and India) that had Russian advisors who are reluctant to account for all of their exploration efforts. The Soviets reluctantly admit to drilling dry holes, thus unsuccessful wildcat wells become "scientific/stratigraphic tests" and accordingly are not listed (if any wells are reported at all). Consequently, DI rates are always suspect and may be overly optimistic for any country during its guidance by Communist advisors. No communist country was included in the DI study because of the incomplete drilling statistics.

Exploratory footage missing in any AAPG table can be estimated for many nations from the text of the annual

(1) Year	(2) Year end reserves (million bbl)	(3) Net new reserves (million bbl)	(4) Annual production (million bbl)	(5) Gross new reserves (million bbl)	(6) Exploration footage (million ft)	(7) DI Rate
	(adjusted)	(2) minus (3)		(3) + (4) = (5)		(5) ÷ (6) = (7)
1959	200 MMB					
1960	150 MMB	−50 MMB	120 MMB	+70 MMB	95 Mft	737

AAPG reports. In those rare cases where exploration drilling footage is completely omitted, other references (such as *Oil and Gas Journal*) or estimates can be used to provide realistic numbers. Cumulative drilling footage is often used (rather than years) in studies of North American discovery trends, but such refinements were not justified for this post-1945 foreign project.

Alaska and the 48 conterminous states of the United States were treated as separate regions in this DI study. Failure to separate Alaska from the conterminous United States skews the exploration statistics for both. Alaskan operating problems, lack of infrastructure, government-owned mineral lands, and other factors are much more typical of major oil companies' foreign operations in the LDCs than of the small independents in the rest of the United States. Furthermore, Alaskan oil is overseas oil; it is shipped to the conterminous United States by tankers from Valdez, which are subject to interruption by terrorists (Ivanhoe, 1983, 1984d).

A significant trend cannot be established in 2 years. Isolated drops in exploration drilling and DIs are often due to temporary national political and/or economic factors (e.g., Iraq/Iran war) rather than a major change in the geologic merit of a country. High and low DI spike years on DI graphs are not particularly meaningful. (An analogy is the daily weather versus the annual climate.) Erratic DI spikes are often the result of irregular time lags between the dates of new discoveries and the time when their reserves are finally reported. If enough countries are combined, many DI spikes are eliminated, but one must adjust all numbers for the individual nations before they are added together.

Each nation must be evaluated separately, not only as a part of a continent or region. Politics and laws, rather than geology, determine where and when oil companies may work (e.g., People's Republic of China). Two countries sharing common oil basins may have very different exploration politics, problems, and success (e.g., northeastern Mexico versus southern Texas).

Graphs of Annual Exploratory Drilling and New Oil Reserves

Figures 7 and 8 summarize the basic information collected and analyzed during the DI studies. These graphs summarize the actual (adjusted) quantities involved, rather than the DI rates that were derived from them (Ivanhoe, 1984c,d). Figure 7 shows the total exploratory drilling in millions of feet, and Figure 8 gives the total (adjusted) new recoverable oil reserves booked per year from 1937 to 1982. Dates of relevant political and technological developments are indicated.

Both of the figures show curves of three separate areas: (1) the conterminous United States (excluding Alaska); (2) the remaining noncommunist ("free world") countries (including Alaska and the Persian Gulf nations); and (3) the Persian Gulf nations.

This three-way breakdown helps put the noncommunist countries' oil supply into proper perspective. There are two abnormal regions that tend to skew all of these statistics: (1) the conterminous United States with its large amount of drilling and low discovery rates, and (2) the Persian Gulf with its small amount of drilling and huge oil reserves. Statistical details from either of these unusual regions may have little direct bearing on the rest of the world.

Exploratory Drilling

It is apparent by visual inspection of the three curves on Figure 7 that the conterminous United States is still the oil industry's favorite country for exploration drilling. This is because the farmers usually own the onshore mineral rights, thus small oil deals are feasible for independent entrepreneurs. In the period from 1973 to 1982, the exploratory drilling footage in the conterminous United States ranged from 2.1 to 3.0 (average, 2.5) times that of all of the rest of the noncommunist world combined, and 85 to 155 (average, 120) times that of the Persian Gulf nations. Before the 1973 Arab oil embargo, the exploration spread was even greater, reaching a maximum of 7 times at the first U.S. drilling peak in 1956, which did not occur in the rest of the noncommunist world. After the 1973 and 1979 oil price shocks, the U.S. exploratory drilling doubled to a peak in 1982, with a corresponding but less pronounced increase elsewhere. As might be expected, the quality of the oil prospects and finding rates decreased during these drilling surges.

About one-third of the rest of the noncommunist world's exploration drilling effort is in Canada. A change in Canada's federal energy policies reduced exploration tax subsidies after 1980, which resulted in a corresponding decline in drilling.

New Recoverable Oil Reserves

The DeGolyer and MacNaughton (D&M) 1945–1982 engineering reports do not show which (new or old) oil fields are responsible for any reserves, nor what time lag may be involved in the bookings (see Figure 8). This is a basic difference between the D&M reserves, which include annual changes resulting from improved recoveries, and the AAPG procedures, which qualitatively attempt to back-date, after 3- and 6-year intervals, any reserve changes for specific fields per AAPG field size categories. In the long run, the final amount of oil discovered must be the same regardless of whether the D&M or the AAPG procedure is used. The quantitative D&M engineering reserves, however, are more useful for economic planning purposes than the AAPG qualitative field sizes. No attempt was made in this study to correlate any D&M reserves with specific fields.

DISCOVERY INDEX: RESULTS AND DISCUSSION

General

Two distinct foreign categories exist: countries with commercial oil discoveries and those without. In 1981, there were 150 noncommunist countries and 18 communist countries. This study shows that 74 countries produced commercial oil, including 31 exporters, 13 of which are in OPEC. For the 94 less prospective nations where commercial oil has not yet been found, the DI rate is zero.

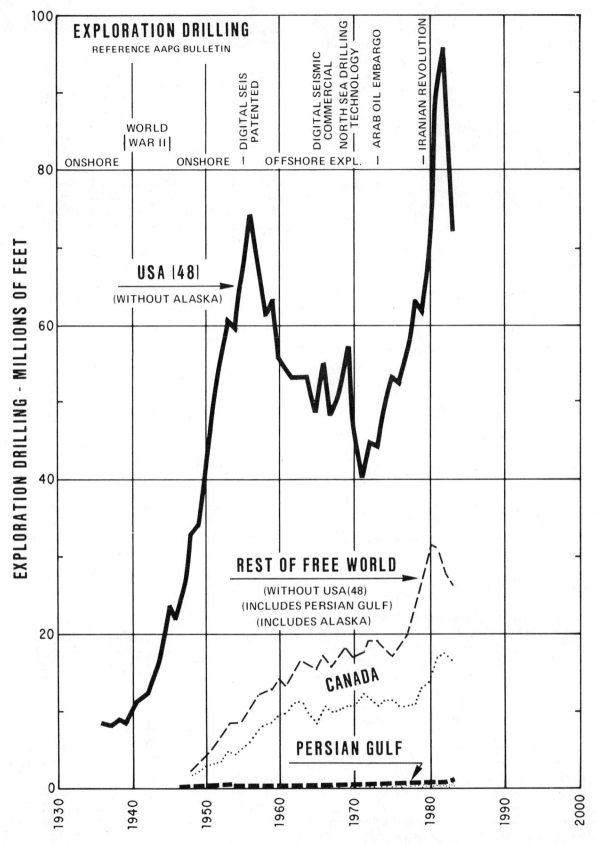

Figure 7. Exploratory drilling of noncommunist world from 1937 to 1983. The three curves distinguish the conterminous United States, the remaining noncommunist countries including Alaska, and the Persian Gulf. About one-third of the noncommunist world's drilling (excluding the United States) is in Canada. (From American Association of Petroleum Geologists, 1945–1983.)

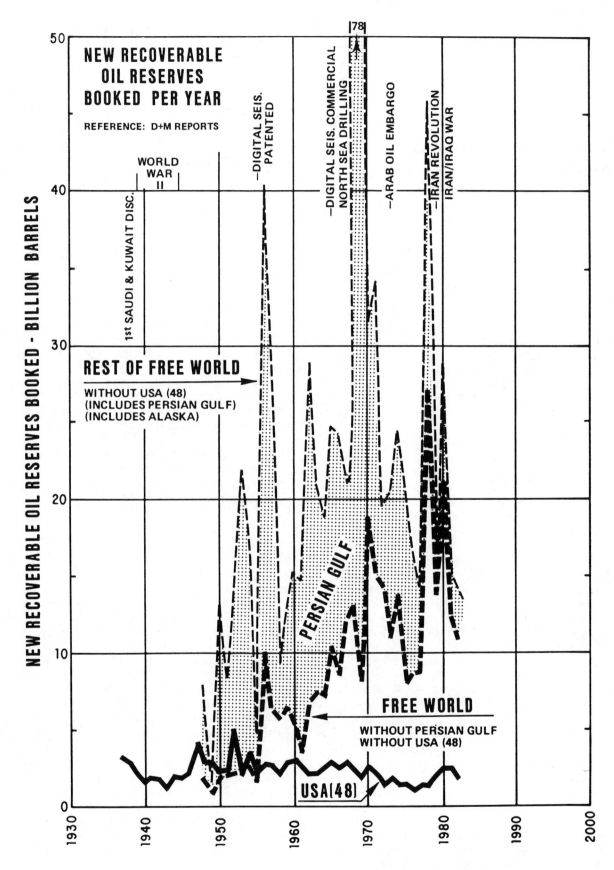

Figure 8. New recoverable oil reserves booked per year between 1937 and 1982. The three curves distinguish the conterminous United States, the remaining noncommunist countries including Alaska, and the Persian Gulf. (From DeGolyer & MacNaughton, 1945–1982.)

The potential reserves of undiscovered fields determine how attractive any country's remaining oil prospects may be, as well as the justified exploration investments, risks, and rewards. On a world scale, any nation's prospects decrease logarithmically in the following manner:

World grade	Oil discovery index (DI) rate (bbl/exploration ft)
Stupendous	>10,000
Excellent	1,000–10,000
Good	100–1,000
Fair	10–100
Poor	1–10

The computed geologic potential of any country represents the "average" grade to date including all past dry holes, and it may not necessarily indicate the quality of any remaining individual drilling prospects. The average remaining prospects in the conterminous United States (1982 DI = 19), however, are approaching the "poor" (DI = 10) category. The oil prospects here are being steadily depleted (Figures 3 and 9) (Ivanhoe, 1983, 1984a,c,e).

Digital seismic and common depth point (CDP) technologies were developed in the United States during the late 1950s and by 1965 were commercially available everywhere. This was the most important exploration breakthrough since the introduction of onshore analog reflection seismic surveys in the early 1930s, because digital seismic surveys made oil prospecting effective at sea and through permafrost. Digital seismic surveys have now been run across every politically accessible continental shelf in the world. Offshore China is the most recent area to be surveyed, thus the U.S.S.R. is the only remaining "nondigital seismic" nation. In many countries (e.g. United Kingdom, Mexico, etc.), there was a pronounced upward jump (digital shift) in the DIs when digital seismic technology opened up virgin offshore basins in the late 1960s and the new reserves were reported soon after 1970 (Figure 4). The digital shift is often very distinct if it is offshore, as in the United Kingdom, but it is sometimes less spectacular for countries, such as the United States and Libya, where digital seismic surveys were first introduced to onshore areas that already had considerable oil production. In the latter cases, the digital shift may appear as a flat or climbing plateau rather than as two offset, steeply declining DI trends (Figure 9). A possible exception to this two-phase digital shift is in the Persian Gulf region where the decline has been fairly steady since 1956, partly because many offshore reserve increases there resulted from extensions of onshore fields, in other words, by development rather than exploration drilling (Figure 4).

Three recognizable "DI periods" can be identified on most national and regional DI graphs: (1) onshore DI decline; pre-1960s; (2) onshore/offshore DI increase due to digital seismic surveys; 1960–70; and (3) offshore DI decline; post-1970. (Figs. 5, 6, and 9). The third (postdigital offshore) DI decline period is almost 15 years old and is well advanced in all regions of the noncommunist world. Oil discovery rates are definitely decreasing everywhere (Ivanhoe, 1983, 1984a,b,c,d).

The DI rates in noncommunist countries peaked in 1969. All 150 of these nations, including the 90 non-oil-producers are now politically accessible and have had at least reconnaissance digital seismic surveys run across their continental shelves. We can now drill to any prospective depth on land or in any ice-free ocean. The best structures were drilled and tested during the last 15 years. No new major petroleum provinces are expected unless the deep ocean areas away from the slopes produce sizable commercial oil reserves, which would happen far in the future, if ever. Effective exploration in arctic Alaska with its permafrost was also delayed until digital seismic technology arrived. The National Petroleum Reserve-Alaska (NPR-A) has been surveyed and drilled by the federal government with poor results. Some parts of the government-owned mineral lands of the conterminous United States and Alaska, however, are still not politically accessible and thus are untested.

No major technological breakthroughs are in sight, although all technologies are being constantly improved to enable us to find smaller and smaller traps and to recover more oil by enhanced oil recoveries (EOR) from those fields that are already known. Governmental geologic research jobs often increase as field sizes go down. National political and military considerations will prolong the search for local oil in many nations long after economic considerations indicate that it should be stopped. Normal technical improvements are already included in projections of the DI rates. What we need are new virgin areas to explore that might contain giant oil fields. Unfortunately none are known. The only completely untested regions of the globe are now those in deep ocean basins and under pack ice. Deep-water, open-ocean completions far from shelf areas may never compete economically with alternate fuel sources (Rozendal, this volume). The stormy, icepack-covered, deep oceans of the Arctic or Antarctica are no more likely prospects than the deep-water basins right next to the U.S. markets. (St. John, 1980; Ivanhoe, 1980e, 1981a,b).

Discovery Index: Decline Rate

A significant trend that came out of this study is the average DI decline rate of 7% per year (Figs. 4, 5, 6, and 9). This decline rate is equivalent to a drop of one order of magnitude per 30 years. It can be used to help project DI trends into the future as oil provinces are depleted. This 7% per year DI decline rate was derived empirically from the DI graphs of the six major oil-producing regions (Figs. 4, 5, and 6), which include North America (1940–1955 and 1967–1977), Western Europe (1958–1969), Latin America (1948–1970), Persian Gulf (1950–1979), Africa (1967–1981), and the Far East (1970–1981). This is a conservative DI decline rate, and a case can be made for a more rapid DI decline for several important regions including the Persian Gulf. After 30 years at a rate of 7% per year, the amount of new oil that will be discovered per unit of drilling effort in the noncommunist world may be only 10% of the current rate. The economic and political consequences of such an oil discovery decline are very alarming for the world's people (Ivanhoe, 1983, 1984a,b,c,d).

All effective new petroleum exploration or production techniques have been quickly introduced everywhere in the noncommunist world since World War II. Consequently, all

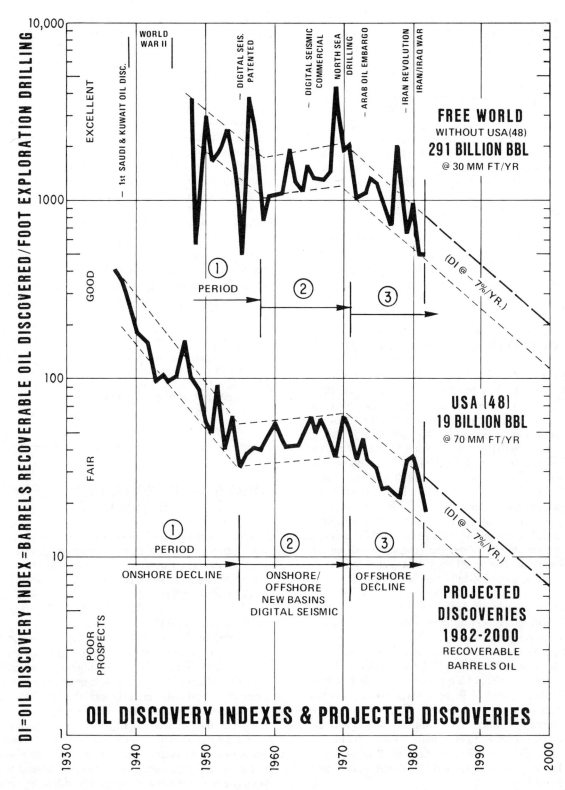

Figure 9. Semilogarithmic plot of oil discovery indexes (DI) and projected discoveries for the conterminous United States and the remaining noncommunist countries. Composite DI numbers versus years for 1945–1981. Projected oil discoveries for 1982–2000 are given at 1980 exploratory drilling rates: 19 billion bbl for the conterminous United States and 291 billion bbl of recoverable oil for the remaining noncommunist world. Empirical high (and low) dashed lines bracket the trends. Graph for noncommunist countries shows three periods: (1) is erratically high due to several new supergiant oil fields in the Persian Gulf region; (2) includes new onshore and offshore basins in Africa, Nigeria, Libya, western Europe (North Sea), and Saudi Arabia; and (3) includes positive spikes from Mexican offshore fields and decreases resulting from the Iraq/Iran war. (From Ivanhoe, 1984a.)

nations' oil discovery rates show similar technology related surges with little time lag, and their DI curves are roughly subparallel, differing only in the richness of oil in that nation's basins.

Discovery Index Costs and Investment Risks

The DIs and graphs can be used to roughly compare annual exploration costs per effort between countries. To do so, drilling expenses must be comparable and the total cost must include not only the contracted drilling price of the wells, but also all direct and indirect exploration expenses such as overhead, land, legal, bonuses, accounting, seismic surveys, geology, engineering, roads, and well logging. The cost per barrel found is infinite before the first oil discovery (Ivanhoe, 1984d).

The average 1958–1980 spread between the subparallel conterminous United States and noncommunist world DI curves indicates that the oil industry has consistently discounted gross oversea oil prospects versus U.S. long-term profit potentials by ∼ 30 times (Figure 9).

WORLD PRODUCTION TRENDS

Unfortunately, the world's petroleum prospects (excluding communist countries) are much worse than they were 10 years ago, but the good news is many people have made a major conservation effort and have accordingly postponed the time of the next major energy crisis (Figure 10). The question is not whether but rather when the next crunch will occur (Hubbert, 1979; Ivanhoe, 1984e). Major new reserves to replace the globe's oil production are simply not being found. Only three giant (> 500 million bbl) oil fields were discovered worldwide from 1979 to 1983. In 30 years (at −7% per year) the oil discovery rate will be only 10% of the present total. The current world "oil glut" will be as fleeting as a desert community's "water glut" caused by one winter storm. (Some glut! The total U.S. 1983 expenditure for imported oil was about $55 billion or 90% of our international trade deficit of $61 billion.) The world's production oversupply is a short-term imbalance caused by a sudden drop in total oil consumption resulting from long overdue conservation in the United States and elsewhere. (The U.S. usage of crude oil, natural gas liquids, and their products decreased from 19.0 to 15.2 million bbl per day between 1979 and 1983, which is analogous to dieting from 190 to 152 lbs in 4 years. A European-type tax on gasoline would reduce our usage even more.) Additional conservation will be increasingly difficult to accomplish, thus long-term consumption is expected to eventually stabilize and to balance production. At this point the world price of crude oil will quickly rise again—to the public's surprise and dismay. (Figures 9 and 10).

The continual addition of new petroleum provinces is critical to maintain the world's long-range oil supplies. The world total will drop as soon as no significant new basins (or nations) can be brought into production. The question now is where the international oil industry can turn to find new virgin provinces. The non-U.S. regions seem to be our best hope. The Persian Gulf is in a class by itself—no other areas in the world comes close (Figures 4, 7, and 8).

The communist nations were not included in this study because of a lack of statistical information. We know, however, that the small socialist nations have prospected for and produce some oil for many years and that they are oil importers today. Only giant Soviet Union and China are net petroleum exporters and will undoubtedly discover more fields in their extensive virgin or semivirgin onshore and offshore basins. The introduction of long-delayed modern digital seismic oil exploration in newly politically accessible communist nations such as China should help keep up the world's discoveries and production. Let us hope that these nations will add substantially to the free world supplies (Ivanhoe, 1984e).

World's Reserve to Production Ratios

At the world's present semimature stage of exploration, the current reserve to: production (R/P) ratios indicate where future oil discoveries are most likely. The 1982 R/P ratios for the globe's four major petroleum, economic, and political regions (Ivanhoe, 1984e) are as shown below

Region	R/P
Conterminous United States	8 years
Middle East	75
Remaining noncommunist world	27
Communist nations	20
Total world	34 years

DISCOVERY INDEX PROJECTIONS OF OIL TO BE DISCOVERED

All projections are limited by optimistic or pessimistic assumptions, which can be best verified on graphs. The DI graphs summarize the noncommunist world's historical discovery rate since 1945. Recognizable discovery rate trends place limits on the probable order of magnitude of oil that may be discovered until the end of the next decade (1982–2000 inclusive).

The information on the regional graphs shown in Figures 3, 4, 5, and 6 was recombined into two summary curves in Figure 9, one showing the conterminous United States and the other, the remaining noncommunist countires.The estimates for oil discoveries through the year 2000 are based on extrapolations of the post-1970 DI curves, which have both been declining at roughly the normal rate of 7% per year. Exploratory drilling for future years was assumed to be about that of the near-record year of 1980, which was 70 million ft in the conterminous United States and 30 million ft for the rest of the noncommunist world. More or less drilling should find proportionally more or less oil. The annual DI, when multiplied by the estimated exploratory footage, gives the amount of new oil that may be found each year. The sum of the projected annual discoveries provides the gross amount of oil that can be expected through the year 2000. High (and low) projections, to bracket the top (and bottom) of the "band" of the discovery trend, are both indicated in Figure 9.

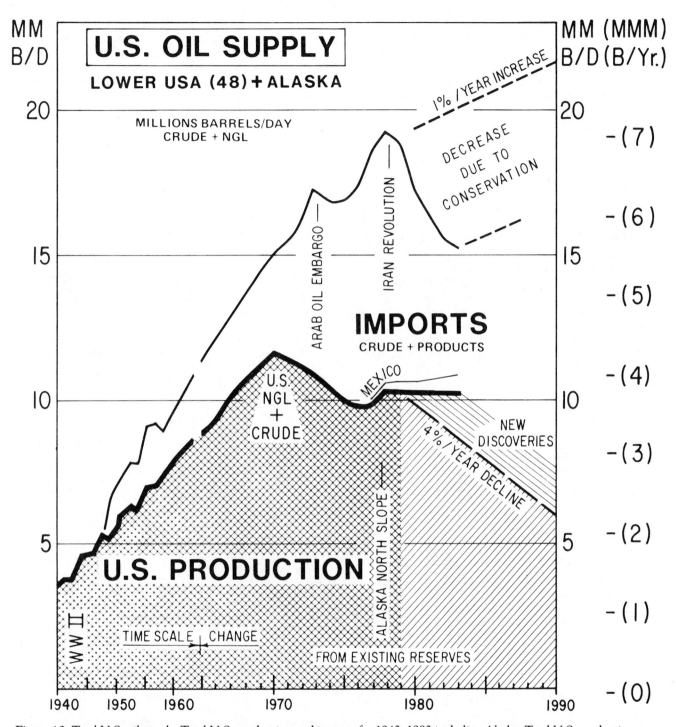

Figure 10. Total U.S. oil supply. Total U.S. production and imports for 1940–1983 including Alaska. Total U.S. production includes conterminous U.S. and Alaskan crude oil and natural gas liquids; imports include crude oil and its products. The cost of 1983 imports is calculated as 5 million bbl per day times 365 days at $30 per bbl equals $55 billion. This was 90% of the 1983 U.S. trade deficit of $61 billion. The major decrease in total oil consumption and imports between 1980 and 1983 was due to conservation. Note Mexico's minor share of U.S. imports. (From Ivanhoe, 1980d.)

The total low and high projections of new oil reserves to be discovered or developed during the period from 1982 to 2000 are 12 and 19 billion bbl for the conterminous United States and 162 and 291 billion bbl for the remaining noncommunist countries. In summary, it appears that conventional crude oil discoveries in the rest of the noncommunist world may keep pace with production through the year 2000, but that U.S. discoveries (excluding Alaska) definitely will not. In 1982, the United States consumed 5.7 billion bbl of oil, natural gas liquids, and refined products (3.8 billion bbl produced here and 1.9 billion bbl imported). We will consume over 100 billion bbl of oil during the next 18 years at the 1982 rate. But my high projection of U.S. oil discoveries is for only 19 billion bbl of new oil (slightly over 1 billion bbl per year) during this period. By 1999, the conventional crude oil finding rate of the lower conterminous United States may be only one-third of today's rate, or about 0.5 billion bbl per year.

CONCLUSIONS

My final oil projections are regrettably lower than most estimates made by others. This discrepancy is mainly due to the lack of time limits on most oil resource studies. My high U.S. estimate through the year 2000, however, is close to the 90% cumulative probability (i.e., low ultimate production) number of Nehring's (1981) U.S. report (my estimate is 19 billion bbl versus Nehring's 21 billion bbl). Comparing the rest of the noncommunist world's estimate with Nehring's (1982) world paper gives similar magnitudes (my estimate is 291 billion bbl versus Nehring's 475 billion bbl). My studies project that our entire globe's ultimate conventional crude oil recovery may be close to 1,700 billion bbl, which falls within the range of 1,600–2,000 billion bbl predicted by Nehring (Nehring, 1980; Ivanhoe, 1984e). My figure is calculated from the following (all values in billion bbl):

World Discoveries to 1982		Projected Noncommunist Countries to 2000		Estimated Communist to 2000		Estimated World after 2000		Total
1170	+	310	+	70	+	150	=	1170

I reluctantly agree with Hubbert's prediction that the globe's oil age will effectively end during the lifetimes of many alive today. By 1999 we will have spent most of our national petroleum heritage by public consent, and the end of our oil age will be in sight. We will thereafter become one of the oil-poor nations. The social and political consequences of this foreseeable decline in our country's economic base will be staggering for the entire world, particularly when combined with the concurrent population explosion in the LDCs (Fox, 1984; Ivanhoe, 1984e).

ACKNOWLEDGMENTS

The writer thanks R. H. Vaughan, retired executive vice president of worldwide exploration of Occidental Exploration and Production Co. (Oxy), for instigating this study, Oxy's John A. Carner for authorizing its publication, Dr. M. King Hubbert for very pertinent suggestions and corrections, plus all of the critics, typists, and draftsmen who patiently revised the manuscript and figures.

This paper is a synthesis of several earlier studies on evaluations of foreign petroleum basins by L. F. Ivanhoe for Oxy. To ensure consistency in the final results, Oxy had all data analyzed by a single experienced and objective petroleum explorationist. Summaries of several of Oxy's previous studies were published in the Oil and Gas Journal 1976–1984. All published opinions are Ivanhoe's and not necessarily those of Oxy. Persons interested in more information should write directly to the author.

SELECTED REFERENCES

American Association of Petroleum Geologists, 1945–1983, (AAPG), Annual, Domestic/Foreign Bulletins (drilling statistics for each country). Tulsa, Oklahoma.

Baker, R. A., et al., 1984, Geologic field number and size assessments of oil and gas plays: AAPG Bulletin, v. 68, n.4, p. 426–432.

Chevron Corporation Economics Department, 1984, World Energy Outlook — Forecast through the year 2000. San Francisco, California, 17 p.

DeGolyer & MacNaugton, Inc., 1945–1982, Annual, Twentieth Century Petroleum Statistics: Annual reports (production and reserves statistics for each country). Dallas, Texas.

Deitzman, W. D., et al., U.S. Department of Energy/Energy Information Administration/FESAS, 1979, Report on the petroleum resources of the Federal Republic of Nigeria: DOE/IA-0008, 132 p.

———, 1982, Libya, Algeria, and Egypt — Crude oil potential from known deposits: DOE/EIA-0338, 105 p.

———, 1983a, The petroleum resources of the North Sea: DOE/EIA-0381, 97 p.

———, 1983b, The petroleum resources of Middle East: DOE/EIA-0395, 169 p.

———, 1983c, The petroleum resources of Venezuela and Trinidad and Tobago: DOE/EIA-0398, 53 p.

———, 1983d, The petroleum resources of Mexico: DOE/EIA-0423, 107 p.

———, 1984a, The petroleum resources of Libya, Algeria, and Egypt: DOE/EIA-0435, 189 p.

———, 1984b, The petroleum resources of Indonesia, Malaysia, Brunei, and Thailand: DOE/EIA-0447, 183 p.

Dolton, G. L., et al., 1981, Estimates of undiscovered recoverable conventional resources of oil and gas in the United States: U.S. Geological Survey Circular 860, 87 p.

Fox, R. W., 1984, The world's urban explosion: National Geographic Magazine, August 1984, p. 176–185.

Grenon, M., 1982, A review of world hydrocarbon resource assessments: Report EA 2658, Contract TPS 80-763 by International Institute for Applied Systems Analysis for Electric Power Research Institute, Palo Alto, California.

Grossling, B., 1976, Window on oil—A survey of world petroleum sources: London, Financial Times, Ltd., 140 p.

Grove, N., 1974, Oil—The dwindling treasure: National Geographic Magazine, July 1974, p. 792–825.

Halbouty, M. T., 1970, Geology of Giant Oil Fields: AAPG Memoir 14, 575 p.

———, 1982, The Deliberate Search for the Subtle Trap: AAPG Memoir 31, 410 p.

Hall C. A. S., and C. J. Cleveland, 1981, Petroleum drilling and production in the U.S.—Yield per effort and net energy analysis: Science, Feb. 6, p. 576–579

Haun, J. D., 1975, Methods of Estimating the Undiscovered Volume of Oil and Gas Reserves: AAPG Studies in Geology n. 1, 206 p.

———, 1981, Future of Petroleum Exploration in the United States, AAPG Bulletin, v. 65, n. 10, p. 1720–1727.

Hubbert, M. K., 1950, Energy from Fossil fuels: American Association for the Advancement of Science, Centennial, Washington, D.C., p. 171–177.

———, 1956, Nuclear Energy and the Fossil Fuels: American Petroleum Institute, Drilling and Production Practice, p. 7–25.

———, 1979, Measurement of energy resources: Transactions of the ASME Journal of Dynamic Systems, Measurement, and Control, v. 101, p. 16–20.

———, 1982, Techniques of prediction as applied to the production of oil and gas, in S. I. Gass, ed., Oil and Gas Supply Modeling: National Bureau of Standards Special Publication 631, p. 16–141.

Ivanhoe, L. F., 1976a, Oil/gas potential in basins estimated: Oil & Gas Journal, Dec. 6, p. 154–155.

———, 1976b, Foreign prospective basins evaluated: Oil & Gas Journal, Dec. 13, p. 108–110.

———, 1976c, Economic feasibility appraised for petroleum search in remote regions: Oil & Gas Journal, Dec. 20, p. 82–84.

———, 1979, Petroleum prospects of non-OPEC LDCs' Oil & Gas Journal, Aug. 27, p. 144–157.

———, 1980a, World's prospective petroleum areas: Oil & Gas Journal, April 28, p. 146–148.

———, 1980b, Evaluating petroleum exploration activities in foreign countries: Oil & Gas Journal, May 5, p. 342–349.

———, 1980c, World's giant petroleum provinces: Oil & Gas Journal, June 30, p. 146–148.

———, 1980d, Will there be enough oil in the 1980s?: World Oil, Oct., p. 175–184.

———, 1980e, Antarctica—Operating conditions and petroleum prospects: Oil & Gas Journal, Dec. 29, p. 212–220.

———, 1981a, Superstructure icing will hike Alaska's Bering Sea operating costs: Oil & Gas Journal, Feb. 9, p. 158.

———, 1981b, Sea ice will increase operating costs in Alaska's Bering Sea waters: Oil & Gas Journal, Oct. 26, p. 267–268.

———, 1983, Free world oil discovery indexes—1945–81: Oil & Gas Journal, Nov. 21, p. 88–90.

———, 1984a, Projected free world oil finds through 2000: Oil & Gas Journal, Jan. 16, p. 138–139.

———, 1984b, Oil discovery indexes of Africa and Far East—1945–1981: Oil & Gas Journal, March 12, p. 115–116.

———, 1984c, Free world exploratory drilling 1937–82 (and new recoverable oil reserves): Oil & Gas Journal, July 9, p. 126–128.

———, 1984d, U.S. vs. foreign oil deals: Oil & Gas Juornal, Nov. 19, p. 176–178.

———, 1984e, World crude output, reserves by regions: Oil & Gas Journal, Dec. 24, p. 65–68.

———, 1984f, Advantages and limitations of geological consensus estimates of undiscovered petroleum resources: 27th International Geological Congress, Moscow, August 1984; Proceedings, Oil and Gas Fields, v. 13, p. 227–285: Utrecht, the Netherlands, VNU Science Press.

Klemme, H. D., 1971a, Trends in basin development—Possible economic implications: World Oil, Oct., p. 47–56.

———, 1971b, What giants and their basins have in common: Oil & Gas Journal, March 1, p. 85–90.

———, 1971c, to find a giant, find the right basin: Oil & Gas Journal, March 8, p. 103–110.

———, 1971d, Look in Permian reservoirs or younger to find super giants: Oil & Gas Journal, March 15, p. 96–100.

———, 1983, Field size distribution related to basin characteristics: Oil & Gas Journal, Dec. 26, p. 169–176.

Lahee, F. H., 1946, Exploratory drilling in the United States in 1945: AAPG Bulletin, v. 30, n. 6, p. 813–828.

McKelvey, V. W., 1973, Mineral resource estimates and public policy: U.S. Geological Survey Professional Paper 820, p. 9–19.

Mast, R. F. et al., 1980, Resource appraisal of undiscovered oil and gas resources in the William O. Douglas Arctic wildlife range: U.S. Geological Survey Open File report 80–916, 36 p. plus Appendices.

Masters, D. C. et al., 1981–1984, U.S. Geological Survey World Energy Resources Program: Open file reports/circulars on assessments of conventionally recoverable petroleum resources: Open-file reports: U.S. Geological Survey Circulars (USGS), Argonne National Laboratory (ANL), and the Department of Energy Energy Information Administration and Foreign Energy Supply Assessment Program (collectively abbreviated here as DOE):
(DOE) 81-986, Persian Gulf Basin and Zagros fold belt
(USGS) 81-1027, Volga-Urals basin, U.S.S.R.
(DOE) 81-1142, Indonesia
(DOE) 81-1143, Northeastern Mexico
(USGS) 81-1144, SE Mexico, N. Guatemala, Belize
(DOE) 81-1145, Trinidad
(DOE) 81-1146, Venezuela
(USGS) 81-1147, West Siberian basin and Kara Sea Basin, U.S.S.R.
(ANL) 82-0296, Middle Caspian basin, U.S.S.R.
(USGS) 82-1027, East Siberia basin, U.S.S.R.
(DOE) 82-1056, North Africa
(ANL) 82-1057, Timan-Pechora basin and Barents—N. Kara shelf, U.S.S.R.
(USGS) 83-0598, NW, Central, and NE Africa
83-0801, Onshore China
83-0728, Distribution and quantitive assessment of world crude oil reserves and resources
(DOE) 84-0094, NW European region
West Africa
China Offshore
(DOE) Malaysia, Brunei
(DOE) Thailand
(USGS) South Asia
(USGS) Australia
(USGS) New Zealand

Megill, R. E., 1984, Exploration outlook: Is the past a key to the future?: World Oil, June, p. 159–168.

Meyerhoff, A. A., 1976, Comparisons of national and private petroleum company exploration techniques, philosophy, and success: Bulletin Canadian Petroleum Geology, v. 24, p. 282–304.

Miller, B. M. et al., 1975, Geological estimates of undiscovered recoverable oil and gas resources in the United States: U.S. Geological Survey Circular 725, 78 p.

———, 1982, Application of exploration play-analysis techniques to the assessment of conventional petroleum resources by the USGS: Journal of Petroleum Technology, Jan., p. 55–64.

Moody, J. D., 1978, Where oil and gas stand in the energy and ecology dilemmas: Oil & Gas Journal, Aug. 28, p. 185–190.

National Geographic Society, 1981: Energy—A special report in the public interest: Feb. 1981, 115 p.

Nehring, R., 1978, Giant oil fields and world oil resources: Report R-2282-CIA, Rand Corp, Santa Monica, California, 162 p.

———, 1980, The outlook for world oil resources: Oil & Gas Journal, Oct. 27, p. 170–175.

———, 1982, Prospects for conventional world oil resources: Annual Review Energy 1982, v. 7, p. 175–200, Annual Reviews Inc., Palo Alto, California.

———, and E. R. Van Driest II, 1981, The discovery of significant oil and gas fields in the United States: Report R-2654-1 and 2-USGS/DOE, Rand Corp, Santa Monica, California, 713 p.

Odell, P. R., and K. E. Rosing, 1975, Estimating world oil discoveries up to 1999—The question of method: Petroleum Times (London), v. 79, p. 26–29.

Roadifer, R., 1975, A probability approach to estimate volumes of undiscovered oil and gas in J. C. Davis et al., eds., Probability methods in oil exploration: AAPG Research Symposium, Standford University, California.

Root, D. H., and L. J. Drew, 1979, The pattern of petroleum discovery ratio: American Scientist, v. 67, n. 6, p. 648–652.

St. John, B., 1980, Sedimentary basins of the world and giant hydrocarbon accumulations: AAPG Special Publication, 23 p., 1 map.

Wood, P. W. J., 1979, There's a trillion barrels of oil awaiting discovery: World Oil, June, p. 141–148.

World Oil, 1983, Forecast—Review issue, Feb. 15.

World Crude Oil Resources: U.S. Geological Survey Estimate and Procedures

Charles D. Masters
U.S. Geological Survey, Oil and Gas Branch
National Center, Reston, Virginia

The World Energy Resources Program of the U.S. Geological Survey applies an assessment technique to international petroleum provinces that was established initially for purposes of domestic analysis. Additions to the procedures that permit quasi quantitative evaluations of exploration maturity are being evolved for the assessment process, as are methods of disaggregating the assessment into a field size distribution. Program results to date indicate that while oil reserves are high, they are also declining steadily because of continued production in excess of discovery such that a deficit of ~70 billion bbl has accumulated in the 10-year period between 1971 and 1981. More disturbing than the discovery/production deficit, however, is the assessment that the distribution of world crude oil has been established, and no new areas with undiscovered resources of > 20 billion bbl of oil have a most likely probability of being discovered.

INTRODUCTION

I do not intend in this paper to present new assessment methodology; rather, my purpose is to rationalize the assessment methodology used by the U.S. Geological Survey (USGS) in the World Energy Resources Program (WERP) and to report some of the results derived from these program studies. A number of program publications are available through the USGS, and in particular, the summary paper for the program to date, on world crude oil, has been published elsewhere (Masters et al., 1983).

THE ASSESSMENT PROCESS

Resource assessment is a people-oriented endeavor. At every stage of the exercise, good judgment is essential to good results. There is no single procedure that can guarantee absolute truth, but clearly one can select procedures and techniques from within the context of the problem one is trying to solve that can lessen subjectivity in the final outcome.

As a first step, the assessor must clearly define what is being assessed; this demands a rigorous resource classification, the necessary elements of which are included in the USGS resource classification diagram shown in Figure 1 and described, as applied to petroleum, in USGS Circular 860 (Dolton et al., 1981). Classification elements that the USGS focuses on for resource assessment are the economically recoverable conventional resources shown in the upper part of Figure 1. The remainder of the in-place oil resources are considered to be marginally economical or subeconomical: (1) marginally economical if there is some concept of future recovery envisioned, for example, for enhanced oil recovery application, or (2) subeconomical if there is no expectation of future recovery.

In the USGS assessment program, with respect to both domestic and foreign activities, the problem we face is the necessity of deriving responsible assessments for hundreds of basins in a relatively short period of time with a limited staff. The job is to assess basin potential, not to design an exploration program. Only those data, relevant to basin analysis are critical to us; in fact, detailed exploration data are commonly not available anyway. We must be generally aware of the nature of the possible plays and their compatibility with the regional geology, but the answers sought in the assessment process do not require that we measure the closures, count the structures, or quantitatively evaluate the source rock.

There are exceptions in which particular downstream analyses require the assessment to be presented in a particular form, such as the assessments for the North Slope of Alaska (Mast, et al., 1980; Miller, 1981; Miller, 1982;

CUMULATIVE PRODUCTION 445	IDENTIFIED RESOURCES		UNDISCOVERED RESOURCES		
	DEMONSTRATED (MEASURED, INDICATED)	INFERRED	PROBABILITY RANGE 95%	MODE	5%
ECONOMIC	RESERVES 723	INFERRED RESERVES (NOT ASSESSED)	321	550	1,417
MARGINALLY ECONOMIC	MARGINAL RESERVES EACH 1% WORLD AVERAGE RECOVERY INCREASE EQUALS 34	INFERRED MARGINAL RESERVES (NOT ASSESSED)	—	—	—
SUB-ECONOMIC	DEMONSTRATED SUBECONOMIC RESOURCES 2,267	INFERRED SUBECONOMIC RESOURCES (NOT ASSESSED)	623	1,068	2,750
OTHER OCCUR-RENCES	EXTRA-HEAVY OIL AND BITUMEN				

Figure 1. Classification of world ultimate crude oil resources (units in billion bbl). Identified resources and production are as of January 1, 1981, and undiscovered resources as of March 1983. From Masters et al. (1983).

Bird, 1984). For this area, data and manpower-intensive Monte Carlo play analysis assessment procedures were followed. Most commonly, however, assessments are made by what we call a modified Delphi process (see Dolton, 1984, for a more complete discussion of this process). It is modified from the commonly understood Delphi method in the sense that the experts are presented with a common body of data that is subject to discussion and modification on the basis of their personal experiences. An individual assessment is subjectively produced by each expert considering his or her interpretation of the petroleum geology and his or her weighing of the scaling factors derived from various number-generating processes. The reported assessment is a consensus average of all the experts' opinions.

THE ASSESSMENT METHODS

To be scientific about basin assessment work, one must organize data impartially, synthesize hypotheses, and test results. Assessors do those things to varying degrees. In our program at the USGS, organization of data done to achieve an understanding of resource quantities at the scale of basins (or other large areas) varies to a considerable degree, because the available data and professional talent needed to conduct the assessment process vary considerably. Commonly, we utilize only two broad number-generating techniques: one is an areal or volumetric analog and the other is a statistical projection. The ultimate disaggregation technique in volumetric analysis is, play analysis, but the need for measured structure sizes and their distribution, or their analogs, commonly precludes our use of that method. The USGS, however, has done a play analysis and assessment of South Asia (Kingston, 1986). Extensive professional experience in this area and a generalized non-Monte Carlo processing approach made possible a form of play analysis that produced satisfactory results.

With respect to quantitative data for general volumetric analogs, we have compiled volumetric yield data from all U.S. basins (which are on file, at the USGS Resource Appraisal Group, Denver, Colorado), and we have ranges of volumetric yield data from the work of Klemme on his worldwide basin classification system (Klemme, 1980). We also use volumetric data from internal system analogs wherever appropriate. A problem with the analog approach is the difficulty of quantifying the analog relationship. Ulmishek and Harrison (1984) experimented with a technique for objectively determining the analog relationship, but to date we continue to subjectively determine that relationship.

Statistical projection, another means of applying data and making assessments, provides both a check on companion methods and a strong short-term extrapolation of basin potential. Previously we at the USGS have considered our assessments to be timeless and have thus been bothered by our inability to express the meaning of the undiscovered resource assessment in economic terms. We now have finding-rate data reasonably available for most countries and regions and are able to project a short-term assessment—for approximately one to two decades or for a given number of exploratory wells—that is anchored to past exploration efficiencies. For the short term, an assessor desiring to alter the assessment projection will be able to clearly select and measure the elements of the equation he or she chooses to change; for example, one would be able to alter the number of exploratory wells per year, increase the discovery efficiency value, or improve the recoverability factor. The geologic assessment deriving from the modified Delphi process, however, (described above) will continue to represent WERP's hypothesis for the total undiscovered resource potential.

This short-term finding-rate projection adds a new dimension of analytical potential to our work. For example, as shown in Figure 2, an estimate of 4 billion bbl of oil (based on a finding-rate projection for Nigeria) closely approximates the modal value of 6 billion bbl determined in the geologic assessment (DOE/USGS, 1979). The finding-rate projection shown in Figure 2 simply estimates a possible amount for discovery given a doubling of the number of wildcat wells; no date for such an occurrence is given here. The similarity of the two assessments reflects that Nigeria is in a very mature stage of exploration for oil, as indicated in Figure 3. On this map a dot shows every location where there is an exploration well density equal to or greater than one well per 8 sq mi. A similar map (not shown) for wells drilled deeper than 15,000 ft (4,500 m) also shows a high density of wells.

In contrast, in North Africa (which includes Egypt, Libya, Tunisia, Algeria, and Morocco), the finding-rate projection based on the same doubling of cumulative wildcats suggests a much smaller resource potential (5 billion bbl of oil) than does the modal value of the geologic assessment (16 billion bbl) (Figure 4). In this case, the geologic assessment reflects the presumed exploration immaturity of a vast area and the prospects for success in untested plays. The immaturity is demonstrated in Figure 5, which shows a map of Algeria and Morocco (only a part of North Africa), like the one for Nigeria; again, the dots show where the exploration well density is greater than or equal to one well per 8 sq mi. The

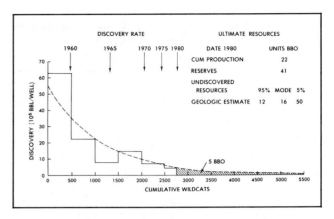

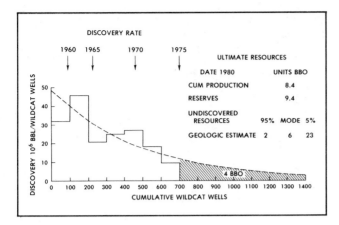

Figure 2. Discovery efficiency through time for Nigerian crude oil; shows projection of possible discovery quantities given the doubling of wildcat exploratory wells from 700 to 1400 wells. The geologic resource assessment shown is very similar to the projection assessment, suggesting a mature stage of exploration in the province (see Figure 3).

Figure 4. Discovery efficiency through time for North African crude oil; shows projection of possible discovery quantities given the doubling of wildcat exploratory wells from 2750 to 5500 wells. The geologic resource assessment is significantly larger than the projection assessment, suggesting an immature stage of exploration in the province (see Figure 5).

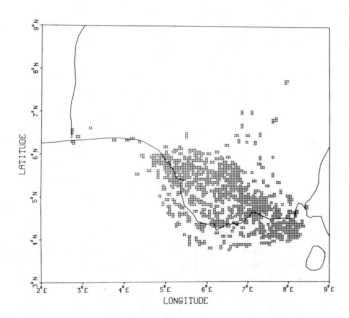

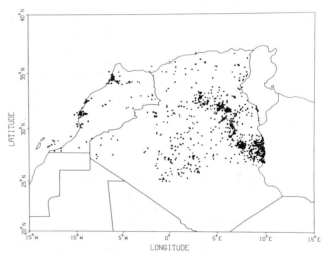

Figure 3. Nigerian exploration density map. The black dots represent areas where well density is equal to or greater than one well per 8 sq mi.

Figure 5. Algeria and Morocco (part of North Africa) exploration density map. The black dots represent areas where well density is equal to or greater than one well per 8 sq mi.

extensive white areas on the map are apparently explored only minimally.

The responsible use of a wide variety of data and number-generating techniques produces what in effect are scaling factors, for our modified Delphi assessment procedures. Because the USGS does not quantitatively compare analogs and because the probabilities for the certainty of occurrence for the different plays vary with the individual assessors and with their personal experiences, we ask the assessors to mentally synthesize the geology and the scaling factors presented to them and to derive their own

judgments of the undiscovered recoverable resource potential of the area. These judgments are expressed in terms of a most likely value, a low value (a 95% probability of more than that amount), and a high value (a 5% probability of more than that amount). (The high and the low percent probabilities were selected to subjectively constrain both the optimist and the pessimist; clearly we cannot know the exact probability.) These values are for the assessors' view, and differing opinions are often debated, especially by those searching for giant fields, which assessed occurrence can become dominant at any probability level of the assessment. This may lead to one or more reassessments until a general consensus is achieved. This process is the hypothesis stage in our scientific endeavor, and it is eventually supported by publications in which the petroleum geology characteristics and the resource occurrences as we know them are described.

To continue with the scientific method analogy, the next stage would be hypothesis testing. This commonly occurs through the analysis of exploration data accumulated over time, but hypothesis testing also occurs by comparison of our data with published results of investigators not with the USGS who may have approached the problem with different data and different techniques. As new exploration data evolve, we not only gain a measure of the plausibility of the assessment hypothesis, but we also begin to see whether the geologic synthesis on which the assessment was based is being sustained. The degree to which we have disaggregated the assessment (geographically, stratigraphically, by field size distribution, or by play) determines the sensitivity with which an analyst can evaluate the new exploration data relative to the assessment. Regardless of the assessment techniques, however, one must evaluate new data and reevaluate old data, because any assessment at any point in time is just an hypothesis that must always remain under review. In fact, a necessary criteria in selecting a methodology is its ability to be reviewed; the use of a detailed assessment technique without comparable level testing over time can become misleading. We can derive a significant improvement in testing ability, and hence assessment, from being able to subdivide the amount of the undiscovered resource into numbers of fields and sizes based on existing or analogue distributions. Such research activities are a part of the present USGS program.

WORLD CRUDE OIL ASSESSMENT

Through world petroleum assessment studies, the USGS has completed its preliminary examination of the occurrence of world crude oil. We have not yet completed our gas analysis and assessments, but the completion of a report is anticipated for 1986. The overall conclusions of our crude oil study are presented in Figure 6, which shows our estimates of the ultimate resources of conventionally recoverable crude oil for the world. The lower bar shows the cumulative production of 445 billion bbl of oil and reserves of 723 billion bbl as of January 1, 1981, for a total original reserve of 1,168 billion bbl. These numbers can be updated readily by considering a world annual production of 20 billion bbl of oil and a discovery rate of 10–15 billion bbl per year. The reserves are higher than commonly reported (see, e.g., the annual reports of the *Oil and Gas Journal*), because of an expanded definition for reserves used in the World Petroleum Congress report (Masters et al., 1983). We are not, however, including any projection of what eventually might be derived from enhanced recovery operations beyond water flood, nor are we including reserves associated with extra heavy oil or tar sands. World production from these two sources is < 200 million bbl per year, and recovery potential and production rates are not sufficiently well understood to derive comparable estimates.

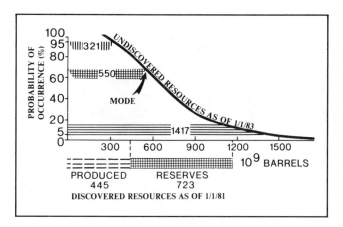

Figure 6. The ultimate resources of conventional crude oil for the world (numbers given are in billion bbl). Discovered oil is shown on the lowest horizontal bar; estimates for undiscovered oil are shown along curved line for selected probabilities, as determined in 1983.

The undiscovered resources are shown as a range from a 95% probability of there being more than 321 billion bbl to a 5% probability of there being more than 1,417 billion bbl. The most likely estimate of the occurrence of undiscovered recoverable resources is 550 billion bbl, an amount equal to about one-half of what has already been discovered. Should the more optimistic resource perspective occur, we can look forward to the possible discovery of a quantity slightly more than that already discovered.

This is a large amount of oil by anyone's standards; the known reserves alone are 36 times the present annual production. Three significant facts, however, warrant attention and concern. First, the present annual production (represented by the cross-cutting thick line on Figure 7) may be abnormally low because of worldwide recession, and it is likely that in the future we can expect much greater draw-down pressure (by increased annual production) on the reserves. Second, we must be aware of the significant decline in discovery rates over the past 15 years. (Discovery rates are shown on the bar graph in Figure 7 as per year quantities averaged over 5-year increments.) The differential between production and discovery rates over the past 10 years has been about 70–80 billion bbl; this deficit has developed even though exploratory effort has been constant or has slightly increased. The third point of concern is the distribution of world crude oil. Figure 8 shows estimates of the ultimate resources by major world regions in declining order of resource endowment. The diagram shows that the distribution of world crude oil is already established and that future discoveries are not likely to change the overall proportions. Through time, the position of the Middle East will become more dominant, and the market competition for the fuel of choice (oil) will soon force most users in the world toward other sources of energy.

To add another element to this concern about distribution, it is my opinion that the continuing declining exploration success rate in the United States since the last USGS assessment work in 1980, (as reported in USGS Circular 860,

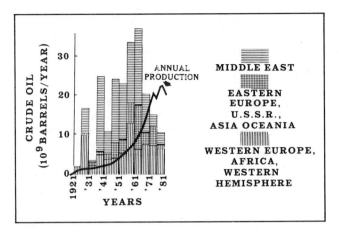

Figure 7. World crude oil discovery rate showing contribution of major geographic regions and a comparison to annual world production. Note that annual production has exceeded the crude oil discovered by some 70-80 billion bbl during the past 10 years. The patterns indicate contributions from various parts of the world, as follows: horizontal lines, Middle East; cross-hatched, eastern Europe, Soviet Union, and Asia Oceania; and vertical lines, western Europe, Africa, and Western Hemisphere. Modified from Masters et al. (1983).

Dolton et al., 1981) calls into serious question the high value of the most likely estimate of the USGS Circular 860 assessment. Exploration has not been as successful area-by-area as had been expected, and we must realistically accept that the value of the undiscovered resource for North America (Figure 8), which reflects the modal value of the assessment, might be less than shown. In other words, exploration outcomes are beginning to suggest that the actual U.S. oil endowment may be on the low side of the assessment range.

Not only is the distribution of crude oil highly skewed by region, it is also highly skewed within regions. Figure 9 shows a map of the undiscovered resource distribution by major regions of the world. The large numbers are the modal values of the assessment, and the dots with the accompanying small numbers show roughly where and how much of that undiscovered oil is still to be found in particular localities within the major regions. Most of the new oil is expected to be found north of the equator; also, 70% of the crude oil assessed occurs in three of the seven regions: the Soviet Union, the Middle East, and North America. Within the various regions, most of the oil is assessed to occur in only a few localities. For example, in North America, most of the new oil will most likely be found in the greater Gulf Coast region (including Mexico) and in the Arctic regions of Alaska and Canada. Venezuela, Peru, and Columbia are the dominant regions in South America for future potential. In Europe, the oil potential lies in the North Sea. In the Soviet Union, West Siberia accounts for almost one-half of the potential. And in the Middle East, about one-half of the assessed amount of undiscovered oil is in Iraq.

What is critical about the crude oil distribution analysis is that it takes the discovery of large new petroleum provinces

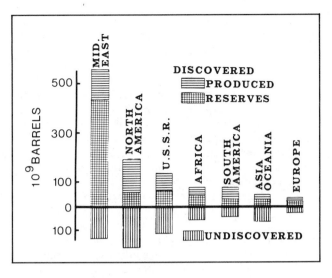

Figure 8. Regional distribution of ultimate resources of world crude oil. Note that the ranking, by major region, of the quantitative distribution of ultimate amounts of crude oil is already established. Modified from Masters et al. (1983).

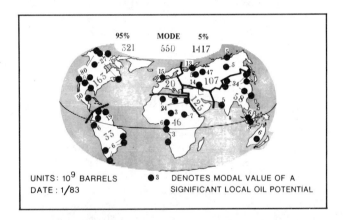

Figure 9. Distribution of undiscovered recoverable crude oil resources by major world geographic regions, showing local areas within the major regions from which most of the oil will be derived.

to affect the distribution of world oil. In these studies, however, we can find no new provinces where we would most likely anticipate a development of as much as 20 billion bbl of oil, which is equal to 1 year of world crude oil production. For example, the Barents Sea has long been considered to be an attractive frontier province. Our studies (Ulmishek, 1982), however, suggest that the favorable Jurassic source rock of the North Sea is mostly inadequately buried in the Barents Sea and thus unfavorable for petroleum occurrence. Instead of the highly favorable Devonian source rock of the adjoining Timan Pechora-Volgs Urals region, facies similar to the Old Red Sandstone (an unlikely source rock) are present in the Barents Sea. Furthermore, organic matter in the otherwise very attractive Triassic dark shales may be overmatured to gas in some parts of this basin. The

Barents Sea is so large that it may yet prove to be productive in giant proportions, but at this time, we have no confident geologic basis for such a prediction. The large but only marginally explored regions of China also must be accorded significant resource potential, but the dominant nonmarine depositional environment and the generally inferior reservoir rock conditions here call for some reservations in the estimation of their ultimate resources. Likewise, the large but minimally explored regions of Africa are not geologically favorable for the occurrence large resources.

Another area of resource controversy and of great interest is Antarctica. Not much is presently known, but work by the USGS and other groups is underway. For the present, however, we note that its former Gondwana neighbors—South America, South Africa, India, and Australia—have received significant exploration attention, with modest results to date. Furthermore, the published works of Hinz (1983) of the German Geological Survey do not present data encouraging to petroleum exploration in the sense of sediment volume or potential source or reservoir rock. The reader should keep in mind also that our assessments are for economically recoverable resources, which for Antarctica, would demand world class, super giant field discoveries. Such discovery sizes have a low probability of occurrence anywhere, and the geology of Antarctica that we have observed to date does not help to increase that certainty.

SUMMARY

Resource assessment is a multifaceted analytical procedure. Unknowns abound and purposes vary such that there is no definitive procedure suitable to all occasions and no answers known to be absolutely correct. Given common U.S. Government objectives and the availability of data and personnel, we believe the modified Delphi procedures offer the most practical assessment approach. This procedure has two main weaknesses. The first is present inability to quantify analogs. This can be improved by the development of better analog models and possibly by following procedures being developed by Ulmishek and Harrison (1984). The second weakness is the inability to subdivide the value of the assessment into expected field sizes so that judgments on actual unexplored areas available can be made relative to assessment amounts and necessary field sizes. This problem can be solved through statistical derivation. Regardless of what assessment techniques are used, it is essential to recognize that the calculations are only hypotheses that must be subjected to constant testing. This testing can best be accomplished in a scientific setting where significant data are available and where responsible analyses of extant assessment hypotheses can be made.

REFERENCES CITED

Bird, K. J., 1984, A comparison of the play-analysis technique as applied in hydrocarbon resource assessments of the National Petroleum Reserve in Alaska and of the Arctic National Wildlife Refuge: Episodes, International Union of Geological Sciences, Publication No. 17, Ottawa, Canada, p. 63-79.

Department of Energy, and U.S. Geological Survey. 1979, Report on the petroleum resources of the Federal Republic of Nigeria: Report DOE/IA-0008, 63 p.

Dolton, G. L., 1984. Basin assessment methods and approaches in the U.S. Geological Survey: Episodes, International Union of Geological Sciences, Publication No. 17, Ottawa, Canada, p. 4-23.

Dolton, G. L., Carlson, K. H., Charpentier, R. R., Coury, A. B., Corvelli, R. A., Frezon, S. E., Khan, A. S., Lister, J. H., McMullin, R. H., Pike, R. S., Powers, R. B., Scott, E. W., and Varners, K. L., 1981, Estimates of undiscovered recoverable conventional resources of oil and gas in the United States: U.S. Geological Survey Circular 860, 87 p.

Hinz, K., 1983, Results of geophysical investigations in the Weddell Sea and in the Ross Sea, Antarctica: Exploration in New Regions, Preprint of the Eleventh World Petroleum Congress, John Wiley and Sons, London, England.

Kingston, J., 1986, Undiscovered petroleum resources of South Asia: U.S. Geological Survey Open-File Report, 86-80, 177 p.

Klemme, H. D., 1980, The geology of future petroleum resources: Revue de l'Institut Francais du Petrole, Mars-Avril, 1980, v. XXXV, n. 2, p. 337-349.

Mast, R. H., McMullin, et al., 1980, Resource appraisal of undiscovered oil and gas resources in the William O. Douglas Arctic Wildlife Range: U.S. Geological Survey Open-File Report 80-916, 62 p.

Masters, C. D., Root, D. H., and Dietzman, W. D., 1983, Distribution and quantitative assessment of world crude-oil reserves and resources: Proceedings of the Eleventh World Petroleum Congress, London, England, v. 2, p. 229-237 (in Tonnes). [Also available as U.S. Geological Survey Open-File Report 83-728.]

Miller, B. M., 1981, Methods of estimating potential hydrocarbon resources by the U.S. Geological Survey—Case studies in resource assessment in the National Petroleum Reserve in Alaska and the William O. Douglas Arctic Wildlife Range: Proceedings of the Southwestern Legal Foundation, Exploration and Economics of the Petroleum Industry, v. 19, p. 57-95.

——— , 1982, Application of exploration play-analysis techniques to the assessment of conventional petroleum resources by the U.S. Geological Survey: Journal of Petroleum Technology, v. 34, n. 1, p. 55-64.

Ulmishek, G., 1982, Petroleum geology and resource assessment of the Timan-Pechora basin, U.S.S.R, and the Adjacent Barents-Northern Kara Shelf: U.S. Department of Energy Report, ANL/EES-TM-199, 197 p.

Ulmishek, G., and W. Harrison, 1984, A quantitative technique for assessment of petroleum resources in poorly known basins: Episodes, International Union of Geological Sciences, Publication No. 17, Ottawa, Canada, p. 80-94.

Some Practical Approaches to World Petroleum Resource Assessment

David H. Root, Emil D. Attanasi, and Charles D. Masters
U.S. Geological Survey
Reston, Virginia

Uncertainty in petroleum resource estimation can be mitigated by using different approaches in making resource estimates for a given area. We divide methods and data into two broad categories. The first is based on geologic data, which aim at estimating the resources of a basin by gaining an understanding of the processes of petroleum formation, migration, and trapping. The second is based on statistical methods, which estimate the resources of a basin by extrapolating the industry's past experience in drilling and discovery to forecast future discoveries. Results of these approaches are compared for Nigeria, North Africa, and many of the world's offshore areas. Undiscovered offshore petroleum resources outside the United States, Canada, and communist countries were estimated by both methods to be ~130 billion bbl. For Nigeria the two methods agree at about 4–6 billion bbl of undiscovered oil. In contrast, for North Africa, where the geologic estimate of undiscovered petroleum is 16 billion bbl and the statistical estimate is 5 billion bbl, the two methods are not in close agreement.

INTRODUCTION

The World Energy Resources Program (WERP) at the U.S. Geological Survey is concerned with all forms of energy resources but has devoted most of its efforts to conventional petroleum, a fuel that possesses all the virtues one could want except plenitude in all locations. How much oil is there? Where is it? Which estimates are the best and how were they made? This volume provides an opportunity for a number of groups to describe the variety of methods by which resource estimates are made, the type of data that are used, and the problems that arise in making these estimates.

RESOURCE ASSESSMENT APPROACHES AND DATA REQUIRED

In 1983, WERP prepared for the World Petroleum Congress an assessment of the world's petroleum resources, both discovered and undiscovered (see Masters et al., 1983). The estimates were prepared by a staff that included representatives of several disciplines: twelve geologists, one reservoir engineer, and two statisticians. Various methods of resource estimation, depending on the geologic situation and the intensity of exploration, were used to assess a wide variety of basins. The geologic approach to basin assessment concentrated on identifying critical factors controlling the

formation and trapping of petroleum. The data used in the assessments varied from detailed cross sections and extensive drilling and discovery histories in some basins to only regional geologic information in others. Whatever the level of detail of the available data, it was the task of the assessors to arrive at a basin-level understanding of the petroleum potential.

The two main classes of evidence used in resource estimation are geologic and statistical. Geologic evidence pertains to the geologic history of the basin and the properties of the rocks. The statistical evidence pertains to the history of industry operations in the basin and includes the drilling and discovery record. For many basins, both classes of evidence should be analyzed separately to obtain an estimate of remaining resources.

Whether one is using a geologic or a statistical approach to estimating undiscovered petroleum, it is necessary to know how much petroleum has already been found and with how much exploratory effort. There is a considerable range among oil-producing countries in the quality and quantity of information about the amount of oil that has already been discovered and the amount of drilling that has been done. For this study, Masters et al. (1983) assembled a list of the world's oil fields that includes the discovery date, the past production, and the estimated ultimate recovery for each. The sizes of many fields were estimated by petroleum engineers at the Department of Energy, Energy Information Administration (DOE/EIA office in Dallas, where there is an international program to identify the world's reserves of crude oil, which complements the WERP efforts of estimating the world's undiscovered resources. Well counts were compiled from the Foreign Developments issues of the AAPG Bulletin and from data from Petroconsultants SA, Geneva.

Even after consulting all these data sources, Masters et al. (1983) still found important fields for which no estimate of recoverable oil had been made. For these fields, an estimate was made of the amount of recoverable oil that was originally present on the basis of the field's production history or whatever other information was at hand. It was not necessary that the estimate for each field be precise as long as the basin totals were reasonably accurate. Whether the data under consideration are statistical or geologic, there are always gaps that must be filled in by ad hoc methods.

The methods used to make an estimate of undiscovered resources are varied but can be broadly categorized. Estimates based on geologic data are usually derived from an analogy between better known areas and lesser known areas (see Miller, this volume). Estimates based on statistical methods generally extrapolate historical discovery rates into the future. A useful survey of such statistical methods was presented in invited papers and was published along with a set of comments by the *Journal of the American Statistical Association* (see Wiorkowski, 1981; Hartigan, 1981; Kaufman, 1981; Mayer, 1981; Schuenemeyer, 1981). If possible, any given geologic assessment should be cross-checked with a statistical estimate, and vice-versa. Indeed, cross-checking for consistency is the central theme in resource estimation. Because none of the data or arguments are absolutely certain, there must always be an effort to find conclusions supported by a preponderance of the evidence. A few examples comparing these methods are discussed in the next section. First, the offshore areas of the world excluding the United States, Canada, and the communist countries are discussed. Following this, geologic and statistical estimates for Nigeria and North Africa are compared with one another and with the North Sea.

EXAMPLES OF APPRAISALS

Offshore Areas Excluding the United States, Canada, and Communist Countries

When conducting a geologic assessment of offshore sedimentary basins, one first evaluates the extent and likelihood of favorable geologic conditions for petroleum formation. The specific geologic factors evaluated include the presence of source rocks, reservoir rocks, seals, and proper timing. Timing must be considered because there is the possibility that the oil was formed and migrated out of the source rock before the seals and traps were present to hold it, as apparently occurred in the western part of the Maricaibo basin. Alternatively, if the oil arrived too late, diagenetic processes might have already destroyed the porosity of potential reservoir rock. If a basin is completely lacking any one of the necessary characteristics, then its resources are considered to be zero. Such basins will most likely not have enough drilling history to extrapolate and so can only be evaluated by geologic methods.

Other considerations that are important in a geologic evaluation can be seen from the following examples. The Barents Sea is thought to have poor oil potential because of a lack of source rock. The Jurassic rocks that are a petroleum source in the North Sea are commonly not buried deeply enough for maturity in the Barents Sea. Also, the Devonian source rock of the Volga–Urals changes its character to the north and loses its source rock qualities, becoming red beds in the Barents Sea. The Triassic may be a source rock in the Barents Sea but little data on its lithologic variation with respect to geography and associated reservoir rock are yet known. In another example, the Bay of Bengal off Bangladesh was given a low rating because the potential source rocks here are overlain by an overpressured zone. This condition suggests the presence of seals that would have prevented the migration of the oil into traps. Mexico, in contrast, received a high rating because of the presence of many undrilled structures in known petroliferous areas.

The assessors also studied the tectonic histories of each basin with the idea of fitting the basins into Klemme's (1980) basin classification scheme. Klemme has demonstrated (see Klemme, this volume) that there is a strong relationship between the type of basin and the likelihood of finding significant petroleum resources. Also, the distribution of field sizes in a basin is apparently related to basin type. Deltas, for example, have a much smaller percentage of their resources in their five largest fields than do rift basins. The variation in distribution of field size has an important effect on the estimate of undiscovered resources relative to the resources that have already been found.

After establishing the essential elements of the basin geology, the next step in the assessment process is the determination of and comparison with a more thoroughly explored analog basin. Resources of the basin under investigation are assigned by a team of geologists as a fraction or multiple of the resources in the better known basin according to relative size and other comparative qualities. These assigned values, representing the collective understanding of the assessors, are identified with the mode (most likely) and endpoints (5 and 95%) of the probability distribution. The process is described in Dolton et al. (1981). As an example of the results of this process, Table 1 presents estimates of the offshore crude oil resources in noncommunist areas excluding the United States and Canada. These estimates are in the form of probabilistic ranges for the different offshore regions.

Before estimates for small areas can be combined to give an estimate of a large area, the question of dependency among estimates must be resolved. The source of the dependence among the estimates is believed to lie in the thinking of the geologists doing the assessment rather than in the actual geology of the basin. Not only were all the appraisers in WERP working with approximately the same information, but they also shared the theories with which the data were organized and interpreted. In addition to these objective similarities among estimators, there is also the subjective element of a common mood of optimism or pessimism. Such moods are usually contagious among people of similar professions working in the same organization.

The manner in which statistical dependence among estimates for different areas can affect the total estimate is

Table 1. Geologic estimates of offshore petroleum undiscovered resources as of January 1, 1981, in billions of barrels of crude oil.

	95%	Mode	Mean	5%	
Europe					
North Sea	8.7	14.5	18.9	34.4	
Mediterranean/Adriatic, etc.	0.3	0.5	0.7	1.2	
Africa					
Cyrenaica shelf/W. Desert	0.3	0.5	1.1	2.6	
Nile Delta	0.2	0.4	1.0	2.6	
Suez/Sinai	0.7	1.3	2.1	4.5	
Nigeria	0.3	0.7	1.5	4.0	
Africa, west coast	1.0	1.9	4.0	10.0	
Red Sea basin	0.1	0.2	0.4	1.1	
Somali basin	0.3	0.5	1.3	3.4	
Asia Oceania					
Malaysia/Brunei	3.1	5.4	7.8	15.4	
Thailand	0.3	0.4	0.4	0.6	
Indonesia	3.3	6.1	10.4	23.0	
Australia	1.2	1.9	2.2	3.5	
New Zealand	.05	.10	0.2	0.5	
India	0.8	1.3	1.7	3.0	
North America					
Mexico, SE	20.0	36.7	58.7	126.0	
Greenland	1.0	1.9	3.5	8.0	
South America					
Venezuela, marg., Tobago	0.3	0.6	1.5	4.0	
Venezuela, Gulf Falcon	0.9	1.8	5.0	14.0	
Trinidad	0.7	1.3	2.0	4.0	
Other	1.6	2.5	3.1	5.2	
Middle East					
Arabian-Iranian	18.0	31.2	43.4	84.0	
Three aggregations for the offshore[a]					Sample Median
Degree of dependence 0.0	117	162	172	250	165
0.5	80	133	171	310	157
1.0	63	114	171	353	149

[a]Statistical totals were computed by Monte Carlo simulations under different degrees of dependence.

shown by three examples in Table 1. The 22 estimates, representing the various offshore regions, were summed by a mean preserving Monte Carlo procedure under three assumptions: no dependence (complete independence), partial dependence, and complete dependence. Only positive dependencies were considered, thus large values tend to cause other large values and small values tend to cause other small values. The 5%, 95%, median (50%), and mean numbers refer to samples of 4999 repetitions in which each small area estimate was taken to be a lognormal distribution. The mode was calculated from a lognormal distribution that was fitted to the 5% and 95% values. Although a sum of lognormal distributions is not lognormal, it can be approximated as such if there are only a few summands. Table 1 shows that the difference between the 5 and 95% estimates and the mode are particularly sensitive to the degree of dependence and the median less so. Complete independence results in a range of uncertainty between the 5 and 95% estimate that seems to us to be too narrow. Some dependence should be introduced in aggregating estimates, although we are not yet prepared to say how much dependence is appropriate.

A statistical analysis (Drew and Root, 1982) of exploration effort and discoveries for the same offshore area was performed independently of the geologic analysis that produced the estimates presented in Table 1. The entire offshore area was treated as a single exploration unit. This is a simplification because there are political and economic barriers among various parts of the offshore area as well as geologic differences. Nonetheless, there is a strong tendency for explorationists to explore the areas where the highest fees are to be earned, and usually those are the places where the best prospects are. Equipment, technology, and consultants can move across boundaries that are impassable to companies, whether private or government; that is, the industry as a whole is not greatly inhibited by political boundaries. Moreover, with respect to large fields (100 million bbl or more), oil fields are commercial almost any place in the world. Economic barriers separating basins are only significant for fields that are small, in very deep water, or affected by extreme climatic conditions.

The first part of the statistical analysis was to compile a tabulation of offshore discoveries that have already been made (see Table 2). It immediately became apparent from

Table 2. A statistical projection of future oil and gas discoveries in the studied offshore areas by size class.[a]

Field Size Class	Field Size (10^6 BOE)	BOE			
		Cumulative Number of Fields Discovered[b]	Expected Ultimate Number of Fields (N_A)	Expected Number of Fields Remaining	Remaining Oil Equivalent (10^9 BOE)
1–11	<6.07[c]	279	no estimate	—	—
12	6.07–12.14	56	4,353	4,297	38.6
13	2.14–24.3	65	2,199	2,134	38.0
14	24.3–48.6	73	1,112	1,039	35.3
15	48.6–97.2	99	565	566	31.0
16	97.2–194.3	97	284	187	25.2
17	194.3–388.6	74	142	68	18.7
18	388.6–777.2	46	62	16	9.0
19	777.2–1,554.4	21	23	2	2.2
20	1,554.4–3,108.8	13	14	1	2.0
21	3,108.8–6,217.6	8	8	0	0.0
22	6,217.6–12,435.2	4	4	0	0.0
23	12,435.2–24,870.4	3	3	0	0.0
24	24,870.4–49,740.8	1	1	0	0.0
Total		839	8,770	8,210	200.0

[a]From Drew and Root (1982).
[b]Sources: Most data from Petroconsultants; fields-size estimates for the Middle East, Nigeria, Venezuela, and Trinidad from William Dietzman, Energy Information Administration/DOE, Dallas, Texas.
[c]Includes discoveries for which no current sizes were available.

this list that something was wrong: there were not enough small fields. In field size classes 15 and smaller the number of fields decreased as class size decreased. These smaller classes onshore have many more fields than the larger classes. A possible explanation is that offshore fields having < 100 million bbl of oil are often judged to be uneconomical because of water depth or distance from market and are thus not reported as discoveries. Even in the Denver basin, Arps and Roberts (1958) recognized the problem of economic truncation.

Because of the effect of economic truncation, the discovery rate analysis was divided into two parts: classes 16 and larger, and classes 15 and smaller. For each size class in the 16 and larger category, the cumulative number of discoveries was plotted versus the cumulative number of wells. This curve was then approximated by a curve having the form

$$F(w) = F_o(1 - e^{-aw})$$

where F_o is the number of fields originally present in the size class offshore, w is the cumulative number of wildcat wells, $F(w)$ is the number of fields in the size class that were discovered by w wells, and a is a parameter measuring the rate of decline of the discovery rate within the size class. Results of these calculations are shown for classes 16 and larger in Table 2. We estimated for this category that there were more fields originally present in each class than in the next larger class. Overall for the large classes, there were about 1.98 times as many fields in each class as in the next larger class. The number of fields originally present in the smaller classes was estimated by assuming that the ratio of 1.98 held for classes 12 through 15.

Values for barrels of oil equivalent (BOE) in Table 2 include gas on an energy equivalent basis. Of the discovered BOE, 66% was oil and 34% was gas (gas had been underreported

probably for the same reason that small oil fields had been underreported). Assuming the same ratio holds for future discoveries as for past, undiscovered oil offshore is calculated to be 130 billion bbl, a figure well within the range determined by geologic methods. Thus, the geologic and statistical estimates reenforce one another. The conclusion that there is much less oil offshore than has already been produced onshore is not surprising. The prospective offshore area is in most places a narrow band around the continents, and based on our knowledge of the continental shelfs, we can expect a lesser quality of reservoir rock with increasing distance from the continent.

Some differences between the estimates produced by geologic and statistical methods of analysis are worth pointing out. Because it was done by field size class, the statistical analysis yielded an estimate in the form of a size distribution and a rate of discovery, but it did not give much information on where the oil might be expected to be found over a very large area (Table 2). The regional geologic estimate (Table 1), however, was less precise about the size distribution of undiscovered fields and rate of discovery but was more informative about the geographic distribution of the undiscovered resources. Because it is more specific as to location, the geologic estimate can be more quickly modified in the case of an exploratory surprise. Because it is more specific as to discovery rate and field size, the statistical estimate is more useful for broad economic projections.

Nigeria and North Africa

Undiscovered resources of Nigeria and North Africa were estimated by geologic methods and extrapolations of the historical discovery rates (Figures 1 and 2). The geologic appraisals were prepared as described earlier. Rather than making a detailed estimate of the number of fields by size class, we fitted an aggregate discovery rate function (bbl of oil

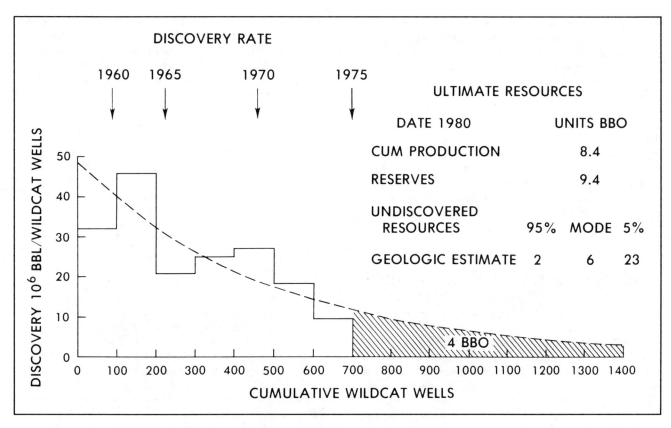

Figure 1. Exponential discovery rate extrapolation and geologic and discovery rate estimates of undiscovered crude oil in Nigeria.

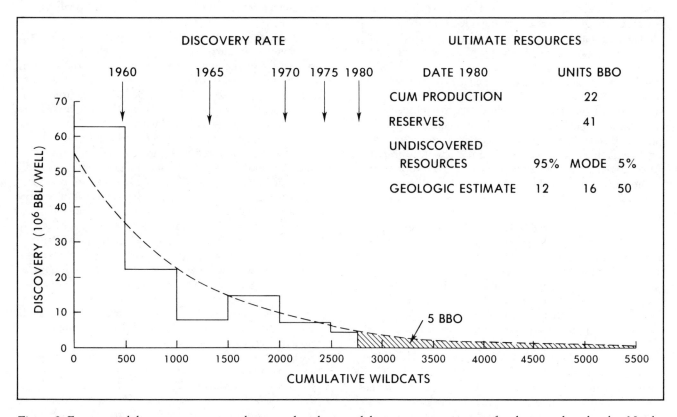

Figure 2. Exponential discovery rate extrapolation and geologic and discovery rate estimates of undiscovered crude oil in North Africa.

per wildcat well) to the historical drilling and discovery data and extrapolated from it. The fitted declining exponential functions are shown in Figures 1 and 2. The geologic estimate and statistical extrapolation are closer to one another for Nigeria than for North Africa. Discovery rate methods of appraisal work best in homogeneous areas that have been systematically explored. The North Africa discovery rate is based on data from several distinct areas having widely different levels of exploration maturity and petroleum richness. The disagreement between the discovery rate assessment and the geologic assessment of North Africa indicates that additional reviews of the data and assumptions of both methods are needed. Alternatively, the difference may derive from the generally low maturity of exploration over this vast area.

MEASUREMENT OF PROSPECTIVE AREA AND ITS APPLICATION TO RESOURCE APPRAISAL

Usually only a small part of any given country or even a small part of the sedimentary area in a country is of interest to petroleum explorationists. For this reason the number of wells relative to the area being explored is more informative than number of wells relative to the area inside political boundaries. The prospective area delineated by drilling can be generally described as all points that are reasonably close to wells and inside to the drilled area. More accurately, we define a point as being in the delineated prospective area if, and only if, there is a well in each of the four quarters of a square centered at the point having sides 40 mi (64 km) long and being parallel to longitude and latitude grid lines.

Figure 3A shows the delineated perspective area in Nigeria. Although Nigeria's total land area is > 350,000 sq mi (910,000 sq km) and the combined area of its onshore and offshore sedimentary basins is 74,000 sq mi (192,400 sq km) (Department of Energy, 1979), the delineated prospective area shown in Figure 3A is only 34,000 sq mi (88,400 sq km). Figure 3B shows the part of the delineated prospective area in Nigeria that is within 2 mi (3.2 km) of a well. Within this smaller area, drilling density is sufficient to nearly preclude future discovery of a field larger than ~ 100 million bbl (8 sq mi or 21 sq km), considering extant reservoir properties and pay thicknesses.

When exploration begins in a country, the prospective area delineated by drilling expands rapidly and then later stabilizes. Exploration following stabilization consists of more intense drilling of an already defined area rather than the opening of new territory. This pattern has been followed in Nigeria where the area delineated by drilling expanded rapidly at first and then slowed down. The last 200 wells added almost nothing to the size of the area being explored. During this interval, the delineated prospective area increased from 32,000 to 34,000 sq mi (83,200 to 88,400 sq km). A measure of the intensity of exploration in the delineated prospective area can be derived from the amount of that area that is within, arbitrarily, 5 mi (8 km) or 2 mi (3.2 km) of a well. These measures of drilling intensity in Nigeria are shown in Figure 4. The area in Nigeria explored most

intensely (within 2 mi [3.2 km] of a well) continues to grow at a steady rate. Thus, on the basis of 2-mi spacing the wells are not crowding each other, whereas, on the basis of 5-m (8-km) or 20-m (32-km) spacing they are. The area of interest in Nigeria is restricted to the Niger Delta and will probably not increase greatly over the present 34,000 sq mi (88,400 sq km). It appears that further exploration will be restricted to the more intense drilling of areas already defined.

In North Africa, exploratory drilling has delineated a prospective area of 278,000 sq mi (723,000 sq km) (Figure 5). This is more than 8 times the delineated area in Nigeria, although North Africa has only 4.5 times as many wildcat wells. For the most recent 200 wildcats, the delineated prospective area in Nigeria grew 10 sq mi (26 sq km) per wildcat, while for the same number of wildcats in North Africa the delineated prospective area grew 44 sq mi (114 sq km) per wildcat. This more rapid growth indicates that exploration is not as far advanced in North Africa as it is in Nigeria. The intensity of drilling is also less in North Africa than it is in Nigeria: in Nigeria 23% of the delineated prospective area is within 2 mi (3.2 km) of a wildcat well, whereas in North Africa only 12% of the delineated prospective area is within 2 mi of a wildcat.

There is more uncertainty about to the size of the prospective area in North Africa than in Nigeria. Even within the already delineated prospective area in North Africa there has been less intense drilling than in Nigeria. This indicates a greater uncertainty for the future as to what plays will develop and how large the prospective area will be. The disparity between the statistical and geologic estimates (discussed earlier) is a natural consequence of uncertainty, and the difference can be expected to narrow as drilling proceeds.

The principle of measuring the stage of exploration by the rate of growth of the prospective area, as delineated by drilling, and by the intensity of drilling within that area can also be applied to the North Sea (Figure 6). These data show that 1,245 wildcat wells drilled in the North Sea from 1961 to 1983 delineate a prospective area of 91,677 sq mi (238,360 sq km). For the most recent 193 wildcats, the prospective area has been expanding by 33 sq mi (86 sq km) per well. Thus the area delineated by drilling in the North Sea is expanding more rapidly than that in Nigeria and less rapidly than that in North Africa. The stage of exploration in the North Sea is probably intermediate between that of North Africa and Nigeria. The part of the delineated area in the North Sea that is within 2 mi (3.2 km) of a well is 14,055 sq mi (36,543 sq km) or 15.3% of the delineated prospective area. Using this measure, we can see that the North Sea is more intensely drilled than North Africa and less intensely drilled than Nigeria. A newly discovered play underlying previously drilled plays could cause an increase in drilling without any increase in delineated prospective area. Thus, a flattening of the curves in Figures 4, 5, or 6 does not in itself indicate there are no new plays.

The modal value of the estimated remaining resources in the North Sea is 14.5 billion bbl (Table 1). Adding past North Sea discoveries of 25.5 billion bbl gives a total resource of 40 billion bbl, of which 63% has been discovered. The corresponding percentage for Nigeria is 75%, which is higher because Nigeria has been more intensely explored. In

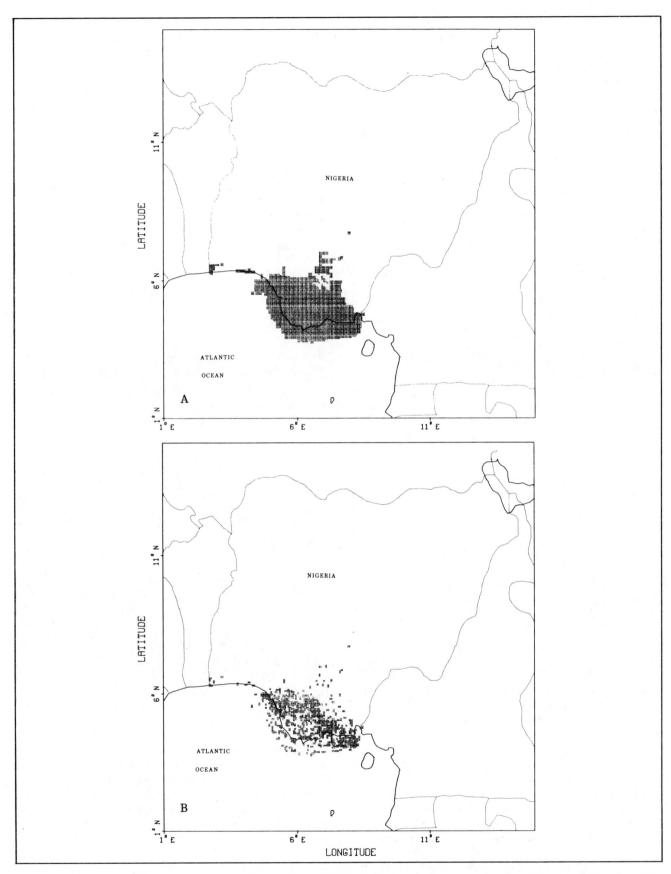

Figure 3. A. Delineated prospective area in Nigeria through 1982. B. Delineated prospective area in Nigeria within a 2 mi (3.2 km) radius of wells drilled before 1983.

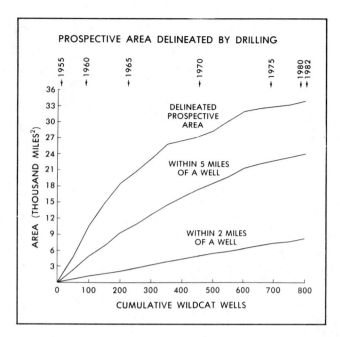

Figure 4. Time profiles of delineated prospective area through 1982 and areas within 5 and 2 mi (8 and 3.2 km) radii of a well in Nigeria.

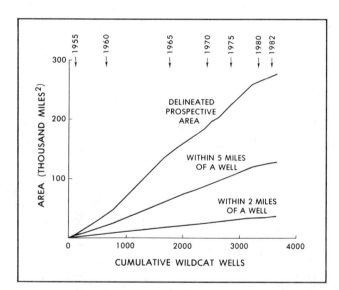

Figure 5. Time profiles of delineated prospective area through 1982 and areas within 5 and 2 mi (8 and 3.2 km) radii of a well in North Africa.

North Africa past discoveries are 63 billion bbl and the geologic estimate of undiscovered resources is 16 billion bbl for a total of 79 billion bbl, of which 80% has been discovered. The high percentage of discovered resources in North Africa seems to result from oil resources in North Africa being concentrated in large fields to a greater extent than in Nigeria or the North Sea. North Africa contains Africa's only two supergiant fields—Hassi Messauod (7

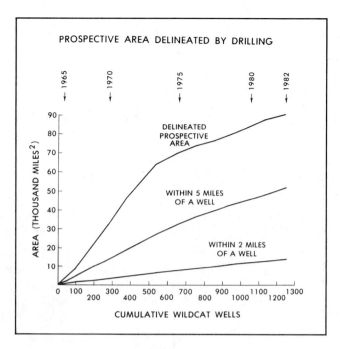

Figure 6. Time profiles of delineated prospective area through 1982 and areas within 5 and 2 mi (8 and 3.2 km) radii of a well in the North Sea.

billion bbl) and Sarir (7 billion bbl)—and near supergiants—Amal (4.3 billion bbl), Nasser (4.2 billion bbl), and Gialo (4.0 billion bbl) (Nehring, 1978).

CONCLUSIONS AND SUMMARY

The primary objective in preparing global petroleum resource estimates is to provide government and corporate planners and the general public with information relevant to long-range decisions. Because we are more concerned with avoiding blunders than with fine tuning assessment techniques, assessors seek alternative methods of cross-checking appraisals. Where different methods produce widely differing estimates, then an explanation should be sought for the differences. In this way, all possible data and expertise are brought to bear on a difficult problem, and the final estimates are as consistent as possible with what is known about the geology and with the success or failure that industry has had in exploration.

Studies of the geology and the industry statistics in Nigeria produced a similar estimates of 4–6 billion bbl of oil remaining to be discovered. The geologic and statistical estimates for North Africa were not in such close agreement. Geologic studies indicate that 16 billion bbl remain to be discovered, while an extrapolation of the discovery rate led to an estimate of 5 billion bbl. The disparity of estimates may be at least partly explained by the relative immaturity of exploration in North Africa. Comparison of statistical and geologic estimates of undiscovered petroleum in offshore areas excluding Canada, the United States, and communist countries showed good agreement at 120–130 billion bbl. The geologic method led to an estimate that was more specific as to location of offshore resources, and the statistical method led to an estimate that included more information on the sizes of fields and rates of discovery.

As a practical aid to both approaches, drilling density maps help to delineate areas within basins that industry is willing to explore or considers commercially prospective for exploration. The graph of explored areas versus cumulative wildcat wells offers a perception of the stage of exploration maturity in a basin; the flattening of the curves indicates maturing exploration.

REFERENCES CITED

Arps, J. J., and T. C. Roberts, 1958, Economics of drilling for Cretaceous oil in the east flank of the Denver–Julesburg basin: AAPG Bulletin v. 42, n. 11, p. 2549–2566.

Department of Energy, 1979, Report on the petroleum resources of Nigeria, DOE/IA-0008, 131 p.

Drew, L. J., and D. H. Root, 1982, Statistical estimate of tomorrow's offshore oil and gas fields, Ocean Industry, May, p. 54–66.

Dolton, G. L., K. H. Carlson, R. R. Charpentier, A. B. Coury, R. A. Crovelli, S. E. Brezon, A. S. Khan, J. H. Lister, R. H. McMullin, R. S. Pike, R. B. Powers, E. W. Scott, and K. L. Varnes, 1981, Estimates of undiscovered recoverable conventional resources of oil and gas in the United States: U.S. Geological Survey Circular 860, 87 p.

Hartigan, J. A., 1981, Comment on "Estimating volumes of remaining fossil fuel resources: a critical review": Journal of the American Statistical Association, v. 76, n. 375, p. 548.

Kaufman, G. M., 1981, Comment on "Estimating volumes of remaining fossil fuel resources: a critical review": Journal of the American Statistical Association, v. 76, n. 375, p. 549–550.

Klemme, H. D., 1980, Petroleum basins—classifications and characteristics: Journal of Petroleum Geology, v. 3, n. 2, p. 187–207.

Masters, C. D., D. H. Root, and W. D. Dietzman, 1983, Distribution and quantitative assessment of world crude-oil reserves and resources: U.S. Geological Survey Open-File Report 83-402, 23 p.

Mayer, L. S., 1981, Comment on "Estimating volumes of remaining fossil fuel resources: a critical review": Journal of the American Statistical Association, v. 76, n. 375, p. 551–554.

Nehring, R. F., 1978, Giant oil fields and world oil resources: Report R-2284-CIA, Rand Co., Santa Monica, California, 162 p.

Schuenemeyer, J. H., 1981, Comment on "Estimating volumes of remaining fossil fuel resources: a critical review": Journal of the American Statistical Association, v, 76, n. 375, p. 554–558.

Wiorkowski, J., 1981, Estimating volumes of remaining fossil fuel resources: a critical review: Journal of the American Statistical Association, v. 76, n. 375, p. 534–548.

Are Our Oil and Gas Resource Assessments Realistic?

C. J. Lewis
Sohio Petroleum Company
Houston, Texas

An examination of past domestic resource estimates of conventional hydrocarbons leads to the conclusion that they are generally not realistic. Furthermore, to be of practical use they need to be qualified by time limits. An approach to resource assessment is presented in this paper that would limit it to a time frame (from the present up to the year 2020) that is considered to be a realistic range for planning. Using discovery rates based on a much higher level of success than those being achieved today and projecting to A.D. 2020, "almost zero" probability figures of 54 billion bbl for oil and 440 trillion cu ft (tcf) for gas are obtained. A pessimistic outcome for this same period (8.2 billion bbl for oil and 82 tcf for gas) is then taken, thus establishing limits within which U.S. oil and gas estimates should lie. These limits should be used as a guide when developing mean estimates that are generally made for regions and combined into one total. In these regional estimates more attention should be paid to three factors: (1) the low side of resource distribution, (2) a realistic assessment of risk based where available on the historical record, and (3) the importance of the economic threshold in frontier areas. This approach produces lower estimates than most other current forecasts, but it should not dampen enthusiasm for future oil and gas exploration in the United States because these figures can still accommodate higher finding rates than are experienced today until well past the turn of the century.

INTRODUCTION

Numerous estimates of U.S. ultimately producible hydrocarbon resources have been made over the past 70 years, the first being ascribed to Day in 1909 (cited in McCulloh, 1973). The art of prediction has become more sophisticated as more data have become available and assessment techniques have improved. For the purpose of this paper, only the most recent estimates have been examined, and the scope is limited to future discoveries of recoverable hydrocarbons by conventional means, that is, recovery under considerations of current economic and technological trends.

These recent estimates have been made by national governmental bodies such as the USGS, by private industry (generally by the large multinational companies), and by committees of experts composed of both government and industry personnel purposely established to carry out resource assessment, such as the Potential Gas Committee and National Petroleum Council committees. The USGS has probably assigned more time and effort than other groups to resource estimating in the most recent years, but this may be an impression gained because they provide full details of their methods and results. Some oil companies may well rival them. Much of the work done by private industry, however, is proprietary, and thus it is often difficult to assess the degree of effort and sophistication that has been applied by

individual companies. Certainly the major oil companies are more likely to have better data bases and experience because they have a long history of exploration in the United States from which to judge future potential. They have been cohesive units for long periods of time. They are also able to apply major efforts locally to frontier areas of the United States, and the proprietary information they obtain is not available to the Resource Appraisal Group of the USGS.

Notwithstanding the quality of information and level of experience of the estimators, I am concerned with the general level of resource estimates that I suspect are unrealistically high and do not support the results of recent exploratory trends. For this reason, these trends are examined in some detail to discover if they were compatible with the resource estimates.

The methodology developed here could successfully be applied to estimates of nonconventional hydrocarbons and other "finite" mineral resources. It involves setting a reasonable time limit for significant future discoveries of the particular resource. Furthermore, it establishes upper and lower bounds for the estimate that can clearly be demonstrated to exceed realistic expectations made on the basis of the historical record and the geological assessment of frontier areas, either partially explored or yet to be explored. These bounds are then used to control hydrocarbon resource estimates that are usually made for specific parts of a whole, in this instance, regions, provinces, or states within the

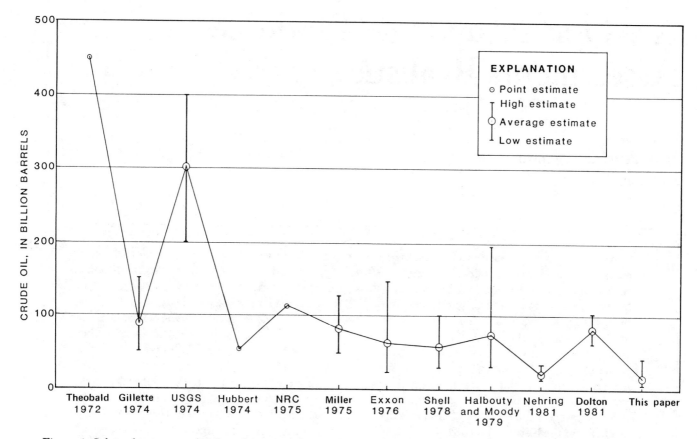

Figure 1. Selected estimates of U.S. undiscovered recoverable resources of crude oil since 1965 (After Dolton et al., 1981.)

United States. It appears to be a natural tendency among geologists to veer toward the optimistic side on both resource distribution and risk. When the results of this optimistic treatment of parts of the country are aggregated into the whole, the problem is then compounded.

The paper will first consider the methodology as applied to oil estimates. The same approach will then be applied to natural gas. In conclusion, some examples will be given of apparent overestimates of significant proportions in selected areas of the United States.

Figure 1 gives various estimates of U.S. undiscovered recoverable conventional resources of oil published since 1972. These estimates should be comparable, although some include natural gas liquids and others do not. The estimates show an overall gradual decrease with time even allowing for declines resulting from production. Especially in the case of the USGS, this may reflect an inadequate data base and inexperience with resource assessment methodology in the earlier years. As knowledge increased and new areas were drilled, such as the Gulf of Alaska and Baltimore Canyon, the assessments tended to decline.

IMPORTANCE OF HISTORICAL RECORD AND TIME LIMIT IN RESOURCE ESTIMATING

The estimates shown in Figure 1 all appear to have no time limit, although they assume recoveries under the economic and technological trends at the time they were made. It

would be more realistic, however, to place a time limit on future oil and gas discoveries that reflects an expectation of their useful life as a major source of energy. It is thus assumed in this paper that by the year 2020, conventional oil and gas supplies will be inadequate to handle their traditional demand and replacement fuels will be available. Oil and gas will probably be used on a different and more restricted basis. In this scenario there is no doubt that hydrocarbons would continue to be discovered and exploited domestically, but for the purposes of our present assessment the amounts discovered after 2020 would not be significant. This assumption may be contentious both in regard to the date of the limit and the availability of replacement fuel. It can be thought of as arbitrary and well beyond the range of accurate prediction. Nevertheless, it is reasonable to assume that it lies close to actuality, within ± 10 years, and thus serves to illustrate the arguments presented in this paper.

This approach to estimating U.S. undiscovered recoverable resources of hydrocarbons is simple and has required a minimum of research because the degree of sophistication that can be reasonably applied to it is limited. The following procedures attempt to illustrate this. First, an "absolute maximum" figure ("almost zero" probability) is developed that is based on discovery rates in the United States during much more prolific years than at present. For this, Table III of the 1980 American Petroleum Institute (API) report "Reserves of crude oil, natural gas liquids, and natural gas in the United States and Canada—1979" was used (Figure 2), and the total of the annual new-field

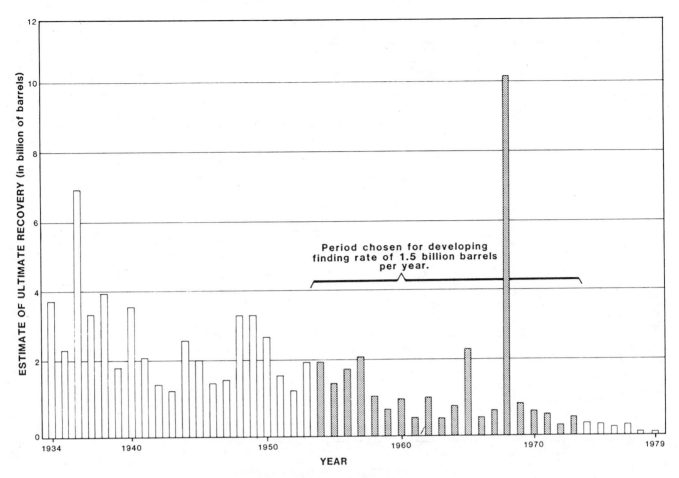

Figure 2. Discovery rates for oil in the United States based on the historical record as assessed in 1979. (From API, 1967–1980.)

discovery rates over the 20 year period from 1954 to 1973, which includes the Prudhoe Bay field, was taken. Subsequent revisions and extensions to these discoveries, as assessed in 1979, have been allocated back in that publication to the date of the original discovery. The growth rate of this 20-year discovery total over the 5-year period from 1975 to 1979 as taken from the API reports for 1976 and 1980 is 6% for oil and 9% for gas.

In-house studies covering growth in the past 5 years (1980–1985) indicate a similar rate of increase. A moderate growth of this order is expected to continue for some years, although not on the scale described by Root in USGS Circular 860 (1981) in which he shows how estimates of the amount of recoverable oil in a given year in the conterminous United States have grown seven-fold over the measured period of 60 years (Figure 3). Many of these large increases occurred at a time when scientific knowledge and technology were improving significantly, and they related predominantly to major discoveries made in the earlier years that were not recognized as such until much later. The average new discovery is smaller, and it is made in the context of a higher level of technology, which enables the resource to be better defined in the early stages. Under these conditions it is anticipated that growth from new discoveries will be less significant in future years.

The 1979 API figures indicate that in the period from

1954 to 1973, discoveries have averaged 1.5 billion bbl of oil per year. (If Prudhoe Bay were excluded, this figure would be approximately 1 billion bbl.) This discovery rate has been projected to the year 2020 (assuming no decline) to a total of ~ 54 billion bbl. Since the present average annual new field discovery rate is significantly lower than the 1.5 billion bbl figure (and is probably closer to about one-third of it, as will be discussed later), 54 billion bbl would seem to be an extremely unlikely estimate to achieve. Yet it is low compared with the 1981 USGS mean of 82.6 billion bbl. A reasonable maximum (5% probability) would be to take the present new field discovery rate as averaging 500 million bbl per year. This is derived by taking the average annual estimate of ultimate recovery (451 million bbl) for the period 1970–1979 from the 1980 API report and rounding it up. This figure is then projected, again with no decline (which is optimistic), to the year 2020 to make a total of 18 billion bbl which should adequately cover the reasonable maximum case of any discoveries made in established oil-producing areas. For frontier provinces let us assume that another field the size of Prudhoe Bay and several multibillion barrel fields are found, making a total of 20 billion bbl. The combined result of established and frontier areas would then be 38 billion bbl, which again is very low compared to the mean of most estimators and most of it falls into the arbitrary 20 billion bbl of the frontier areas.

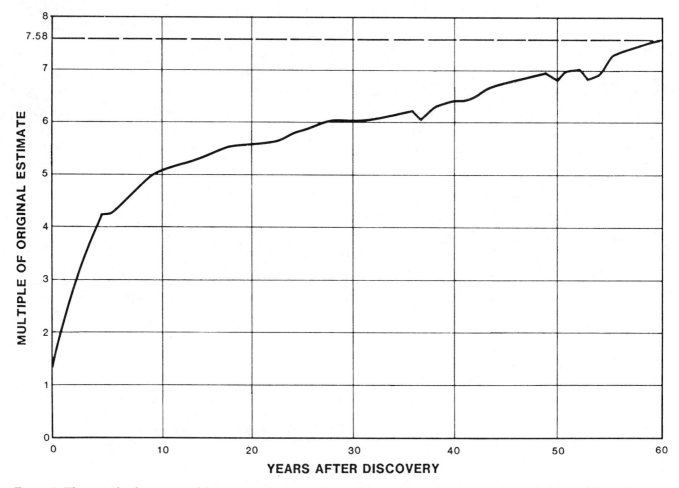

Figure 3. The growth of estimates of the amount of recoverable oil discovered in a given year in the conterminous United States versus the number of years after the year of discovery. (After Root, 1981.)

At the low end of the scale, the present new field discovery rate is taken as 400 million bbl of oil per year, again derived from ultimate recovery estimates for 1970–1979 from the 1980 API report, but in this case rounded down. This rate is then assumed to decline exponentially. An average annual decline to the year 2020 has been set at 2.5%. This is an arbitrary figure, but it does approximate the decline rate in reserves from 1976 to 1983 (U.S. Department of Energy, 1981–1982). This results in a reasonable, low estimate (95% probability) of ~ 10 billion bbl. Yet even in this scenario, we would still be finding ~ 160 million bbl in the year 2020. The "most likely" discovery rate may be closer to this reasonable, low estimate than to the reasonable, high estimate.

On the above basis, a "most likely" estimate has been made of 15 billion bbl. This could assume an annual discovery rate of 400 million bbl of oil per year with no decline to the year 2020 or, depending on preference, a low base estimate of 10 billion bbl to which are added some multibillion discoveries in frontier areas of Alaska and the outer continental shelf.

This procedure may seem somewhat arbitrary, but it uses a logical combination of historical data and generous allowances for future discoveries in frontier areas. Its main purpose is to establish a range within which reasonable estimates of future recoverable resources of conventional hydrocarbons should fall.

GAS RESOURCE ESTIMATES

A similar exercise can be undertaken for natural gas resource estimates. Figure 4 shows selected resource estimates for gas. Those made by the USGS and the Potential Gas Committee (PGC) predominate (see PGC, 1971, 1973, 1977a,b, 1979, 1981, 1982). The upper 5% and lower 95% limits of gas estimates are more difficult to make because potential gas-bearing rocks have been less well explored than oil, which has a shallower depth limitation. It is also more difficult to determine an "arbitrary" cut-off date for gas because there could be continued significant use beyond the year 2020 in local areas where adequate supplies exist. However, future gas exploration drilling in the United States for deeper and more remote high cost objectives could be increasingly challenged by cheaper imports, which could have an adverse impact on domestic discovery rates.

Exploration for gas is extremely price sensitive and one could argue that discovery rates prior to the mid-1970s reflect

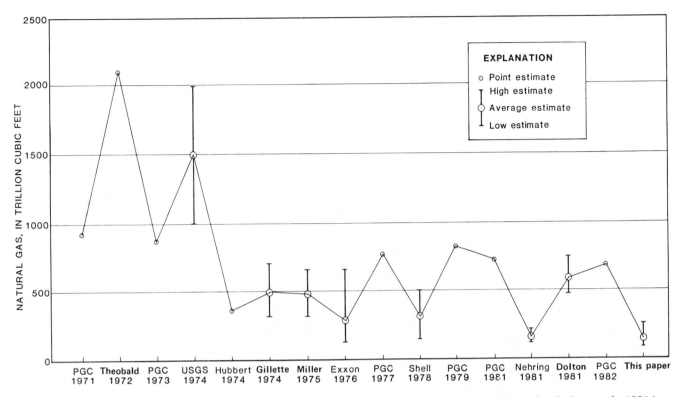

Figure 4. Selected estimates of U.S. undiscovered recoverable resources of natural gas since 1965. (After Dolton et al., 1981.)

the low, controlled price of the commodity. During this period much of the gas was found accidentally by wells drilled on "oil" prospects. Significant increases in discoveries were made as the prices rose in the late 1970s, and it is too early to tell what the recent adverse impact of the "gas bubble" will have on discovery rates. It is thus more difficult to assess representative discovery rates for gas than it is for oil. During the period 1954–1973, which is the same as that used for the oil calculations, the average annual gas finding rate taken from the 1980 API report was 12.2 trillion cubic feet (tcf) (Figure 5). By the mid-1970s this rate had dropped to between 4 and 6 tcf. It has been at the higher end of this range in more recent years, which apparently reflects the drilling boom. The present total annual reserve addition including revisions and extensions from old fields also averages 12.2 tcf (U.S. Department of Energy, 1981–1982). If this figure were projected to the year 2020 with no decline, the result would be 440 tcf. This appears unlikely even under the best economic conditions. The majority of significant discoveries in the United States have been made in the gas-producing provinces of Louisiana, Oklahoma, and Texas. Many of these are now at a mature stage in development, and although several giant discoveries have been made recently, notably off Texas and Alabama in the Gulf of Mexico, it is not anticipated that finding rates overall will increase significantly (i.e., revert to the 12.2 tcf level). The 440 tcf figure is therefore suggested as the "almost zero" probability.

If we assume that the annual discovery rate for new fields and new reservoirs in old fields will continue at the current high end of the range at 6 tcf per year with no decline to

2020, and (as in the case of oil) we assume one supergiant discovery of comparable size to Prudhoe Bay, as well as several other giant discoveries made in frontier areas totaling 60 tcf, the result is 275 tcf. This is considered the high side or 5% probability case. This estimate is very low compared with all other estimates except for Nehring and Van Driest (1981). The 6 tcf annual discovery rate applied to this estimate may be questioned as being too low because of the uncertainties of gas resource estimates (referred to earlier in this paper). This, however, can be considered as compensated by the generous allowance given for the frontier areas where economic thresholds are generally much higher than they are for oil, especially in remote offshore areas and Alaska.

For the low side case, a present average annual discovery rate of 4 tcf is used; it is decreased (as in the case for oil) at a rate of 2.5% per year to A.D. 2020, and makes no allowance for major discoveries in the frontier areas. This totals 100 tcf, which is lower than Nehring's estimate of 143 tcf. The resultant average of the high and low cases is 187 tcf, which I consider to be slightly high. I place the "most likely" estimate at 150 tcf, which assumes that the 4 tcf per year average figure continues without decline to the year 2020.

THE NECESSITY OF A TIME LIMIT FOR RESOURCE ESTIMATES

As mentioned earlier, these determinations may seem somewhat unsophisticated but they are based on present-day discovery rates in established areas with generous allowances for discoveries in frontier areas. Most major areas of high

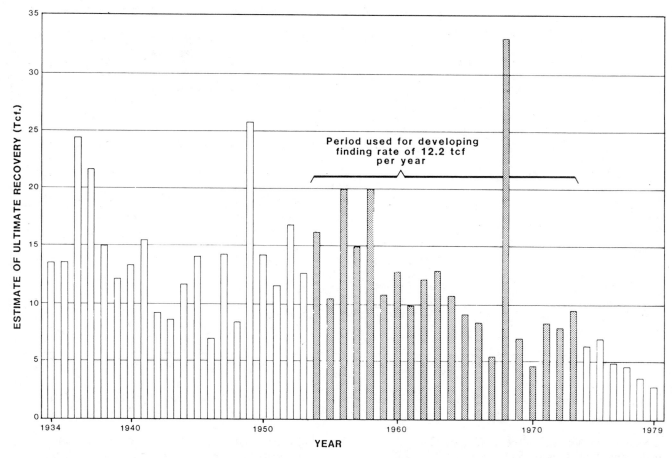

Figure 5. Discovery rates for gas in the United States based on the historical record as assessed in 1979. (From API, 1967–1980.)

geologic promise that have the potential to augment annual discovery rates significantly have already been drilled. The remaining areas lie in deep and/or ice-infested waters, such as the Alaska Beaufort Sea, where exploration and development are severely limited by economic and environmental considerations. The estimates may seem low, but given the circumstances described above, it is difficult to build a case to go much higher even projecting beyond the A.D. 2020 limit. For example, in the low case described above for oil and gas, the finding rates used were 160 million bbl of oil and 1.6 tcf of gas in 2020. If these finding rates are projected with an exponential decline of 2.5% per year for an additional 36 years, this would add only 3.7 billion bbl of oil and 37 tcf of gas. If we assume no decline and that a "low" plateau is reached, the figures are 5.7 billion bbl of oil and 57 tcf of gas. The point is that by the year 2020, the potential in the United States for conventional oil and gas will be extremely well explored. Thus, major additions from these sources cannot be expected beyond that date.

A time limit for resource estimates is essential for planning purposes and should be used as a standard for comparisons. It should, however, be set sufficiently far in the future that there is good certainty that discoveries of conventional oil and gas beyond that date will be insignificant. Beyond 2020 there is a strong likelihood significant quantities of conventional oil and gas will continue to be produced but

outside the United States. Domestic production on a major scale will only be able to be maintained if it is derived from nonconventional sources, which at present are not economically attractive and need major capital investments and long lead times to develop.

REGIONAL ASSESSMENTS WITHIN THE UNITED STATES

My approach to resource estimating has been to consider the United States as a whole. No attempt has been made to analyze in depth any specific regions within the country. Most resource estimators start with regions and build up a resource inventory for the whole. These is nothing wrong with this approach except that historically the regional "means" tend to be on the high side, reflecting that both the 5% and 95% probability cases are too high and the risks too low (examples of this are given later in this paper). This results in the aggregate for the whole being too high to be realistic. It appears to be more accurate if one works within an overall risked mean and then determines how this could be best apportioned to the parts. This would still allow a reasonable range to be established (5–95%) within each region.

Frontier regions that have been little explored, such as offshore Alaska, are where the greatest overestimates are likely to occur. In areas such as these there is usually no established finding rate. A problem also exists with economic thresholds in remote Alaska and in deep-water areas. The adequacy chance (National Petroleum Council, 1981) that at least one 50 million bbl field would be present is of little practical use if economic feasibility is considered in any of the Alaskan Arctic regions. A field of this size might be marginally economic if it were immediately adjacent to the Prudhoe Bay infrastructure, but not if it were in the middle of the Chukchi Sea pack ice, on the Bering Sea Shelf, or in the deep Atlantic. This is not meant to imply that it is practical, considering the major uncertainties facing the estimators, to set up a variety of economic thresholds within a frontier province. Preferably a generalization has to be made, but one that is more characteristic of the province as a whole and takes into account where the most prospective areas might lie. On this basis, the probable minimum economic field size for the Beaufort Sea off Alaska would be ~300 million bbl of recoverable oil.

Because of the large size of frontier areas and the common presence of giant structures indicated by preliminary seismic work, there is the tendency for initial estimates to be on the high side. Most of the regional high side estimates (5% probability) in USGS Circular 860 (Dolton et al., 1981) appear feasible. It is possible, however, that most of these areas will yield no commercial hydrocarbons. The disappointments in the Gulf of Alaska and Lower Cook Inlet well illustrate this possibility. For example, at the time of the Northern Gulf of Alaska sale (OCS No. 39) in April, 1976, the USGS estimated a "minimum" high side (5%) of 2.8 billion bbl of recoverable oil and 9 tcf of gas. Their low side (95%) was 100 million bbl of oil and 300 billion cu ft of gas. The mean for oil was ~850 million bbl. If it were assumed, however, that there was a 25% chance of finding no commercial hydrocarbons (which is a realistic assumption to make in many frontier areas), the resultant mean would have been many times less (~90 million bbl). This illustrates the potential for overestimating resources in frontier areas where the estimator has failed to temper the high side and to recognize the risk that the basin may be completely dry.

Another good example of a USGS (Dolton et al, 1981) mean that appears to be on the high side is in Petroleum Region No. 3, which is the Colorado Plateau and Basin and Range, where the oil estimate is 14.2 billion bbl. One can readily conjecture why this estimate is high. It was made when enthusiasm for the Rocky Mountain overthrust belt was almost euphoric, and it is thus significantly higher than the statistical mean estimate of 4 billion bbl made in Circular 725 (Miller et al., 1975). As more is learned about the limited extent of Cretaceous source rocks having adequate maturity in the subthrust, this enthusiasm has waned. The historical record (API, 1967–1980) shows that the average annual discovery rate of oil in Petroleum Region No. 3 during the period 1954–1973 was ~155 million bbl which assuming no decline, would sum to 5.6 billion bbl by the year 2020. I suggest that this is well on the high side of a "most likely" case, especially since the whole of Colorado, New Mexico, and Wyoming were included in the estimate, and some of this area lies outside of Region No. 3. Even if we

accept it as the mean, it is still 8.6 billion bbl less than the USGS mean of 1981.

The same comments could be made of several other assessments by the USGS including Alaska, the Atlantic Coast, the Rocky Mountains, the northern Great Plains, and even the Gulf of Mexico. The USGS estimates were used here for this discussion because their circulars on resource estimates have the most complete documentation both of assessments and methodology. Industry sources may have access to more current data and more sophisticated technology, but they tend (in the interest of confidentiality) to disclose very little detail of how their figures were obtained.

With regard to the gas estimates, those provided by the Potential Gas Committee, (see PGC estimates in Figure 4) seem to be at odds with most other recent estimates. Their "most likely" figure produced in 1982 for the combined possible and speculative categories was 684 tcf. The Committee has been consistently high in its estimates from 1971 onward, whereas recent industry estimates have been much lower. On the basis of the reasoning in this paper, the PGC estimates lie well above the "almost zero" probability. This may be because the committee depends heavily on aggregating the results of individual basin appraisals, but does not appear to use an overall realistic limit based on the historical record, which usually helps to contain the enthusiasm of the local experts. On one hand, it could be argued that the major oil companies are more interested in oil and tend to undervalue gas resources. On the other hand, these companies probably have more sophisticated appraisal methods at their disposal and represent a more objective approach. It would take over 100 years at current discovery rates assuming no decline, to achieve the 684 tcf figure.

CONCLUSIONS

The main points covered in this paper on oil and gas assessment methodology can be summarized as follows:
1. Resource assessments of future discoveries of conventional hydrocarbons are generally not realistic. They tend to be too optimistic.
2. Growth rates for discoveries made in the last decade are significantly lower that those of earlier discoveries.
3. Long term assessments of conventional hydrocarbon resources are of limited value unless set in a time frame.
4. Estimators should establish realistic upper and lower bounds (5 and 95%) for the whole area that can be assessed on the basis of projections from the historical record and also include realistic representation for frontier areas; this will then act as a guide for control over local assessments within the whole.
5. A knowledge of realistic economic thresholds is critical in the assessment of high cost and remote frontier areas.
6. Assessment of the low side is frequently too optimistic, which also optimistically distorts the resultant conditional mean.
7. Risking must be more stringently applied than in past estimates and must be tempered by the historical record.

The ranges (5–95%) in national resource estimates of

conventional hydrocarbons given in this paper and based on the above conclusions are much lower than most previous estimates. Although sobering, this is not necessarily cause for concern for future oil and gas exploration. Even with the high estimates published in the 1970s (Figures 1 and 4), it was unrealistically optimistic to think that discovery rates were really expected to leap up from the then current 400–600 million bbl of oil per year to 1.5 billion bbl per year in the 1980s and 1990s, which would have been about the level required to see these estimates achieved. According to the AAPG Annual Review of World Energy Developments, only ten oil or gas fields greater than 50 million BOE have been recorded during the 10 years prior to 1983. It is probable that there have been an additional five fields in this category resulting from discoveries in Alaska and offshore California that have not yet been recorded. All the increases during this 10-year period have been in fields of the ⟨ 1 million bbl size category, but there have been decreases in all the larger field sizes, which are the important ones in making significant reserve additions.

Nevertheless, the low resource projections suggested in this paper can still accommodate annual finding rates for hydrocarbons on a level greater than that presently achieved without decline until well past the turn of the century. This should not dampen enthusiasm for exploration in the United States. Large quantities of oil and gas still remain to be found, especially in the frontier areas, but the search will become increasingly difficult. This is a major challenge to geologists in the next decade and will require ingenuity, skill, effective use of new technologies, optimism, aggressiveness, and luck.

REFERENCES CITED

American Petroleum Institute (with the American Gas Association and Canadian Petroleum Association), 1967–1980, Reserves of crude oil, natural gas liquids and natural gas in the United States and Canada (Annual volumes for the years 1966–1979): New York, American Petroleum Institute.

Dolton, G. L., K. H. Carlson, R. R. Charpentier, A. B. Coury, R. A. Crovelli, S. E. Frezon, A. S. Khan, J. H. Lister, R. H. McMullin, R. S. Pike, R. B. Powers, E. W. Scott, and K. L. Varnes, 1981, Estimates of undiscovered recoverable conventional resources of oil and gas in the United States: U.S. Geological Survey Circular 860, 87 p.

Exxon Company, 1976, U.S. Oil and Gas Potential: Exxon Company, U.S.A., Exploration Department, March, 10 p.

Gillette, R., 1974, Oil and gas resources—did the USGS gush too high?: Science, v. 185, n. 4146, p. 127–130.

Halbouty, M. T., and J. D. Moddy, 1980, World ultimate reserves of crude oil: Bucharest, 10th World Petroleum Congress Proceedings, v. 2, p. 291–301.

——— , 1974, U.S. Energy resources, a review as of 1972, Pt. 1, in

A National Fuels and Energy Policy Study: U.S. 93rd. Congress, 2nd. Session, Senate Committee on Interior and Insular Affairs, Committee Print, Serial No. 93-40 (92-75), 267 p.

Johnston, R. R., North American drilling activity in 1982: AAPG Bulletin v. 67, p. 10.

McCulloh, T. H., 1973, Oil and gas, in Brobst, D. A., and W. P. Pratt, eds., United States mineral resources: U.S. Geological Survey Professional Paper 820, p. 477–496.

Miller, B. M., H. L. Thomsen, G. L. Dolton, A. B. Coury, T. A. Hendricks, F. E. Lennartz, R. B. Powers, E. G. Sable, and K. L. Varnes, 1975, Geological estimates of undiscovered recoverable oil and gas resources in the United States: U.S. Geological Survey Circular 725, 78 p.

National Petroleum Council, 1981, U.S. Arctic oil and gas, 17 p.

National Research Council, Committee on Mineral Resources and Environment, 1975, Mineral resources and the environment: National Academy of Science, 348 p.

Nehring, R., and E. R. Van Driest II, 1981, The discovery of significant oil and gas fields in the United States: Report R-2654/1-USGS/DOE (2 vols.), Rand Corp., Santa Monica, California, 236 p.

Potential Gas Committee (PGC), 1971, Potential supply of natural gas in the United States (as of December 31, 1970): Golden, Colorado, Potential Gas Agency, Colorado School of Mines, 41 p.

——— , 1973, Potential supply of natural gas in the United States (as of December 31, 1972): Golden, Colorado, Potential Gas Agency, Colorado School of Mines, 48 p.

——— , 1977a, A comparison of estimates of ultimately recoverable quantities of natural gas in the United States: Golden, Colorado, Potential Gas Agency, Colorado School of Mines, 27 p.

——— , 1977b, Potential supply of natural gas in the United States (as of December 31, 1976): Golden, Colorado, Potential Gas Agency, Colorado School of Mines, 45 p.

——— , 1979, Potential supply of natural gas in the United States (as of December 31, 1978): Golden, Colorado, Potential Gas Agency, Colorado School of Mines, 75 p.

——— , 1981, Potential supply of natural gas in the United States (as of December 31, 1980): Golden, Colorado, Potential Gas Agency, Colorado School of Mines, 119 p.

——— , 1982, Potential supply of natural gas in the United States (as of December 31, 1981): Golden, Colorado, Potential Gas Agency, Colorado School of Mines, p. 28.

Root, D. H., Estimation of inferred plus indicated reserves for the United States: U.S. Geological Survey Circular 860, Appendix F, 5 p.

Shell Oil Company, 1978, Alaska holds 58 percent of future U.S. oil finds: Oil and Gas Journal, v. 76, p. 214.

Theobald, P. K., S. P. Schweinfurth, and D. C. Duncan, 1972, Energy resources of the United States: U.S. Geological Survey Circular 650, p. 27.

U.S. Department of Energy, U.S., 1981 and 1982, Crude oil, natural gas and natural gas liquids reserves: USDOE Annual Reports.

U.S. Geological Survey, 1974, USGS releases revised U.S. oil and gas resource estimates: Washington, D.C., U.S. Department of the Interior Geological Survey, News Release, March 26, 1974.

The Assessment of Heavy Crude Oil and Bitumen Resources

Richard F. Meyer
U.S. Geological Survey, Reston, Virginia
Christopher J. Schenk
U.S. Geological Survey, Denver Federal Center, Denver, Colorado

Heavy oils and bitumens can be separated into three categories for assessment purposes: identified bitumen deposits, identified heavy oil deposits, and undiscovered bitumen and heavy oil deposits. Bitumen-bearing rocks and many heavy oil reservoirs are devoid of natural reservoir energy that would enable primary recovery, thus assessments are essentially made for oil in place. Resource estimates are volumetric calculations based on extent and thickness of the reservoirs and bitumen or heavy oil saturations. Primary recovery, however, enables the calculation of recovery factors and reserves. The economics of many phases of heavy oil recovery, transportation, and refining ultimately effect the reserve estimates.

Most heavy oils and bitumens probably represent the degraded remnants of conventional petroleum deposits that were 50–90% larger, a concept important in assessing undiscovered heavy oil and bitumen deposits. To predict accurately the occurrence of bitumen and heavy oil, assessments need to involve the formulation of models for oil generation, expulsion, migration, and degradation. Some assessments of undiscovered or poorly known heavy oil deposits simply use the geologic characteristics of known deposits to predict the presence and amount of undiscovered heavy oil. Others use the proportion of cumulative conventional oil production that was heavy to make rough estimates of reserves and undiscovered resources. The amount of undiscovered bitumen in most countries is probably extremely low, because most bitumen deposits crop out or were located during the search for conventional petroleum.

INTRODUCTION

Bitumen and heavy crude oil have similar physical properties, and related chemical properties, and in some cases, parallel genetic histories; thus, they share problems associated with their extraction and processing. Collectively, heavy crude oil and bitumen deposits represent an enormous hydrocarbon resource. The technically recoverable amount worldwide has been estimated to be as large as 600 billion bbl (Meyer et al., 1984); the resource-in-place in Alberta, Canada, and eastern Venezuela alone is about 2,500 billion bbl, and the world total is probably about 1.5 times this amount.

The assessment of in-place heavy oil and bitumen in presently known deposits follows conventional petroleum and mining engineering procedures. Expanding the assessments to include the undiscovered resource and to deal with the factor of recoverability introduces complications to the assessment process.

DEFINITIONS

For purposes of this report, the definitions of heavy crude oil, extra-heavy crude oil, and bitumen as proposed by Martinez (1984) are followed. Heavy and extra-heavy crude oils and bitumens are naturally occurring petroleum or petroleumlike liquids or semisolids in porous and fractured media. Bitumen deposits are also called tar sands, oil sands, oil-impregnated rocks, and bituminous sands. Viscosity should be used first to differentiate between crude oils and bitumens. Subsequently, density should be used to differentiate among extra-heavy crude oils, heavy crude oils, and other crude oils.

Bitumens have viscosities of > 10,000 centipoise (cP). Crude oils have viscosities of < 10,000 cP. These viscosities are gas free as measured and referenced to original reservoir temperature.

Extra-heavy crude oils have densities greater than 1,000 kg/m^3 (American Petroleum Institute [API] gravities < 10°).

203

Table 1. Soviet classification of natural bitumens and oils of high viscosity.[a]

Class	Constituents (%)		Viscosity (Pa · s)	Density	
	Oil	Asphaltenes and Resins		(g/cm³)	(°API gravity)
High-viscosity oil	>75	<25	0.05–2	0.935–0.965	20–15
Maltha	75–40	25–60	2–20	0.096–1.030	15–5
Asphalt	40–25	60–75	20–1,000	1.030–1.100	5
Asphaltite	<25	>75	>1,000	1.050–1.200	5
Kerite	3–5	>90	[b]	1.070–1.350	—
	5–10 (rare)	—	—	—	—
Anthraxolite	—	100	[c]	1.300–2.000	—
Ozokerite	—	<50	[d]	0.850–0.970	—
Pure	100	—	—	—	—
Impure	0–50	—	—	—	—

[a]From Khalimov et al. (1983).
[b]Hard, partly soluble in chloroform
[c]Hard, insoluble in chloroform
[d]Semihard, paraffinic

Heavy crude oils have densities from 934 to 1,000 kg/m³ (API gravities 10–20°) inclusive. These densities (API gravities) are referenced to 15.6°C (60°F) and atmospheric pressure.

Crude oils with densities <934 kg/m³ (API gravities >20°) may be classified as medium, light, or other crude oils.

A similar terminology is employed by Soviet geologists (Gol'dberg and Iûdin, 1979; Khalimov et al., 1983; Lavrushko, 1984), who use the terms *natural bitumen* or *natural petroleum bitumen* and *high-viscous oils*. Table 1 clarifies this terminology; the category of extra-heavy oil (API gravity 10°) thus falls within the class of malthas (API gravity 15–5°). The break at 10° (API gravity) is convenient in that it is 1.00 g/cm³ which is the specific gravity of water. Table 1 makes an important point in distinguishing the high-viscosity oils (API gravity 15–20°) from the bitumens (API gravity <15°). For resource assessment purposes, the hydrocarbons above 15° (API gravity) are conventional oils.

If an oil field is producing oil that is <10° (API gravity) and viscosity data are not available (the usual case), it is assumed to be extra-heavy crude oil and not bitumen. The term *bitumen* generally has a broader connotation than this definition assigns to it. Excluded from it are the mineral waxes, asphaltites, and organic matter insoluble in carbon disulfide.

The definitions of resources and reserves of the U.S. Bureau of Mines and U.S. Geological Survey (1980) are used in this paper.

ASSESSMENT METHODOLOGIES FOR HEAVY OIL AND BITUMEN

For assessment purposes, heavy oil and bitumen resources are divided into three categories: identified bitumen deposits, identified heavy oil deposits, and undiscovered bitumen and heavy oil deposits. The boundaries of the three are also gradational, and they are gradational with conventional oil deposit. Most assessments of petroleum resources are directed to conventional oil and gas and specifically exclude heavy crude oil and bitumen (e.g., Dolton et al., 1981). The reasons for doing so are perfectly valid: different assessment techniques are required.

In general, the most significant differences between assessing heavy oils and bitumen and assessing conventional oils result from the greater density and viscosity of the former. Only minor amounts of natural gas are associated with heavy oil and bitumen, which results in a lack of reservoir energy. Almost invariably the resource requires thermally enhanced recovery, although in some heavy oil deposits, water flooding may be effective. Primary and secondary methods of recovery alone seldom are sufficient for materials below ~17° (API gravity); this is why most of the produced heavy oil falls between 17° and 20° (API gravity). The severity of recovery problems of heavy oil and bitumen imposes an economic penalty that, to date, has only been overcome under special circumstances.

Identified Bitumen Deposits

In assessing bitumen deposits, the grade or richness is critically important, whether expressed as weight percent, barrels per acre-foot, or gallons per ton, as are net pay thickness and area of the deposit. A chart by Herkenhoff (1972) expresses the relationships between grade and barrels per acre-foot for oils of two densities, 10 and 20° (API gravity) (Figure 1A). A second chart by Herkenhoff (1972) (Figure 1B) demonstrates the combinations of area, pay thickness, and weight percent of hydrocarbon needed for a reserve of 250 million bbl.

Surface Deposits

Palacio (1957) evaluated the bitumen quarry on Leyte Island, in the Philippines, as a source of road material. The quarry was an openpit mine about 90 × 140 m (300 × 460 ft) in size, but the extent of the bitumen-bearing sand lenses was defined by tunnels, test pits, trenches, and 32 drill holes. The bitumen content varied from 1 to 21 wt. %, with a market requirement for paving of greater than 8 wt. %. Palacio (1957) calculated the total in-place resource dividing the area

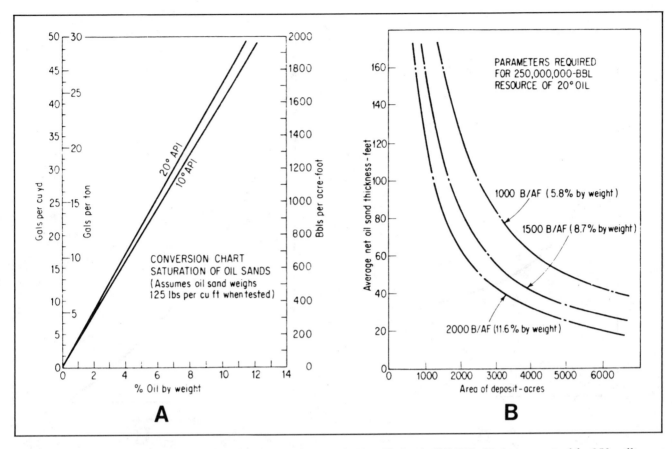

Figure 1. (A) Oil content for 10° and 20° (API gravity) heavy oil. From Herkenhoff (1972). (B) Areas required for 250 million bbl reserve. From Herkenhoff (1972).

into polygonal blocks on the basis of the percentage of saturation, computing the volumes of rock asphalt, and converting the tonnage of the bitumen-saturated rock at 15 cu ft/ton. Using a cut-off of 6 wt. % (about 1,000 bbl/acre-ft), Palacio allowed 20% for a mining loss for reserves and 30% for inferred reserves, yielding estimates of 350,000 and 175,000 tons, respectively. He then calculated an additional subeconomic resource (3–6 wt. %) of 98,000 tons. If the 525,000 tons of saturated rock averaged 8 wt. % bitumen, then the material represents about 3.4 million bbl.

The largest known bitumen deposit in the world is Athabasca, in Alberta, Canada. The oil sands of the Lower Cretaceous McMurray Formation crop out of Athabasca, whereas they occur in the subsurface at Cold Lake and Peace River. Part of it is surface minable, another part is buried too deeply for mining, and an intermediate part between about 75 and 300 m (250 ft and 1,000 ft) cannot be mined economically by surface methods and is too shallow for *in situ* thermal recovery. The intermediate part may be accessible to tunneling, mine-assisted *in situ* thermal recovery, or slurry mining.

For surface mining and subsequent hot water plant extraction of the bitumen, the grade and amount of fines (clay content) as well as ore, overburden, and reject thicknesses are the critical elements (Pearson, 1981). The amount of feedstock required by economics increases as grade

decreases (Figure 2A). The tailings storage volume increases as grade decreases; this is important because of the economic and environmental costs of tailings ponds (Figure 2B). For a test plot of 6,700 acres with 62 core holes for control, Pearson (1981) determined that 1,240 million cu ft of minable reserves were available. Of this, 24% was below a grade of 4 wt. % and 38% below 6 wt. %. He concluded that production to a minimum grade of 6 wt. % is justifiable on economic and environmental grounds, based on a minimum mining interval thickness of 20 ft. He also found that the economics of a mine were apparently less sensitive to ore thickness than to grade. Figure 3 is plot of waste:ore ratios and grade as interpreted by different operators for defining the economic viability of openpit mining and hot water extraction of bitumen. It is only on the basis of criteria such as these that subeconomic bitumen resources can be defined.

Strom and Dunbar (1981) and Energy Resources Conservation Board (1982) defined the mining and extraction requirements of the Athabasca deposit and related Alberta Cretaceous bitumen deposits (Cold Lake, Peace River, and Buffalo Head Hills), as well as the underlying Devonian Grossmont Limestone. Typically, the bitumen in these deposits is characterized by 4.5 wt. % sulfur, 6–14° (API gravity), with a viscosity of 10^5–10^6 cP. Minimum values are a grade of 3 wt. % and a thickness of 2 m (6.5 ft). Where fines exceed 15–20% of total solids, the bitumen saturation

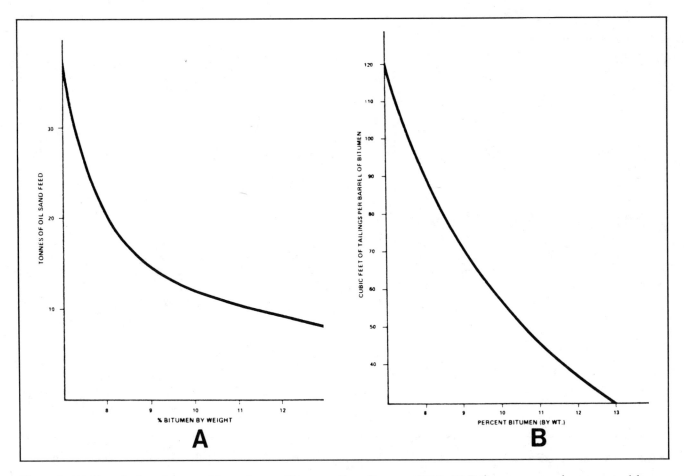

Figure 2. (A) Plant feedstock required for each ton of bitumen. From Pearson (1981). (B) Tailings storage volume required for various grades of oil sand feedstock. From Pearson (1981).

sharply decreases, the water saturation increases, and the grade drops below the economic limit of 6–8 wt. % bitumen. Using logs and cores, researchers determined the sand porosity, shale lens thickness, and net sand thickness by grade. Because of lateral variation in bitumen and fines content, the in-place volume of bitumen was mapped for blocks of 23 km^2 (8.8 sq mi). The surface-minable areas were defined by thickness of overburden and content of fines. In these areas, the overburden is 0–75 m thick, with 20 m (65 ft) being typical, the underlying minable sand being typically 50 m (165 ft) thick and 12 wt. % in grade.

Constraints to surface mining of the Athabasca and related deposits included a low grade of deposit, the presence of many shale lenses, a thick overburden, a high content of fines, and areal losses for overburden and tailings storage and for plant sites. *In situ* recovery with steam at 150–200° C reduced viscosity to 10^2 cP and resulted in 20–30% recovery in favorable areas. Problems with *in situ* recovery included heat loss, difficulty in attaining flow normal to formation partings or high-permeability sand lenses, the presence of viscous interfingering, and complex multiphase flow affecting displacement to producing wells. In the surface mining of these deposits, in addition to the areal losses, there was a mining loss of 25%. Also, the bitumen extracted by either surface mining or *in situ* recovery by steam entailed a 25% loss in carbon-rejection upgrading.

Pearson (1981) described a prismatic calculation of bitumen reserves in which it is assumed that all holes are vertical and that the ground surface and the top and bottom of the deposit are planes. Thus, in his calculations the waste and ore volumes are represented by prisms (Figure 4). It is assumed that ore thickness times grade at each hole is proportional to an equivalent column of bitumen. Therefore, quality and quantity vary linearly between points. The calculations are as follows:

$$V_w = X(W_1 + W_2 + W_3)/3$$
$$V_o = X(O_1 + O_2 + O_3)/3 \quad \text{and}$$
$$G = X(O_1G_1 + O_2G_2 + O_3G_3)/3/V_o$$

where V_w is waste volume, V_o is ore volume, G is average grade, X is area, W_n is waste thickness (overburden and reject) at hole number n, O_n is ore thickness at hole n, and G_n is average grade (bitumen content) in O_n. The plan area (X) used in the calculation is derived from the north (N) and east (E) coordinates for each hole:

$$X = N_1(E_2 - E_3) + N_2(E_3 - E_1) + N_3(E_1 - E_2)/2$$

This method is independent of planimetry errors and map distortion and permits rapid calculation and simple data revision. It can be applied to any stratified deposit.

VARIOUS ECONOMIC CRITERIA FOR MINEABLE OIL SANDS

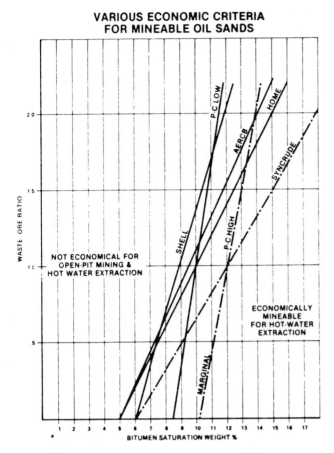

Figure 3. Various economic criteria for mineable oil sands. From Pearson (1981).

PRISMATIC CALCULATION OF ORE RESERVES

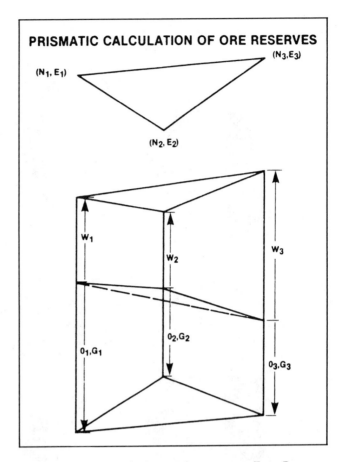

Figure 4. Prismatic calculation of ore reserves. From Pearson (1981).

Isopach mapping is more commonly used in areas of sparse well control. Lewin and Associates (1984) assessed most of the major bitumen deposits of the United States. For each well they used cores and wireline logs to determine the depth to the top of the bitumen saturation, the thickness of the saturated interval, the bitumen saturation, and the porosity and permeability. From these data, they prepared richness (grade) maps based on bbl per acre from which resources were calculated. They separately computed measured resources, which are those deduced from core analysis, and speculative resources, those presumed to exist from showings in drilling logs or from interpretation, commonly based on extrapolation from wells having shows of bitumen.

Subsurface Deposits

An example of an assessment of subsurface bitumen resources is given by Minken (1974) for the Cold Lake deposit in Alberta, Canada, which contains four oil sands: Minken's cut-off was a grade of 3 wt. % or greater and a sand thickness of 1.5 m (5 ft) or more. The sands were classified into three saturation categories: rich, with an oil saturation of ≥10 wt. %; intermediate, >5 wt. % but <10 wt. %; and lean, <3 wt. % but <5 wt. %. He mapped the oil sand thicknesses for 53 cored wells and 359 wells not cored. These volumes were then converted to oil volumes in place by conversion factors based on saturations determined from core analysis. Core analysis yielded the bitumen saturation values

for the grade categories, whereas resistivity values were related to saturations in the uncored wells. Resistivities were determined by the 16-in. normal log and the LL8 laterolog. He then compared core analysis and resistivity data to determine saturation values for the uncored wells. The comparative data for two key wells are illustrated in Figure 5.

The Cretaceous bitumen-bearing sands in Alberta truncate the upturned eroded edges of bitumen-bearing carbonate rocks ranging in age from Devonian to Permian over an area of 70,000 km² (27,000 sq mi). Outtrim and Evans (1977) attempted to assess the resource potential of these Paleozoic rocks using the few available drill holes penetrating the section. The bitumen is similar to that of the Cretaceous oil sands and is apparently derived from them or from a common source bed. From available sources (literature and log and core analyses) they estimated bitumen saturations and porosity values, dismissing from calculations all values of saturation of 20% pore volume or less. For each formation, they then estimated the strike length, average subcrop width, saturated sand thickness from the unconformity to the downdip oil/water contact, and porosity and saturation ranges. They were then able to compute the rock volume and reduce it by the fraction of the rock having a saturation below 20% pore volume. The remaining rock volume multiplied by the porosity and saturation yielded an estimate of the in-place bitumen resource (actually a demonstrated subeconomic resource). A range of conditional probability

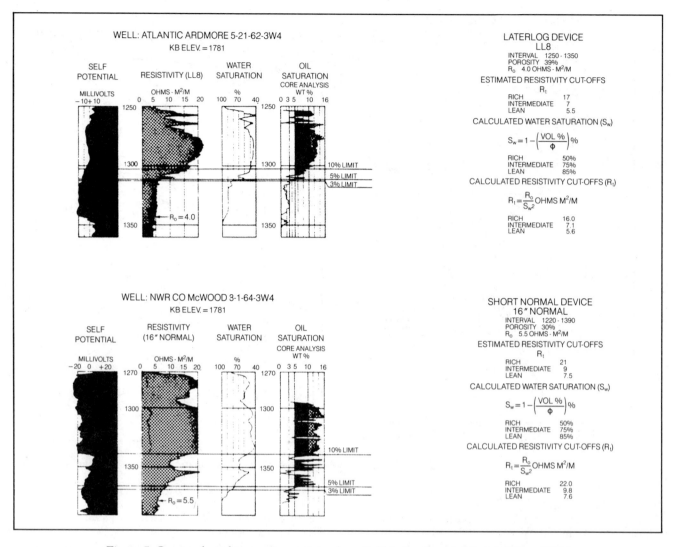

Figure 5. Core analysis data versus resistivity data, Cold Lake oil sands. From Minken (1974).

values could have been obtained by Monte Carlo simulation of the estimated data inputs, which would have better reflected the uncertainty in the estimates.

Methods for assessing resources of Permian bitumen deposits in the Tatar Republic of the Soviet Union were outlined by Troepol'skii (1976). The main accumulations are in sand lenses in the Lingula shales. The deposits, identified in sparse wells, are associated with structural uplifts. The demonstrated subeconomic resources (geologic reserve) are estimated from a volumetric formula:

$$Q = F \times h \times m \times B \times g$$

where Q is the bitumen resources (tons), F is the area of field from the structure map (m²), h is the bitumen-saturated thickness of formation (m), m is the fracture porosity determined from maps of porosity distribution, B is the bitumen saturation from core and wireline logs that is greater than 1 wt. %, and g is the density of bitumen (g/cm³). The average values for these inputs were determined from 785 samples from 14 wells. The average bitumen saturation of

51 wt. % is considered to be conservative because the unconsolidated core samples yielded poor laboratory results.

The difficulties with obtaining good core results from unconsolidated sands are described by Dusseault and Scott (1984), who point out that core damage is pervasive in uncemented formations, commonly the case with heavy oil and bitumen deposits. The total resource and the grain mineralogy are not affected; the effect on bulk density and microstructure is small. However, critical reservoir parameters, such as gas saturation, relative permeability, and compressibility, may be greatly affected. For a relatively small stress change, the potential for core expansion is large. The damage may involve volumetric expansion, core intrusion, radial planes opening within the core sleeves, shear failure structures, and fine pervasive fracture networks in fine-grained and clayey cores.

Britton's (1984) work on the South Texas San Miguel deposit, which produces from the Upper Cretaceous San Miguel "D" Sand, shows that measurements of the reservoir properties of bitumen deposits in unconsolidated sands can lead to standard core analyses of porosity that are as much as

20% too high and bitumen saturations that are 20% too low. Wireline logs give better porosity values but overly optimistic bitumen saturations. As a result, the in-place oil from cores tends to be too low and from logs, too high.

Identified Heavy Oil Deposits

Calculation

Reserves and subeconomic resources in heavy and extra-heavy crude oil reservoirs are calculated in much the same way as in conventional oil fields. A particularly lucid explanation of the volumetric calculation of oil reserves is given by Webber (1961):

$$\text{recoverable oil} = \frac{A \times t \ (\text{or } V_o) \times 7{,}758 \times \theta(1 - S_w) \times R}{FVF}$$

where A is the area of reservoir (acres), t is the net thickness (ft), V_o is the bulk reservoir volume (acre-ft), 7,758 is the number of bbl in 1 acre-ft, θ is the effective porosity (fractional), S_w is the interstitial water saturation (fractional), R is the recovery factor, and FVF is the formation volume factor (ratio of in-place oil to recovered oil).

Some special considerations apply to the calculation of heavy crude oil, particularly if the oil is below about 15° (API gravity). It is likely that the gas:oil ratio will be very low, in the range of 100–200 cu ft/bbl. This means that FVF will be very close to unity, there will be no effective gas drive, and, by disregarding R, the formula will essentially measure oil in place. Unless the geothermal gradient is exceptionally high, as is the case of the Orinoco oil belt, the high viscosity will work against an effective water drive. As a result, most heavy oil reservoirs have low natural energy and primary recoveries are very low to absent.

Connan and Coustau (1984) state additional considerations regarding the hydrodynamic environment of heavy crude oil. First, the heavier the oil, the more likely that the oil/water contact will be tilted and the greater will be the movement of the oil. Second, if degradation preceded accumulation or if the oil is unaltered but immature, the oil can only saturate the highly permeable and most porous zones. Because these are the zones preferentially swept in a steam flood, the sweep will be efficient. Finally, if alteration of conventional oil occurs after entrapment because of inspissation by water washing or by evaporation through a poor cap, oil in the less permeable zones in the reservoir will be modified to heavy oil. A steam flood then will be inefficient because the steam follows highly permeable zones already depleted by water washing.

Other mechanisms may interfere with steam flooding. One way is through the development of a "tar mat" at the oil/water interface caused by precipitation of asphaltenes from natural gas, either original or introduced after entrapment.

The principal difficulty in estimating reserves of heavy oil fields lies in estimating the recovery factor R. In many fields (e.g., in California) steam flooding of the reservoirs is successful, if not very energy efficient. In other places, such as the Bati Raman field in Turkey, although the resource is very large, the recovery by known methods is only about 2%.

In the Bolivar Coastal fields of Lake Maracaibo in Venezuela, subsidence plays an important role in improved recovery. Given existing technologies, about 10% recovery from heavy oil fields is an overall average. We should recognize, however, that conditions in heavy oil fields vary enormously, so that generalizations may not be accurate.

Problems of Heavy Oil Assessment and Recovery

Because heavy oil is viscous to very viscous and because it is typically contained in deltaic and fluvial sandstones of erratic occurrence and is associated with weak or no natural gas drive, natural reservoir energy is thus very low. As a result, production capacities are low and production decline curves are much flatter than in the case of higher quality crude oils. Thus, for a developed field, reserves can safely be estimated on the basis of a 20-year decline. Although this is very imprecise, if it is added to cumulative production, it gives a figure for ultimate recovery useful for geologic (but not economic) planning purposes. This figure presumably would represent about 10% of the original heavy oil in place.

Commonly, heavy oil fields are borderline in classification, containing oil in the reserve and inferred reserve categories and in the subeconomic resources category, which with changing economics could become reserves. Heavy oil reservoirs are particularly sensitive to changes in economics. A fine example of such a field is Lloydminster in Alberta and Saskatchewan, Canada, for which a number of studies are available. Jha and Berma (1984) succinctly state the problems involved with Lloydminster heavy oil reservoirs: (1) low primary recovery due to high viscosity; (2) production of sand with oil, leading to abrasion and plugging of equipment; (3) corrosion; (4) difficulty in separating sand from oil; (5) need for heated storage and transportation facilities; (6) need for dilution with light crude for transportation by pipeline; and (7) difficulty in marketing the oil, which is low quality because of low API gravity, high viscosity, and high content of asphaltenes, sulfur, nitrogen, and metals.

McCallum (1981) and Raicar and Procter (1984) have described the geology of the producing reservoirs in Lloydminster and the difficulties associated with producing them. There are eight principal producing zones in the sands interpreted as being deposited in a nearshore environment. These are cut by a set of anastomosing channel sands whose location is governed by leaching of salt in the Devonian substrate. Tilting subsequent to sedimentation has distorted the fluid interfaces. The oil, emplaced normally, has been degraded to near immobility. Older wells lack porosity logs, but the newer ones have full wireline log suites. Porosities are on the order of 28–33%, and where they are <28%, the oil saturations do not exceed 30%. Evidently a permeability threshold is involved in the entrapment of the oil. Formation water resistivities are consistent at 0.07 ohms (Ω), thus readings above 2 Ω indicate oil-saturated clean sands. Dirty sands with low oil saturation, however, show readings of 5 Ω, which causes some difficulty in identification of the saturated zones. Because the Alberta portion of the field involves 600 pools and 11,180 km² (4,300 sq mi), it was divided into 9,000 segments on the basis of oil saturation, thickness, extent of bottom water, gas cap, and area. The segments were then screened to exclude channel sand segments and those with less than 1,500 bbl/acre-ft of saturation, extensive gas cap, or < 1.5 m (< 5 ft) of net saturation. Because the viscosity decreases as temperature increases (Figure 6), wet combustion is considered to be the most viable recovery method.

Hickerty et al. (1984) have made use of a computerized data base to assess the resources of the Lloydminster deposit. They have addressed three levels of activity: exploration, pool extension and development, and enhanced oil recovery pilot projects. All data are stored in four compatible computer files: LOCATIONS, with spatial coordinates, kelly-bushing elevation, and total depth; TOPS, with stratigraphic data; LITHOLOGY, with wireline log and core data; and DIGITIZED LOG, with digitized borehole log traces. Logs are used to generate net pay thickness values. Hickerty and co-workers base reservoir quality on SP response, and such logs are commonly available. They then encode four lithologic types and their corresponding reservoir grades on the SP log: good, for clean sand; fair, for silty sand; poor, for shaley siltstone; and very poor, for shale. All zones thicker than 0.6 m (2 ft) are recorded, and individual stratigraphic units are mapped at three levels: regional (total area), local (one field, such as Aberfeldy), or detailed (an enhanced oil recovery pilot). This approach is very useful in dealing with fields such as Lloydminster, which is actually a play.

Undiscovered Bitumen and Heavy Oil Resources

General Considerations

Assessing the resource potential of undiscovered bitumen and heavy oil deposits is difficult. One method commonly used to assess undiscovered hydrocarbons is play analysis, which requires identification by some means (usually subsurface geology or geophysics) of prospects that can be assembled into prospect-related plays (Energy, Mines and Resources Canada, 1977; Proctor et al., 1982). In essence, the play is reduced to a series of inputs (Figure 7), and an estimate for each is made at various probability levels that a particular parameter is greater than a certain value. We have inserted an array of numbers in Figure 7 for illustration. The numbers ordinarily reflect the observations of a panel of experts, providing a Delphi approach. If specific prospects within the play, such as known occurrences, can be identified, they are considered separately and the entire data set is subjected to a Monte Carlo simulation, which provides unrisked frequency distribution curves of conditional probabilities. Because other factors, such as the presence of adequate source beds, govern whether oil is present, these are assigned marginal probabilities (Figure 7). The product of the marginal probabilities is used to reduce the unrisked frequency distribution curves to risked distributions. Examples of unrisked curves reduced to risked curves are given in Figure 8.

Some of the difficulties in applying the play analysis method to heavy oil and bitumen deposits (but not limited to play analysis) involve the amount of hydrocarbons originally generated, the amount actually lost through alteration, the fate of the lost hydrocarbons, and the lack of precision possible in estimating recoverability. An exhaustive treatment of the latter is given by Outtrim and Evans (1977).

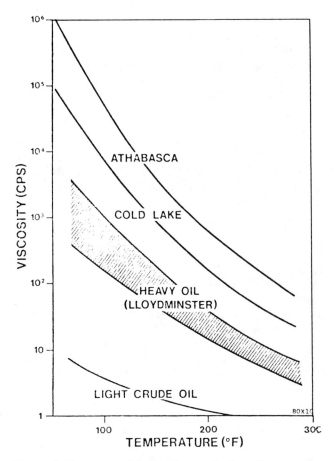

Figure 6. Viscosities of some Alberta oils. From Raicar and Procter (1984).

The recovery estimation process is complicated by the various deposits differing widely in reservoir conditions because of their complicated geological histories.

It would be of great value in assessing petroleum resources to know the exact amount of hydrocarbons generated. We can quantify only the amount known to be presently trapped. The difference between the two amounts is either lost or yet to be discovered.

Claypool et al. (1978) followed the concept of McDowell (1975) in evaluating hydrocarbon generation and entrapment in the black shale of the Phosphoria Formation. They estimated liquid hydrocarbons generated on the basis of the amount and distribution of organic carbon and the thermal history of the basin. They then estimated the amount of heavy hydrocarbons still present in the thermally mature source beds by applying observed hydrocarbon to organic carbon ratios to the total organic carbon. They next converted the total organic carbon to heavy oil on the basis of pyrolitic yield. By comparing this yield to the amount of oil known to be in the reservoir and estimated to have been lost after expulsion, they determined that about 8% of the heavy hydrocarbons generated were actually expelled from the source rocks and about 2% formed recoverable oil deposits. Their results are given in Figure 9.

Other more complex ways of approaching the value of hydrocarbons expelled have been devised. Bishop et al.

PLAY/PROSPECT									DATE	
POTENTIAL EQUATION VARIABLES								"OIL" OCCURRENCE FACTORS		REMARKS
	Conditional probability Per cent GT.							Presence or adequacy	Marginal probability	
	100	95	75	50	25	5	0			
Area of closure	1	10	18	23	26	33	50	Geometric closure	8	
Reservoir thk.	10			40			90	Lithofacies	1	
Porosity	.08			.12			.16	Porosity	8	
Trap fill	.05			.4			.7	Seal	1	
Recovery				.35				Timing	5	
Water sat				.25				Source	.5	
Shrinkage				.7				Preservation	1	
Gas fraction				.5				Recovery	1	
No. of prospects	3	8	15	20	25	40	60			
										6,000 ft.
										average depth
								Product	.16	

Figure 7. Data form used in estimating play parameters. From Energy, Mines and Resources Canada (1977).

(1983) prepared a comprehensive procedure for estimating hydrocarbon accumulation and dispersion. Their method begins with an estimation of the volume of expelled hydrocarbons and an integration of this with the trap volume. They conclude with estimates of recoverable hydrocarbons. The model involved is data intensive, as indicated by the required sequence of equations (Table 2). The variables are assigned ranges of values, and, after many iterations are generated through Monte Carlo sampling, a range of results is obtained. Bishop et al. (1983) emphasize the importance not only of oil and gas expulsion from source beds, but of trap volume. Because heavy crude oil and bitumen deposits ordinarily contain very little gas and commonly have been altered severely, the list of parameters included in the calculation process requires amplification.

The problem with using material balance approaches to assess heavy oil and bitumen is that the existing deposit in most cases is altered or degraded. If, as is the case with the Monterey Formation diatomite, the oil is not degraded but is immature (Orr, 1978; Petersen and Hickey, 1984), then the method of Bishop et al. (1983) would work without modification. If the oil is severely degraded, however, then volume and composition of the oil prior to degradation should be estimated before one attempts to calculate the amounts of oil originally generated and expelled.

Orr (1978) approaches such a reconstitution of the original oil volume through studies of sulfur content. Water washing and microbial degradation are two principal means by which

oil deposits are degraded. Criteria for determining the presence of microbial degradation include a relative decrease in n-alkanes, an increase in oil density, and an increase in cycloalkanes, aromatics, and N—S—O compounds. The isoprenoids, such as pristane and phytane, show an initial relative increase because they are less desirable as bacterial food, but a later decrease relative to cycloalkanes and other compounds. This pristane–phytane effect is characteristic on gas chromatographs. Orr (1978) contends that the two- to threefold increase in sulfur in the heavy oils and bitumens is a concentration effect of sulfur compounds already present in the original oil and is not a result of sulfur addition from exogenous sources. If this is so, he concludes that 50–70% of the original oil must have been destroyed. Also, because selective removal of n-alkanes alone is volumetrically insufficient, losses of iso- and cycloalkanes and aromatics must also be involved.

Tannenbaum et al. (1984) made a detailed study of light oils and bitumens from the Dead Sea area and compared estimates of removal of light constituents with laboratory degradation experiments. On the basis of material balance equations, they concluded that bacterial degradation and water washing have led to removal of about 75% of the original oil constituents in the C_{15+} range and that the surface asphalts represent a residue of 10–20% of the original oil. They also found an apparent fourfold sulfur enrichment, which is in line with the observations of Orr (1978). Tannenbaum et al. (1984) assumed that there was removal of

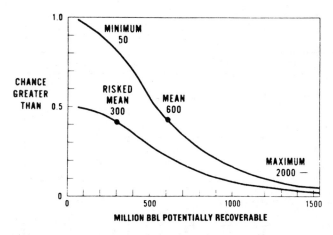

Figure 8. "Risked" and "unrisked" assessment probability curves. From Baker et al. (1984).

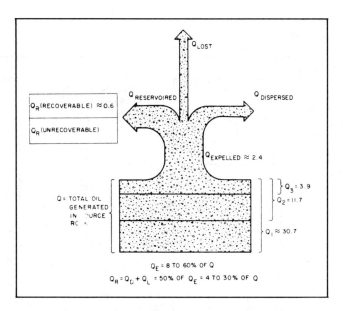

Figure 9. Diagram illustrating, for the Phosphoria Formation, quantities (in units of 10^9 metric tons) of heavy hydrocarbons generated, expelled, lost, and pooled. Q is the total oil generated, Q_e is the expelled oil, Q_d is the dispersed oil, Q_l is the lost oil, and Q_q is the pooled oil. From Claypool et al. (1978).

saturates and aromatics in varying amounts, with resins and asphaltenes in an inert phase.

An idea of the magnitude that a natural petroleum system can attain is demonstrated by Moshier and Waples (1985) in a material balance study of the Mannville Group of western Canada. They concluded that the assumed source rocks could not have generated the amount of oil reservoir in the Alberta bitumen deposit complex plus the 20% assumed to have been lost through degradation. Therefore, there must have been additional Mannville equivalent source beds or undocumented multiple sources. Usually the reverse case is true for bitumen deposits: identified source beds have generated far more oil than can be reasonably accounted for by expulsion, diffusion, and migration into discovered traps.

Alteration of Hydrocarbons

Gol'dberg and Iûdin (1979) state that knowledge of the amount of hydrocarbons lost during alteration is important to an understanding of (1) the alteration process (supergene transformation in the zone of oxidation), (2) conditions of formation of the heavy oil or bitumen deposits, and (3) reconstruction of the original oil resources. They cite model experiments by Korenkova and Arkhangel'skaya indicating that biodegradation of an *n*-alkane naphthene petroleum is accompanied by a loss of 18–21 wt. %, of *n*-naphthene and naphthene aromatic petroleum by 16–19 wt. %, and of a high *n*-alkane petroleum by as much as 50 wt. %. Based on earlier studies by Uspenskiy and Gol'dberg, Gol'dberg and Iûdin (1979) prepared a table (Table 3) to estimate the possible concentration of various bitumens and crude oils in reservoirs of different porosities. In Table 3, the data characterize bitumen deposits that are the product of petroleum alteration under conditions of one-time formation of the primary petroleum accumulation. Gol'dberg and Iûdin (1979), however, recognize that deposits may go through more than one cycle of alteration, as reflected by bitumens of differing compositions. Such relationships between porosity (or perhaps another physical reservoir parameter) and bitumen composition are useful for interpreting resource quantities; they have the further value of being accessible from borehole data.

Resins and asphaltenes, as well as low-volatility oils, are important components of kerogen and of the oil generated from that source material. Tissot (1984) has described the resins and asphaltenes as normal, commonly subordinate components of all crude oils, with the heteroatoms providing important links. Pyrolysis of asphaltenes shows that they follow a similar thermal history to that of kerogen, that is, the hydrogen:carbon, oxygen:carbon, and sulfur:carbon ratios decrease with increasing maturity. According to Tissot (1984), the ratio of asphaltenes plus resins to hydrocarbons in source rocks is about 1:1, whereas this ratio is about 1:6 in oil originally in the reservoir. He suggests that the asphaltenes may be dissolved by an "atmosphere" of resins surrounded by aromatic hydrocarbons. Thus, an oil migrating from the source rock would carry a quantity of asphaltene relative to the available resins and aromatics. This would then create the original composition of the present heavy oils and bitumens, but intervening alteration would alter the proportion of the light components to the asphaltenes plus resins.

The qualitative effects of the alteration of a pooled oil and some of the mechanisms of alteration are summarized in Figure 10. Thermal alteration in the form of thermal cracking normally results from increased burial with a concomitant rise in temperature. This alteration most likely will result in the destruction of heavy oil and bitumen deposits as part of the natural maturation process.

Deasphalting is a process by which asphaltenes are precipitated from an oil by the addition of natural gas. The most common result is the formation of a so-called tar mat at the oil/water contact of the type observed in the Burgan field, Kuwait.

Table 2. Equations used to calculate fundamental volumes for resource assessment.[a]

Indigenous Oil and Gas Volumes Migrated From Source
1. Kerogen quantity = drainage area × effective source rock thickness × original total organic carbon
2. Oil volume (in reservoir bbl) = kerogen quantity × oil yield × units constant × oil formation volume factor
3. Gas volume (in reservoir bbl) = kerogen quantity × gas yield × units constant × gas formation volume factor
Trap Volume
4. Barrels = trap area × reservoir thickness × porosity × (1 − interstitial water saturation) × trap geometry correction × net/gross × units constant
Gas Volume Dispersed
Formation water volume
5. Barrels = drainage area × [(reservoir thickness × porosity) + (effective source thickness × porosity)] × units constant
Dissolved gas
6. Volume of dissolved gas, if water and oil are both saturated = (gas solubility in water × water volume) + (gas solubility in oil × total oil volume)
7. If oil and water are not saturated with gas, the calculation of dissolved gas volume in Eq. (6) cannot be used.
Diffused gas
8. Volume of diffused gas = (residence time × gas concentration in water × seal diffusion coefficient × drainage area × units constant) ÷ (seal thickness).

[a]From Bishop et al. (1983).

Table 3. Amount of bitumen and highly viscous oil (in wt. %) as a function of porosity.[a]

Porosity (%)	Petroleum 0.85 g/cm³	0.95 g/cm³	Maltha	Asphalt	Asphaltite
25–35	10–15	8.0–12.5	7.5–11.5	5.0–7.5	2.8–4.2
15–25	6–10	5.0–8.0	4.5–7.5	3.0–5.0	1.7–2.8
10–15	4–6	3.5–5.0	3.0–4.5	2.0–3.0	1.2–1.7
5–10	2–4	1.7–3.5	1.5–3.0	1.0–2.0	0.5–1.2

[a]From Gol'dberg and Iûdin (1979).

Formation waters result in three types of alteration, all important in the reduction of an original crude oil accumulation to a heavy oil reservoir or bitumen deposit. The first of these is inorganic oxidation, probably mainly operative in near-surface deposits but continuing in play indefinitely. The process involves large amounts of molecular oxygen. Of much greater significance are two other effects of water-induced alteration: water washing and biodegradation.

Water washing is effective because of the differing solubilities of hydrocarbons in water at various boiling temperatures and at various salinities. Thus, Bailey et al. (1973), in a study of a sequence of Canadian fields, demonstrated that as the formation waters became fresher, the oils showed lower solution gas:oil ratios, lower contents of hydrocarbons in the gasoline range (C_4–C_7), and also lower contents of hydrocarbons boiling below 270°C. Hydrocarbons in the gasoline range decrease by a factor of 9, and those in the boiling range below 270°C by a factor of 3. Thus, the heavier hydrocarbons are less affected than the light ones by water washing.

The other effect of water-induced alteration is bacterial degradation, which occurs because bacteria can metabolize most types of hydrocarbons. Although aromatics and naphthenes are included, the normal paraffins are the most susceptible. In the suite of reservoirs studied by Bailey et al. (1973), the saturates decreased from 47.1 to 19.1% of the C_{15-} fraction in the direction of fresher formation water, but the N—S—O compounds increased 100%, and the asphaltenes, 200%. They also found an increase in percentage of sulfur, caused largely by removal of the sulfur-free saturates. There is a relationship between the increases in sulfur and asphaltene contents: a small fraction of the total sulfur apparently was actually added during the alteration process.

Alteration resulting from incursion of meteoric waters may have an important effect on the assessment of oil resources. This is clearly evident in the reservoirs in the eastern part of the Western Canada basin. The continuity of a geologic play may depend as much on oil-to-oil correlations as on structural or stratigraphic relationships.

Case Studies

A discussion of the assessment of undiscovered heavy oil and bitumen can be approached through a review of work

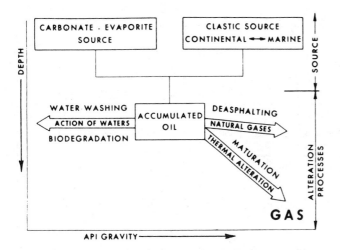

Figure 10. Model for predicting oil quality. Original oil available for migration from source rocks has an API gravity of 18–27° and contains < 1.5% sulfur. Oils outside these limits have been altered in the reservoir, as shown in the model. From Bailey et al. (1973).

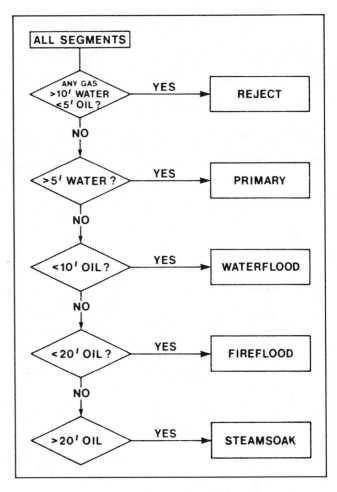

Figure 11. Schematic filter for computer processing of reservoir segments. From McCrossan et al. (1981).

that has been done in several areas. For heavy crude oil, the Western Canada basin and the North Slope of Alaska serve as good examples. The greatest accumulation of extra-heavy crude oil is found in the Orinoco oil belt of the Eastern Venezuela basin. By far the Soviet geologists have given the greatest attention to natural bitumen.

Heavy Crude Oil. More than 1.3 trillion bbl of bitumen and heavy oil are in place in the Cretaceous sandstones of Alberta and Saskatchewan, Canada. Most of the oil has been identified, but where well control is scant, it is as yet undiscovered. The results of the resource appraisal are given in detail by McCrossan et al. (1981). We will briefly review here the procedures used and results produced.

The stratigraphic section was divided into 20 sand–shale couplets, each representing a cycle of transgression and regression. Using well control, McCrossan and co-workers defined areas of richness, with the richest areas having more than 1,750 bbl/acre-ft. They also found that, out of about 1,200 pools in the area, half were delineated by only one well, but totaled, they indicated a homogeneous population. Dealing with the discovered pools, they divided each into segments according to recoverability of the oil by various methods, using the chart shown in Figure 11. This yielded 9,000 segments to which they applied probability distributions to give the identified reserve base and subeconomic resources (Figure 12A). The lower curve, the composite of the others, indicates as much as 30 billion bbl of oil in place with about 2.5 billion bbl, or 10%, recoverable. Finally, the identified oil data were processed with data estimated to fall in the category of the inferred reserve base plus hypothetical data, all based on projected plays, to yield the probability distributions given in **Figure 12B.** In this case, the uncertainty was expressed both by the probability distributions and the bands around the curves associated with the methods, which were not defined by McCrossan et al. (1981), for estimating the undiscovered resources and the use of other recovery combinations. The

result was that there may be about 30–40 billion bbl of oil in place in the area studied by McCrossan et al. (1981) and about 2.5–4.5 billion bbl of recoverable oil.

The work of the Canadian geologists in advancing the techniques of resource assessment has been important. This is especially true for the heavy hydrocarbons. Their work has been important because they account for several complicating factors: pool size, resource estimates based on individual pools or prospects, both technological and economical recoverability, and probabilities.

Extra-Heavy Crude Oil. Extra-heavy crude oil is defined as that oil with an API gravity of < 10° (specific gravity of 1.0) and a viscosity of < 10,000 cP. If a presently producing field contains oil that meets the density criterion, it is assumed that it also meets the viscosity criterion. Studies indicate that very little oil being produced today outside of Venezuela is extra heavy. For those places where it is, the same method for estimating the undiscovered heavy oil resource can be applied to extra heavy oil. The only known extensive resource of extra heavy oil is found in the Orinoco oil belt of the Eastern Venezuela basin. Recent studies have defined the hypothetical resource with far greater precision than has been done in the past.

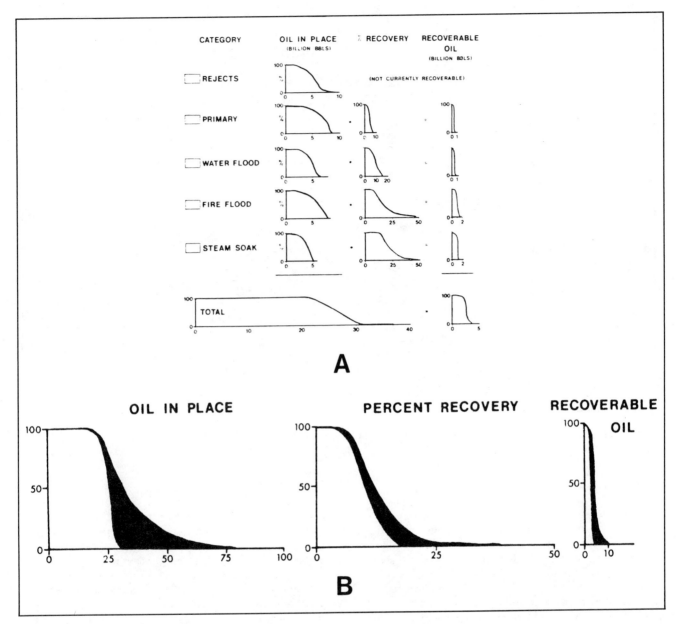

Figure 12. (A) Cumulative probability distributions of estimates of oil occurring in reservoir categories amenable to various recovery techniques showing proportion recoverable. (B) Estimated total heavy oil resource in Lloydminster area of Alberta including unexplored areas. From McCrossan et al. (1981).

The geology of the Orinoco oil belt is explained by Alayeto and Louder (1974) and Valera (1981) and is outlined by Fiorillo (1984a, b). The belt is an area of 54,000 km² (20,800 sq mi) lying along the north side of the Orinoco River. The oil is contained in Miocene sedimentary rocks interpreted to have been deposited in a nearshore environment. The source of the oil lies in the deep basin to the north, and the oil migrates southward and updip toward the craton. Distribution of oil gravity values (Figure 13) indicates the intensity of alteration of the oil through water washing and biodegradation.

The oil varies from 4 to 17° (API gravity) and has 3–4 wt. % sulphur and 500 ppm of vanadium and nickel. The pay zones average 50 m (160 ft) in thickness at depths of 150–1,300 m (500–4,300 ft), with a porosity of 30–34%, a water saturation of 10–25%, an oil saturation of 80%, permeability of 1,000 md, and a reservoir temperature at 450 m (1,500 ft) of 127°F.

For purposes of evaluating the resource, cut-off values were established for the pertinent reservoir parameters to permit machine calculation; these values varied with each of the four major areas. From the reservoir data, a net oil sand thickness map was prepared to show oil in place (Figure 14), yielding for the entire belt an estimate of 1,182 billion bbl.

On the basis of these data, two priority areas were established that emphasized oil-saturated sand thickness and

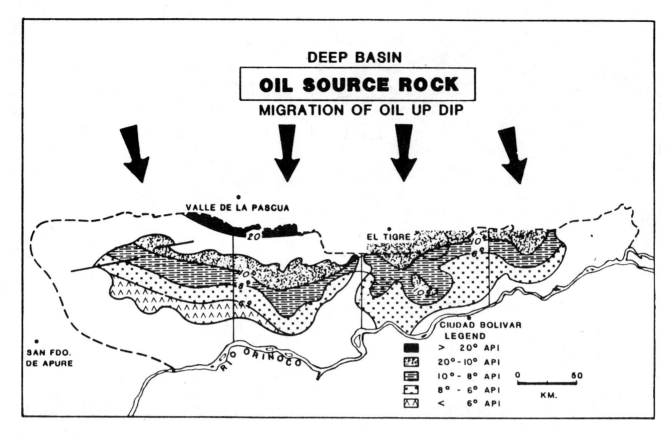

Figure 13. Diagram showing the Orinoco oil belt migration pattern. The oil source rock is located to the north, in the deep basin. From Fiorillo (1984b).

estimated economics of recovery; these are centered on Zuata and Cerro Negro, two of the four exploration areas established by Petroleus de Venezuela for its operating companies. The priority areas contain 700 billion bbl of oil in place, and the nonpriority areas contain only 500 billion bbl because of such factors as high viscosity, intermediate water zones, and structural complexity. Oil in place was converted to a reserve base by estimating an average recovery of 31% in the priority areas and 10% in the other areas, giving 217 billion bbl and 50 billion bbl, respectively. The average recovery for the priority areas was calculated by 7% from solution gas combined with cyclic steam injection, 0–8% from compaction, and another 20% from enhancement by steam drive. The goal was stated to be 500,000 bbl per day by the year 2000.

The experience with the Orinoco belt is instructive because it typifies on a large scale the problems of assessing such deposits. The fluvial deltaic, nearshore, and shallow water clastic depositional environments give rise to widespread but erratic net oil saturations that are most easily assessed as single volumetric reservoirs. As data accumulates, a single reservoir may be reduced to plays and prospects, as was seen in the case of the Lloydminster complex.

A large-scale example is presently unfolding on the North Slope of Alaska, where a heavy oil deposit of Cretaceous and Tertiary age overlying the Kuparuk oil field is being defined (Werner, 1984). First penetrated in 1969, the reservoir has since been logged in 23 exploratory wells over nearly 780 km² (300 sq mi), but only rarely has it been cored and tested.

Preliminary estimates of oil in place are as high as 40 billion bbl. The rocks apparently represent a shallow marine, deltaic complex. The oil is believed to be from the same source rocks as the deeper Kuparuk, Sadlerochit, and Lisburne pay zones, but it has been degraded through loss of straight-chain n-alkanes. The API gravity and the salinity increase in successively deeper reservoirs, reflecting incursion of fresh meteoric waters and bacteria. The eventual development of this resource will almost certainly follow the pattern of similar deposits. Steam-enhanced recovery will pose special problems because of the permafrost and other environmental considerations, but adequate natural gas is available for generating steam. Conventional oil is also available for diluting oil deposits for transporation.

Bitumen. Many bitumen deposits, most prominent of which is the Athabasca deposit in Canada, are already known and have been or are being evaluated. Most of these are listed in Meyer et al. (1984). Although none compare in magnitude with the Athabasca deposit, many are potentially of commercial interest. The Soviet Union appears to be the largest remaining area of poorly understood identified deposits and undiscovered resources. Because of this, Soviet geologists have researched extensively for ways of assessing these resources.

Lavrushko (1984) formulated the general approach to the characterization of bitumen deposits in the Soviet Union, followed by Valeev et al. (1979) and Khalimov et al. (1983). This approach requires information on bitumen genesis and

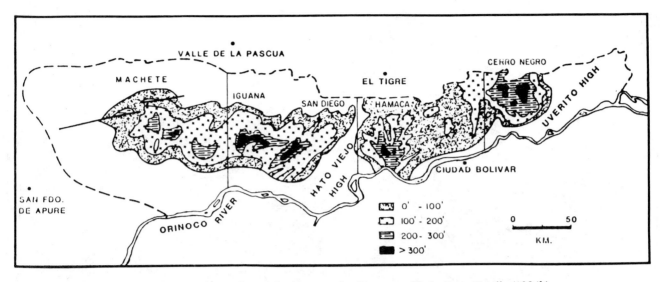

Figure 14. Map of net oil sand thickness in the Orinoco oil belt. From Fiorillo (1984b).

migration, ways in which the deposits are formed, basin classification, regional geology, and observed regularities in the distribution of bitumen deposits, especially with respect to the structure and mode of origin.

The formation of bitumen is a normal phenomenon of the process of oil formation. It is the final stage in the process (not always reached) in which oil pools are destroyed in the zone of oxidation (hypergenic or supergenic zone), giving rise to a suite of new hydrocarbons (Table 1). Thus, bitumen may be associated with both present and ancient hypergenic zones, which are the upper limit of conventional oils. The ratio of degraded oil reserves to the total amount of oil that existed in a basin reflects the degree of hypergeneity of the basin in question (Lavrushko, 1984). Basins, therefore, can be subdivided into predominantly oil–gas bearing (hypergeneity degree < 10%), bitumen–oil–gas bearing (10–50%), oil–bitumen bearing (50–90%), and bitumen bearing (> 90%). The degree of hypergeneity is closely related to the structural style of the various sedimentary basins.

Valeev et al. (1979) list four main types of bitumen–oil–gas basins. In the Soviet Union, most of the bitumen is found in foredeep bitumen-bearing basins (16%), platform basins (78%), and activated (structurally negative) basins, mostly grabens and other fault-related structures. Figure 15 is a map showing the bitumen deposits in the Soviet Union.

The richest areas of bitumen accumulations are associated with ancient platforms. The oil-forming process is most mature in these areas, and the degree of hypergenic destruction is the greatest, whereas younger platforms are mainly oil and gas bearing. Some large accumulations of bitumen are related to basins of the structurally negative, fault-related group, small ones to the orogenic group, and a few very small ones to recent geosynclines, especially along the border of the Pacific Ocean, as at Sakhalin. The platform-related basins are characterized mainly by stratiform deposits. The large Olenekskoye deposit (Figure 16) is an example of a platform-related basin accumulation in East Siberia.

The method used by Lavrushko (1984) for estimating

hypothetical bitumen resources is based on analogy to a known area. His formula is

$$Q = S \times q \times K$$

where Q is the hypothetical bitumen resources, S is the area of the bituminous unit being estimated, q is the density of bitumen reserves of a reference area, and K is the analogy factor that permits consideration of differences between reference areas and areas to be evaluated. The areas selected for determining Q require: (1) bitumen units with at least 1 wt. % bitumen at depths of < 500 m (1,640 ft) and with known geologic structure; (2) determination of thickness of bitumen unit and grade of individual beds; and (3) delineation of a reference area similar in structure and in size.

After calculation of Q, the area is subdivided into prospects. Areas of very fair prospects have a resource density 10^5 tons/km² and a grade of > 5 wt. % bitumen. Fair prospects have densities of 10^3–10^5 tons/km², and those with poor prospects have densities of < 10^3 tons/km². The density and grade values were determined from data from countries where bitumen deposits are being exploited.

Akishev et al. (1974) found it difficult to assess bitumen resources for planning purposes because of the lack of standardized parameters, that is, minimum values of thickness, porosity, and grade. They found that in the Tatar Republic in the Soviet Union, which includes the Melekess depression and the Tatar arch, most bitumen deposits are closely related to present structures, and within a trap there is one bitumen/water contact. Knowledge of the Permian bitumen deposits comes mainly from drilling down to the underlying conventional deposits. The deposits become larger and more numerous between the lower and upper parts of the Permian stratigraphic section. On the basis of known occurrences, study of other structures, and projection of potential deposits, Akishev et al. (1974) prepared a prospective map of Tatar (Figure 17).

Gol'dberg et al. (1981) have approached the prediction of hydrocarbons on the Siberian platform by studying the tectonic development of the platform and the distribution of cap rocks. They have analyzed the composition of the

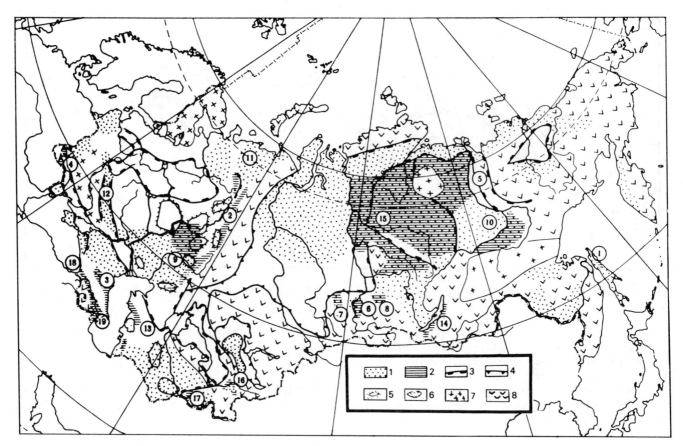

Figure 15. Bitumen–oil–gas basins of the Soviet Union. (1) Regions of oil and gas accumulation; (2) regions of bitumen accumulation; (3) boundaries of principal tectonic elements; (4) boundaries of young platforms; (5) domes, systems of arches; (6) basins; (7) shields; and (8) folded structures. Bitumen–oil–gas basins: (1) Sakhalin–Okhotsk; (2) Cis–Ural; (3) Cis–Caucasus; (4) Cis–Carpathian; (5) Cis–Verkhoyansk; (6) Southern and Northern Minusinsk; (7) Kuznetsk; (8) Rybinsk; (9) Northern Caspian; (10) Vilyuy; (11) Timan–Pechora; (12) Dnepr–Donets; (13) Southern Mangyshlak; (14) Baikal; (15) Tungusska; (16) Ferghana; (17) Afghan–Tadzhik; (18) Eastern Black Sea; and (19) Southern Caspian. From Valecv et al. (1979).

hydrocarbons and bitumens and the evolution of oil in the rocks of various ages. They found that petroleum evolution apparently began in the platform-associated Tatar basins as early as Cambrian time. The areas of present bitumen accumulation are those where the thick salt cap is absent, permitting degradation of the oil.

Preliminary estimates of the bitumen reserves of central Asia have been made, indicating that they are large but not yet defined. Khaimov et al. (1982) observe that the oil and bitumen are found mainly in gravity segregated anticlinal traps. Although the pools of each have attributes in common, they also differ in many respects. This is a result of the bitumen being formed in the zone of oxidation, which results in uneven dissemination and substantial amounts of water saturating lenses and intercalated zones. The deposits are characterized by inclined and uneven bitumen/water contacts. These factors, along with the geologic structure, nature of the reservoir, type of bitumen, and depth of occurrence, govern the recovery method, the percentage recovery, and the category and amount of reserves. Khaimov et al. (1982) point out that little actual prospecting for bitumen has been conducted even though bitumen-saturated rocks are known to occur at economically recoverable

depths. They suggest that in the exploration process, special attention should be paid to the content of rare and disseminated elements in the bitumens, because metals are already being recovered as a by-product with oil in some deposits.

The Soviet geologists generally place strong emphasis on their assessments on the presence of cap rock, either gypsum–anhydrite or shale, and on structural control. The prospect assessments are usually volumetric, and regional totals are based on trends (plays) or on regional analogs. They apparently do not use conditional or marginal probabilities.

RECOVERY

The process of recovery is an integral part of the resource assessment of heavy oil and bitumen, both technologically and economically. The technology must be able to handle both removal of the material from the ground and upgrading to a petroleum liquid by accepted methods. Upgrading of the raw bitumen by a carbon rejection process entails a 25% loss.

Factors related to surface mining profitability of bitumen include exploration cost, waste:ore ratio, grade, and plant

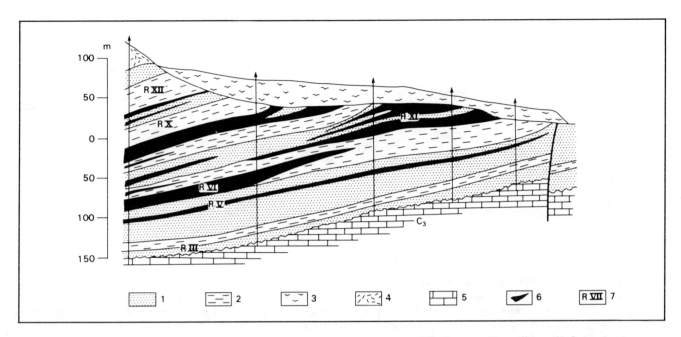

Figure 16. The Olenekskoye oil field (eastern Siberia) geologic cross section: (1) sandstones; (2) argillites; (3) Quaternary deposits; (4) tuffaceous sandstones; (5) carbonate rocks; (6) bituminous rocks; and (7) paybeds of Permian age. From Khalimov et al. (1983).

size (Earley, 1984). If it is assumed that no exploration cost is involved, there remains the core drilling, logging, and sample analysis costs involved in determining the waste:ore ratios and ore grade. Once the deposit is delimited, it is necessary to allow for a 25% mining loss, as well as losses for settling ponds, surface facilities, and spoil piles. Earley (1984) showed that, down to opencast mining depth limits of about 90 m (300 ft), waste:ore ratios are not critical (Figure 18). Grade, however, is very important. Minimum values of about 8 wt. % bitumen or about 1,300 bbl/acre-ft are required, with an extraction plant capable of processing at least 15,000 bbl/day. Figure 19 shows the sensitivity of the rate of return on grade and plant size; small plants, feasible for deposits with small resources, require very rich deposits.

Shallow, abandoned oil reservoirs can also be surface mined. After being produced, such fields have already lost up to one-third of the oil in place. If the oil were heavy, these reservoirs would surely have been subject to enhanced recovery, which would have reduced the ore grade and may have ruined the reservoir. This certainly is the case following fire, flood, or wet combustion.

Reservoirs can also be mined by underground methods in which the bitumen-bearing rock is lifted to the surface for processing as though it were opencast mined. This eliminates consideration of overburden but places a premium on reservoir thickness, continuity, and grade (Resnick et al., 1981). Competency of the rocks in the stratigraphic section and depth are significant; at about 150 m (500 ft) depth, soft shales and mudstones undergo plastic deformation and more indurated shales do so at 300 m (1,000 ft). Although as much as about 90% of the hydrocarbon in place may be recoverable, Dvorets et al. (1984) estimate that lifting rock having a grade of 7–10% is even less profitable than mining low Btu coal.

A third option is mine-assisted, *in situ* recovery. The best example of this is in the Yaregskoye field in the Soviet Union (Dvorets et al., 1984). The sandstone reservoir here is 200 m (660 ft) deep, 17 m (56 ft) thick, and has an average permeability of 3,000 millidarcys and a reservoir temperature of 45–60°F. The oil is 19° (API gravity) and has a viscosity of $1.5–2 \times 10^4$ cP. Attempts to produce the field during the 1930s with conventional wells on a spacing of 1.25–2.5 acres resulted in recovery of only 2% of oil in place. Beginning in 1939, shafts were sunk and mines developed in a tuff deposit above the reservoir, to which vertical and slant wells were drilled. The system of slanting holes reduced the volume of excavation by two-thirds (Baibakov and Garushev, 1981). Oil recovery was then added to the recovery scheme, and the slant holes were then used for recovery and the vertical holes for injection. Recovery of 57–58% of oil in place is presently being achieved from three mines, with a production of about 4 million bbl per year.

Hydraulic mining may provide an alternative approach to recovery of bitumens and heavy oils in sandstone reservoirs at depths of 30–180 m (100–600 ft). In this process, a borehole is drilled from the surface into the reservoir, at which point a hydraulic tool uses high-pressure water jets to create a slurry, which is then pumped to the surface (Wagner and Hodges, 1984). In theory, such a method could recover 100% of the oil in place within the cavern created around the jet tool.

A number of methods produce heavy oils and bitumen *in situ*. Of these, only three are considered to be economically viable at this time: cyclic steam injection, steam flooding, and *in situ* combustion. The intent of these is to improve or induce recovery through viscosity reduction.

Enhanced oil recovery has been summarized by the National Petroleum Council (1984). This report briefly

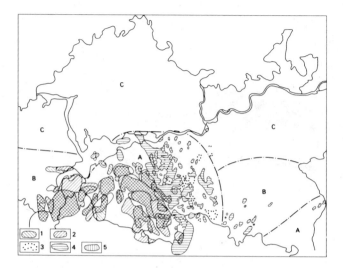

Figure 17. Map of prospects for bitumen occurrences in the Permian deposits of the Tatar Republic, Soviet Union. Bitumen occurrences: (1) in the lower thickness of the upper Kazanian substage; (2) in the upper thickness of the upper Kazanian substage; (3) in the Ufimian deposits of the Upper Permian; (4) in the Sakmarian deposits of the Lower Permian; (5) in the Asselian deposits of the Lower Permian. Areas: (A) highly promising for prospecting for bitumen occurrences in the Permian deposits; (B) promising for prospecting for bitumen deposits; and (C) undetermined prospects. From Akishev et al. (1974).

describes the steam stimulation method. A prospective reservoir is expected to have some natural reservoir energy in the form of solution gas or gravity. Steam (5,000 to 15,000 bbl on a cold water basis) is injected into a well during a 1–4 week period. After a variable period of time, the well is produced for as long as 1 year. This cyclic steam process may recover about 15% of oil in place. Then continuous steaming—the steam drive or flood—is commenced. The hydrocarbons are either pulled or driven to the production wells. Oil recovery, including that from the cyclic steaming, may reach a maximum of 65–70% of oil in place in ideal reservoirs and should average at least 40%. With the steam-enhanced recovery processes, the equivalent of one barrel of produced oil must be burned to generate steam for each 3–4 bbl produced, yielding a net of 2 or 3 bbl. This, therefore, imposes a 25–30% reduction in the size of the recoverable hydrocarbon resource.

CONCLUSIONS

Except for a few accumulations of very immature heavy oil, bitumen and heavy oil deposits are the altered remnants of previously larger conventional oil deposits. The alteration takes place in the zone of oxidation through the action of fresh water, which dissolves light-oil components by water washing, and degrades the oil through introduction of bacteria which can consume hydrocarbons. Other reduction takes place inorganically by deasphalting, the process of

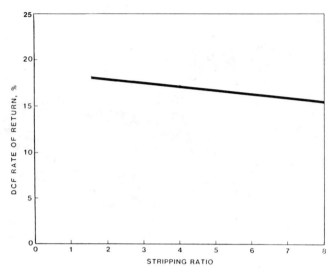

Figure 18. Mineable tar sands—sensitivity of discounted cash flow rate of return (DCF) to stripping ratio. From Earley (1984).

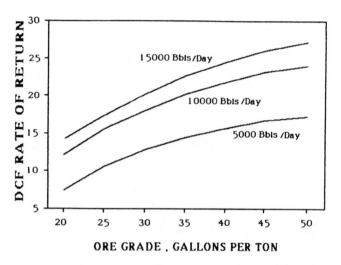

Figure 19. Mineable tar sands—sensitivity of the DCF to plant size. From Earley (1984).

precipitation of asphaltenes by natural gas from solution in the oil. The removal of the light oil components and gas effectively decreases the natural reservoir energy, in many cases to zero.

Identified bitumen and most heavy oil deposits are assessed using calculations similar to those for ore deposits. Saturations, pay thickness, and pay extent are critical parameters for obtaining assessments of oil in place. Nearly all bitumen deposits crop out or are in the shallow subsurface. Thus, a few countries probably still have undiscovered bitumen deposits. Most heavy oil was discovered and bypassed in the search for conventional petroleum deposits. Only recently have geologists focused attention on delineating the undiscovered extent of heavy oil deposits.

REFERENCES CITED

Akishev, I. M., R. K. Muslimov, N. P. Lebedev, and X. X. Troepol'skii, 1974, Bitumnye zalezhi permskikh otlozhenii Tatarii, perspektivy ikh poiskov i razvedki [Bitumen occurrences in the Permian deposits of Tatary, outlook for prospecting for and exploring them]: Geologiia nefti i gaza, n. 3, p. 23–28.

Alayeto, E. M., and L. W. Louder, 1974, The geology and exploration potential of the heavy oil sands of Venezuela (the Orinoco petroleum belt), in L. V. Hills, ed., Oil sands—fuel of the future: Canadian Society of Petroleum Geologist Memoir 3, p. 1–14.

Baibakov, N. K., and A. R. Garushev, 1981, Teplovye metody razrabotki neftianykh mestorozhdenii [Thermal methods of oil field development]: Moscow, Nedra, 286 p.

Bailey, N. J. L., H. R. Krouse, C. R. Evans, and M. A. Rogers, 1973, Alteration of crude oil by waters and bacteria—evidence from geochemical and isotope studies: AAPG Bulletin, v. 57, n. 7, p. 1276–1290.

Baker, R. A., et al., 1984, Geologic field number and size assessments of oil and gas plays: AAPG Bulletin, v. 68, n. 4, p. 426–437.

Bishop, R. S., H. M. Gehman, Jr., and A. Young, 1983, Concepts for estimating hydrocarbon accumulation and dispersion: AAPG Bulletin, v. 67, n. 3, p. 337–348.

Britton, M. W., 1984, Problems frequently encountered in evaluating tar sand resources—example: the South Texas San Miguel deposits: AAPG Research Conference on Exploration for Heavy Crude Oil and Bitumen, Santa Maria, California, 23 p.

Claypool, G. E., A. H. Love, and E. K. Maughan, 1978, Organic geochemistry, incipient metamorphism, and oil generation in black shale members of Phosphoria Formation, western interior United States: AAPG Bulletin, v. 62, n. 1, p. 98–120.

Connan, J. and H. Coustau, 1984, Influence of geological and geochemical characteristics of heavy oils on their recovery: AAPG Research Conference on Exploration for Heavy Crude Oil and Bitumen, Santa Maria, California, 31 p.

Dolton, G. L., et al., 1981, Estimates of undiscovered recoverable conventional resources of oil and gas in the United States: U.S. Geological Survey Circular 860, 87 p.

Dusseault, M. B., and J. D. Scott, 1984, Coring and sampling in heavy oil exploration: difficulties and proposed cures: AAPG Research Conference on Exploration for Heavy Crude Oil and Bitumen, Santa Maria, California, 15 p.

Dvorets, N. L., V. A. Sorokin, and M. L. Surguchev, 1984, Practical aspects of recovering heavy crude, bitumen and oil shale, in R. F. Meyer, J. C. Wynn, and J. C. Olson, eds., The future of heavy crude and tar sands, Second International Conference: New York, McGraw-Hill, p. 265–269.

Earley, J. W., 1984, Economic factors in near-surface heavy oil/tar and mining: AAPG Research Conference on Exploration for Heavy Crude Oil and Bitumen, Santa Maria, California, 12 p.

Energy, Mines and Resources Canada, 1977, Oil and natural gas resources of Canada, 1976: Energy, Mines and Resources Canada Report 77-1, 76 p.

Energy Resources Conservation Board, 1982, Alberta's reserves of crude oil, gas, natural gas liquids, and sulphur: Energy Resources Conservation Board, Alberta, ERCB-82-18, p. 3–1 to 3–4.

Fiorillo, G. J., 1984a, Exploration of the Orinoco Oil Belt: Review and general strategy, in R. F. Meyer, J. C. Wynn, and J. C. Olson, eds., The future of heavy crude and tar sands, Second

International Conference: New York, McGraw-Hill, p. 304–312.

——— , 1984b, Exploration and evaluation of the Orinoco Oil Belt: AAPG Research Conference on Exploration for Heavy Crude Oil and Bitumen, Santa Maria, California, 26 p.

Gol'dberg, I. S., and G. T. Iûdin, 1979, Voprosy klassifikatsii, obrazovaniiâ i razmeshcheniiâ skoplenii bitumov [Problems of the classification, formation and distribution of bitumen accumulations], in V. I. Ditmar, ed., Geologiia bitu mov i bitumovmeshchaiushchikh porod [Geology of Bitumens and bitumen-containing rocks]: Moscow, Nauka, p. 15–20.

Gol'dberg, I. S., B. A. Lebedev, and B. M. Frolov, 1981, Razdel'nyi prognoz razmeshcheniiâ gaza, nefti i bitumov na Sibirskoi platforme [Separate prediction of the distribution of gas, oil and bitumens on the Siberian platform]: Geologiia nefti i gaza, n. 2, p. 22–26.

Herkenhoff, E. C., 1972, When are we going to mine oil?: Engineering and Mining Journal, v. 173, n. 6, p. 132–137.

Hickerty, R. S., G. S. Jones, J. E. Klovan, and P. E. Putnan, 1984, Geological assessment of Lloydminster heavy oil reservoirs using a computerized data base: AAPG Research Conference on Exploration for Heavy Crude Oil and Bitumen, Santa Maria, California, 11 p.

Jha, K. N., and Arun Berma, 1984, Heavy oil development in Saskatchewan, in R. F. Meyer, J. C. Wynn, and J. C. Olson, eds., The future of heavy crude and tar sands, Second International Conference: New York, McGraw-Hill, p. 205–211.

Khaimov, R. N., Smol'nikov, I. R., Pen'kova, et al., 1982, Bitumy Srednei Azzi—vozmozhnii istochnik dopolnitel'nogo uglevodorodnogo syr'ia [Bitumen of central Asia—a possible source of additional hydrocarbon raw material]: Neftegazovaia Geologiia i Geofizika, n. 2, p. 20–26.

Khalimov, E. M., I. M. Klimushin, L. I. Ferdman, and I. S. Gol'dberg, 1983, Geological problems of natural bitumen: Londo, England, 11th World Petroleum Congress Panel Discussion 1, 114 p.

Lavrushko, I. P., 1984, The problem of evaluating resources of bitumen and viscous oils and the state of its study in the U.S.S.R., in R. F. Meyer, J. C., Wynn, and J. C. Olson, eds., The future of heavy crude and tar sands, Second International Conference: New York, McGraw-Hill, p. 259–264.

Lewin and Associates, 1984, Major tar sand and heavy oil deposits of the United States: Oklahoma City, Interstate Oil Compact Commission, 272 p.

Martinez, A. R., 1984, Report of working group on definitions, in R. F. Meyer, J. C. Wynn, and J. C. Olson, eds., The future of heavy crude and tar sands, Second International Conference: New York, McGraw-Hill, p. lxvii–lxviii.

McCallum, R. G., 1981, Geology of Lloydminster play, Alberta, in R. F. Meyer, J. C. Wynn, and J. C. Olson, eds., The future of heavy crude and tar sands, Second International Conference: New York, McGraw-Hill, p. 223–236.

McCrossan, R. G., Procter, R. M., and Ward, W. J., 1981, Studies of heavy oil identity characteristics of divided areas: Oilweek, v. 32, n. 34, p. 34–39.

McDowell, A. N., 1975, What are the problems in estimating the oil potential of a basin?: Oil and Gas Journal, v. 73, n. 23, p. 85–90.

Meyer, R. F., P. A. Fulton, and W. D. Dietzman, 1984, A preliminary estimate of world heavy crude oil and bitumen resources, in R. F. Meyer, J. C. Wynn, and J. C. Olson, eds., The future of heavy crude and tar sands, Second International Conference: New York, McGraw-Hill, p. 97–158.

Minken, D. F., 1974, The Cold Lake oil sands: geology and a reserve estimate, in I. V. Hills, eds., Oil sands—fuel of the future: Canadian Society Petroleum Geologists Memoir 3, p. 84–99.

Moshier, S. O., and D. W. Waples, 1985, Quantitative evaluation of Lower Cretaceous Mannville Group as source rock for Alberta's oil sands: AAPG Bulletin, v. 69, n. 2, p. 161–172.

National Petroleum Council, 1984, Enhanced oil recovery: Washington D.C., National Petroleum Council, 96 p.

Orr, W. L., 1978, Geochemistry of asphaltic Monterey oils from the Santa Maria basin and Santa Barbara Channel area offshore (abs.): American Chemical Society 198th National Meeting, St. Louis, Missouri.

Outtrim, C. P., and R. G. Evans, 1977, Alberta's oil sands reserves and their evaluation: Petroleum Society of Canadian Institute of Mining and Metallurgy 28th Annual Technical Meeting, 41 p.

Palacio, D. N., 1957, Preliminary report on the geology and rock asphalt deposits of Balite, Villaba, Leyte: Philippine Geology, v. 11, n. 3, p. 69–100.

Pearson, M., 1981, Oil sands: reservoir or orebody?, in R. F. Meyer, and C. T. Steele, eds., The future of heavy crude oil and tar sands, First International Conference: New York, McGraw-Hill, p. 295–300.

Petersen, N. F., and P. J. Hickey, 1984, California Plio-Miocene oils: evidence of early generation: AAPG Research Conference on Exploration for Heavy Crude Oil and Bitumen, Santa Maria, California, 13 p.

Procter, R. M., P. J. Lee, and G. L. Taylor, 1982, Methodology of petroleum resource evaluation: in Third Circum-Pacific Energy and Mineral Resources Conference, Petroleum Resource and Assessment Workshop and Symposium: Geological Survey of Canada Manual, 59 p.

Raicar, M., and R. M. Procter, 1984, Economic considerations and potential of heavy oil supply from Lloydminster—Alberta, Canada, in R. F. Meyer, J. C. Wynn, and J. C. Olson, eds., The future of heavy crude and tar sands, Second International Conference: New York, McGraw-Hill, p. 212–219.

Resnick, B. S., D. H. Dike, L. M. English, III, and A. G. Lewis, 1981, Evaluation of tar sand mining; Volume I—An assessment of resources amenable to mine production: U.S. Department of Energy Report DOE/ET/3020-1, 214 p.

Strom, N. A., and R. B. Dunbar, 1981, Bitumen resources of Alberta: converting resources to reserves, in R. F. Meyer, and C. T. Steele, eds., The future of heavy crude oils and tar sands, First International Conference: New York, McGraw-Hill, p. 47–60.

Tannenbaum, E., A. Starinsky, and Z. Aizenshtat, 1984, Light oils transformation to heavy oils and asphalts—assessments of the amounts of hydrocarbons removed and the hydrological–geological control of the process: AAPG Research Conference on Exploration for Heavy Crude Oil and Bitumen, Santa Maria, California, 16 p.

Tissot, B. P., 1984, Recent advances in petroleum geochemistry applied to hydrocarbon exploration: AAPG Bulletin, v. 68, n. 5, p. 545–563.

Troepol'skii, V. I., ed., 1976, Permskie bitumy Tatarii [The Permian bitumens of Tatary]: Kazan, Izdatel'stvo Kazanskogo Universiteta, p. 25–164.

U.S. Bureau of Mines and U.S. Geological Survey, 1980, Principles of a resource–reserve classification for minerals: U. S. Geological Survey Circular 831, 5 p.

Valeev, R. N., G. T. Iûdin, R. M. Gismatullin, and V. L. Shteingol'ts, 1979, Bitumoneftegazonosnye basseiny [Bitumen–oil–gas basins], in V. I. Ditmar, ed., Geologiia bitumov i bitumovmeshchaiushchikh porod [Geology of bitumens and bitumen-containing rocks]: Moscow, Nauka, p. 3–14.

Valera, R., 1981, The geology of the Orinoco heavy oil belt: an integrated interpretation, in R. F. Meyer, and C. T. Steele, eds., The future of heavy crude oils and tar sands, First International Conference: New York, McGraw-Hill, p. 254–263.

Wagner, C. G., and E. L. Hodges, 1984, Downhole hydraulic mining system (abs.): AAPG Research Conference on Exploration for Heavy Crude Oil and Bitumen, Santa Maria, California, 1 p.

Webber, C. E., 1961, Estimation of petroleum reserves discovered, in G. B. Moody, ed., Petroleum exploration handbook: New York, McGraw-Hill, p.25-1–25.12.

Werner, M. R., 1984, Tertiary and Upper Cretaceous heavy oil sands, Kuparuk River Unit area, Alaskan North Slope: AAPG Research Conference on Exploration for Heavy Crude Oil and Bitumen, Santa Maria, California, 20 p.

Assessment of Natural Gas from Coalbeds by Geologic Characterization and Production Evaluation

Raoul Choate and John P. McCord
TRW Exploration and Production
Lakewood, Colorado

Craig T. Rightmire
TRW Exploration and Production
Houston, Texas

The relative attractiveness of coalbeds as source rocks and potential gas reservoirs are functions of the amount of methane generated by coal and the amount of methane retained in coal. The amount of methane gas generated during the coalification process depends primarily on the thermal maturity (rank) of the coal and the volume (thickness and lateral continuity) of coal present. The retention of methane gas in a coalbed reservoir will depend on a combination of factors including depth (i.e., fluid pore pressure), temperature, coal rank, and the hydrogeologic environment. Given a coalbed with the proper combination of methane generation and storage, the final and most critical factor is the production potential of the reservoir. Experience has shown that short duration testing (drill stem testing) does not provide the necessary data to address this issue. The production case histories presented here suggest that long duration production testing is the only way to adequately assess economic potential.

INTRODUCTION

The presence of methane in coalbeds has been recognized for hundreds of years as a hazard to coal mining. Appreciation of coalbeds, however, as both a source rock and a reservoir for pipeline quality methane is a recent development in the United States. The purpose of this paper is threefold:

1. to summarize important elements concerning the generation and storage of methane in coalbeds;
2. to define geologic characteristics critical to resource assessment; and
3. to present several production case histories and comment on factors relating to producibility.

Coalbeds underlie approximately 380,000 sq mi (988,000 km²) of the conterminous United States (Figure 1). Early estimates of the coalbed methane resources of the United States, conducted as part of the Unconventional Gas Studies by the Department of Energy (DOE) and the Gas Research Institute (GRI), show that this resource (gas-in-place) may be as high as 800 trillion cubic feet (tcf) (Table 1). These early estimates of the potential magnitude of the coalbed methane resource generally considered only the estimated coal resource to a depth of 3,000 ft (910 m) and an approximation of the average gas content (cubic feet per ton) of the coal. Because large quantities of coal are known to exist below a depth of 3,000 ft (910 m) and because the gas content of coal generally increases with increasing depth, the existing estimates of the magnitude of the coalbed methane resource of the United States should be considered conservative.

Currently, there is a significant and growing interest in this resource from the oil and gas industry. This interest was initially stimulated by the DOE's Methane Recovery from Coalbeds Program, which was supplemented by additional and continuing funding from the GRI. Commercial production of coalbed methane is taking place today in the Black Warrior and San Juan basins of Alabama and Colorado–New Mexico, respectively. In both these basins, the bulk of the coal resource is relatively shallow (< 3,000 ft or 910 m in depth). As industry gains confidence in its ability

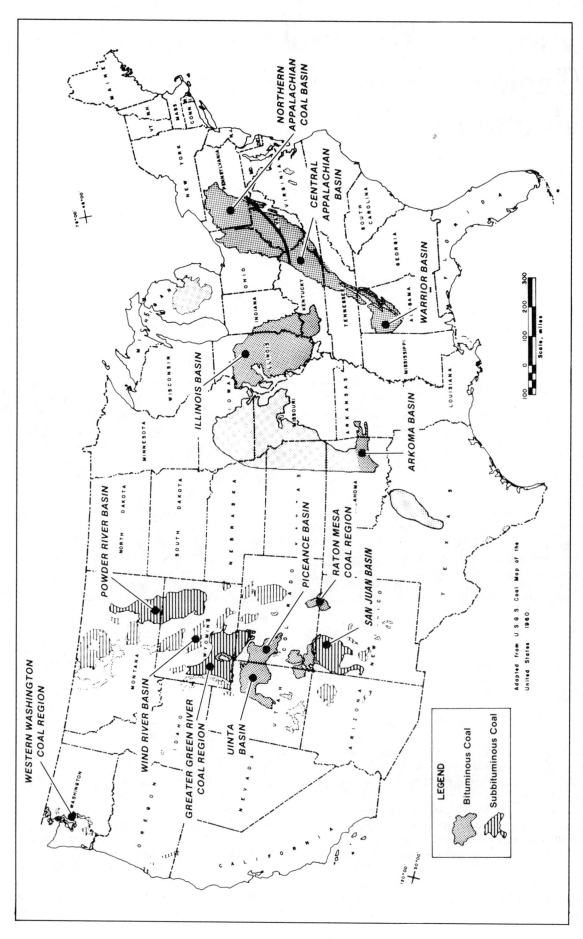

Figure 1. Map of major coal regions in the conterminous United States. From Rightmire (1984).

Table 1. Comparison of estimated total U.S. in-place coalbed methane[a].

Study	Total Resource In-place (tcf)	Recoverable Resource (tcf)
Kuuskraa and Meyer (1980)	550	40–60
National Petroleum Council (1980)	398	45[b]
Sharer (GRI) (1980)	500	10–60
Rosenberg and Sharer (GRI) (1979)	72–860	16–487
National Gas Survey Task Force on Nonconventional Natural Gas Resources Federal Energy Regulatory Commission (1978)	300–850	Not Reported
Deul and Kim (1978)	318–766	Not Reported
MRCP Basin Analysis (13 Basins)	72–400	Not Estimated

[a]From Rightmire (1984); modified from Potential Gas Committee (1981).
[b]Assumes a price of up to $9/mcf and rate of return of 10% (1979 dollars).

to solve the various problems of coalbed methane production in these basins, the trend will be toward exploitation of the potentially large resource at depths > 3,000 ft.

SOURCE ROCK/RESERVOIR CHARACTERISTICS OF COAL

Coalification

The process by which buried plant material evolves from peat to anthracite is termed *coalification*. This process is partly related to thermal maturation and results in changes in calorific value, moisture content or moisture-holding capacity, percent volatile matter, fixed carbon content, and gas content. All of these properties are altered measurably and predictably by increased temperature commonly brought about by increased depth of burial.

The formation of peat from vegetal remains has been termed *biochemical coalification* because of microbial activity. This activity occurs in both oxidizing and reducing environments where bacteria consume a portion of the carbon in the organic matter, leaving a strongly altered residual material. Diagenetic changes occur up to the lignite–subbituminous coal boundary, which is commonly observed at a formation temperature of ~ 50° C (depending on time–temperature relationships). As temperature increases above 50° C because of increased depth of burial or increasing geothermal gradient, the coal rank also increases. Rank increase is not an instantaneous phenomenon, but is related to the time that the coaly material is maintained in a given thermal regime. This time–temperature relationship dictates the level of maturity of the coal.

Most coals are composed of humic organic matter—largely oxygen-rich lignin and cellulose. Humic coal is considered a Type III kerogen. Figure 2 shows a Van Krevelen diagram with the maturation pathways of the different kerogen types, the principal products generated, the related vitrinite reflectance (R_o) value, and thermal alteration index (TAI) value. As humic coal matures (Type III kerogen), it loses oxygen and moves to the left along the Type III branch of the diagram. Because of its chemistry, hydrogen loss in humic coal material will be less than that for sapropelic material. Note that the "oil" product portion of Type III kerogen pathway is very small.

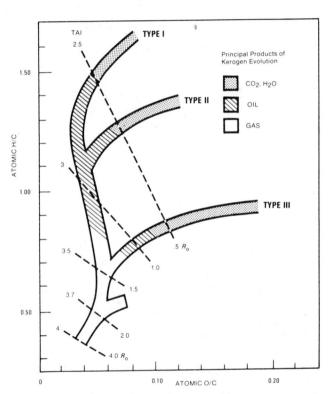

Figure 2. Kerogen maturation and type relative to vitrinite reflectance (R_o). Note long CO_2 and CH_4 windows relative to oil window for Type III kerogen. TAI is thermal alteration index. After Tissot et al. (1974).

Methane Generation by Coal

Major by-products of the coalification process are methane (CH_4), carbon dioxide (CO_2), nitrogen (N_2) and water (H_2O). During the early stages of coalification, at temperatures below 50° C, biogenic methane is produced by microbial decomposition of organic material. This gas is generally referred to as "swamp" or "marsh" gas. Where subsidence and burial are sufficiently rapid, biogenic methane can be trapped in shallow gas reservoirs.

The methane produced at temperatures above 50° C as coal rank increases is termed *thermogenic methane*. In Figure 3, Meissner (1984), using data from Juntgen and Karweil

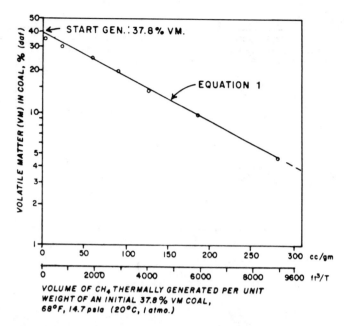

Figure 3. Graphical plot of the log of volatile matter (VM) in coal versus the linear volume of methane generated, based on data points of Karweil and Juntgen shown in Figure 4. Daf is dry ash-free basis. From Meissner (1984).

(1966), has plotted the log of volatile matter in coal versus the linear volume of methane generated. This semi-log plot results in a straight line represented by Equation (1) (Meissner, 1984):

$$\text{volume CH}_4 \text{ (cc/gm)} = -325.6 \, \log\left(\frac{\% \ \text{VM(daf)}}{37.8}\right) \quad (1)$$

It is assumed the CH_4 is at 68°F and 14.7 psia (20°C and 1 atm) and the percentage of volatile matter (VM) was measured on a dry ash-free (daf) basis. This relationship indicates that thermogenic methane generation is initiated when coal reaches 37.8% volatile matter or at the high volatile A–B bituminous boundary. By the time a coal reaches the rank of anthracite, it has generated over 9,600 cu ft/ton (300 cm³g) of methane (Meissner, 1984).

Figure 4 summarizes the volume of methane generated for various levels of volatile matter content and relates these parameters to vitrinite reflectance and coal rank. The volume of methane generated within a given area can be estimated by using either Figure 3 or 4 or Equation (1) and by knowing the coal rank and the lateral extent and thickness of the coal.

The two principal nonhydrocarbon constituents in gas generated by coal are N_2 and CO_2. Both form from the decomposition of organic material and should be expected at some levels as constituents of all coalbed gas. Nitrogen emissions begin as ammonia (NH_3) near the end of the high-volatile A bituminous stage at about 120° C (Hunt, 1979). It is often found as only a minor constituent of the produced gas because of its high solubility and because its small molecular size (3.0 Å) permits it to migrate from the

Volatile Matter VM (daf) (%)	Vitrinite Reflectance R_0 (%)	ASTM Coal Rank	Thermally Generated CH_4 (cc/g)
65±	0.23	Peat	
		Lignite	
53±	0.36	C	
50±	0.43	B	Sub-Bituminous
46.9±	0.47	A	
45.6±	0.49	C	
42±	0.51	B	
38.7	0.69	A (High Volatile Bituminous)	
(37.8)	(0.73)		0
31	1.11	Medium Volatile Bituminous	28.04
22	1.60	Low Volatile Bituminous	76.54
14	2.04	Semi-Anthracite	140.45
8	2.40	Anthracite	219.59
2 0	5.0	Meta-Anthracite	415.62

Figure 4. Summary of thermally generated methane as related to coal rank, volatile matter content on a dry ash-free (daf) basis, and vitrinite reflectance (R_o). ASTM is American Society for Testing and Materials.

system more readily than the gases of larger molecular size with which it is generated. Although a principal constituent of early thermogenic gases, CO_2 is commonly a relatively minor and extremely variable constituent in the produced gas. Although its molecular size inhibits its rapid migration as a gas, its high solubility in water significantly facilitates its mobility. Approximately 1 volume of CO_2 will dissolve in 1 volume of fresh water at conditions of 14.7 psi (1 atm) and 68° F (20° C). At 4,400 psi (300 atm) and 212° F (100° C), conditions that may be equivalent to a coalbed methane reservoir at 10,000 ft (3,050 m), ~ 30 volumes of CO_2 (STP) will dissolve in that same volume of water (Rightmire, 1984).

Hydrogen sulfide (H_2S) has been identified as being generated in trace amounts from humic source material. Because of its very high solubility in water (2.6 times that of CO_2 at 1 atm and 20° C), and because it is the last constituent of natural gas to form (starting at ~ 100° C), only trace amounts of H_2S have ever been identified in coalbed methane reservoirs (Rightmire, 1984).

Analyses of gas produced from coals of various ranks show that with some exceptions, these gases contain in > 90% methane, trace to minor amounts of higher hydrocarbons

Table 2. Desorbed gas composition and heating values for selected coalbed methane gas samples and conventional natural gas[a].

Basin	Sample Source Bed	Gas Composition					Btu/scf
		CH$_4$	C$_2^+$	H$_2$	Inerts[b]	O$_2$	
Central Appalachian	Pocahontas No. 3	96.87	1.40	0.01	2.09	0.17	1059
Northern Appalachian	Pittsburgh	90.75	0.29	—	8.84	0.20	973
Northern Appalachian	Kittanning	97.32	0.01	—	2.44	0.24	1039
Arkoma	Lower Hartshorne	99.22	0.01	—	0.66	0.10	1058
Warrior	Mary Lee	96.05	0.01	—	3.45	0.15	1068
Raton Mesa	Vermejo Fm.	95.06	0.02	—	4.91	—	963
San Juan	Fruitland Fm.	91.63	2.66	—	5.71	—	987
Piceance Creek	Mesaverde Gp.	89.84	4.24	—	5.92	—	984
Conventional Natural Gas		94.40	4.90	—	0.40	—	1068

[a]Modified from Kim (1978).
[b]Inert include N$_2$, CO$_2$ and He.

Table 3. Summary CO$_2$ surface area measurements and corresponding CO$_2$ and CH$_4$ monolayer capacities for 10 coal samples[a].

Sample Number	Coal Rank	CO$_2$ Surface Area (m^2/g) (Dry, Mineral-Containing Basis)	Monolayer Capacity (cm^3/g, STP)	
			CO$_2$	CH$_4$
1	Subbituminous	125	17.5	10.6
2	Subbituminous	107	15.8	8.8
3	Subbituminous	98	14.4	6.7
4	Subbituminous	105	15.4	8.3
5	Subbituminous	117	17.3	9.8
6	Anthracite	339	49.9	14.3
7	Bituminous	138	20.3	8.3
8	Bimuninous	179	26.4	13.9
9	Bituminous	164	21.9	6.8
10	Bituminous	169	24.9	10.9

[a]After Sato (1981).

(ethane, propane, etc.), and < 3% each of N$_2$ and CO$_2$. These coal gases have heating values of approximately 1,000 Btu/standard cubic feet (scf). Table 2 shows representative analyses of gases obtained from various coal samples (modified from Kim, 1978).

Methane Retention In Coalbeds

Methane is retained in coal in four ways: (1) as sorbed molecules on the internal surfaces or within the molecular structure of the coal, (2) as gas held within the matrix porosity (macro- and microporosity), (3) as free gas within the fracture network, or (4) as gas dissolved in groundwater within the coalbed. The first of these retention mechanisms is very important and is the least understood. The conceptual model most often employed considers methane molecules to be adsorbed onto the internal surfaces of coal. In line with this model, much work has been done with adsorption isotherms of various molecular species to calculate internal surface areas and monolayer capacities (Gan et al., 1972; Nandi and Walker, 1971). In a study by Sato (1981), total surface area and CO$_2$ and methane monolayer capacity for 10 different coal samples were calculated. Review of these data (Table 3) indicates that both surface area and monolayer capacity generally increase with increasing coal rank.

An alternate conceptual model considers that methane molecules are absorbed within the free molecular lattice of the coal. Workers who favor this model doubt that the nonpolar CH$_4$ molecule has a strong attractive force to a surface and note that the size relationships suggest that the CH$_4$ molecule will fit inside a benzene ring, which forms the lattice framework of the coal (K. C. Bowman, 1985, personal communication). Although distinction between an adsorbed and an absorbed relationship between the methane molecule and coal may be fundamentally important, much work still needs to be done in this area. Therefore, in this paper we use the generic term *sorbed* when discussing methane retention by coal in the microporosity.

Figure 5 (Meissner, 1984) shows the capacity of coal to both generate and store methane as a function of volatile matter content (rank). The generation volume (A) curve is after Juntgen and Karweil (1966) and is relative to standard temperature and pressures of 68° F (20° C) and 14.7 psi (1 atm). The total storage capacity curve (B) includes the sum of the volume held by sorbtion on molecular surfaces or within molecular space (C) and the volume held within the matrix porosity (D). The total storage capacity curve is relative to temperature and pressure conditions of 212° F (100° C) and 14,700 psi (1,000 atm) as related to the appropriate coal

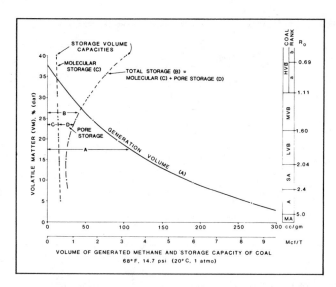

Figure 5. Graph showing volumes of methane generation and storage capacity for coals of various ranks, as characterized by volatile matter content and R_o. The generation volume curve shown and represented by dimension A is defined by Equation 1 (see text). It represents volumes generated from an initial coal at the start of generation with a volatile matter content of 37.8%. The curve characterized by dimension B represents total methane volume storage capacity within a spectrum of coal ranks at 212° F (100° C) and 14,700 psi (1,000 atm). Storage curves are calculated after the scheme of Juntgen and Karweil (1966). The total storage curve represents the sum of volumes held by the molecular absorption (dimension C) and by conventional pore volume storage (dimension D). Coal rank symbols: HVB is high-volatile bituminous, MVB is medium-volatile bituminous, LVB is low-volatile bituminous, SA is semianthracite, and MA is metaanthracite. From Meissner (1984).

rank. Review of the data in this figure indicates the following:

1. Sorptive capacity increases slightly with increasing rank.
2. Pore volume storage is maximum for low rank coals, decreases to a minimum at approximately 12.5% volatile matter content (semianthracite) and then increases slightly with decreasing volatile matter content.
3. Pore volume storage is approximately the same as sorbed volume storage at a volatile matter content of 18% (low volatile bituminous). At higher volatile matter contents, pore volume storage is greater than sorbed volume storage. At lower volatile matter contents, the sorbed volume is greater than the pore volume storage.
4. Methane is expelled from coal when generation volume exceeds total storage capacity (for a given temperature and pressure regime).

If the temperature is increased while the pressure is held constant, the storage capacity curve shown on Figure 5 will move to the left (increasing temperature results in decreasing storage capacity). If the pressure is increased while the temperature is held constant, the storage capacity curve shown on Figure 5 will move to the right (increasing pressure results in increasing storage capacity).

Gas generated in excess of that which can be sorbed on or in the molecular surfaces and contained within the microporosity will initially be present as "free gas." Free gas within coalbeds is contained within fractures. Characteristically, coal contains an orthogonal jointing system called "cleat." This system includes "face cleat," which generally extends relatively great distances, and "butt (or end) cleat," which extends from one face cleat to the next. In addition to the characteristic cleat systems, coalbeds can also be more extensively fractured in response to local structure. Free gas contained within the fracture network is available for dissolution in groundwater, retention in the coalbed as trapped free gas, and free-phase migration away from the area of generation either within the coalbed or into adjacent strata.

Methane is soluble to a limited degree in groundwater at the pressures (normally close to hydrostatic) and temperatures encountered in most coalbed methane reservoirs. For example, at 1,000 psi (68 atm) and 100° F (38° C), a maximum of 8.5 cu ft of methane could dissolve in one barrel of fresh water. Although this is a small volume of gas, in a dynamic groundwater system large total volumes of gas can be removed from a potential reservoir (Rightmire, 1984).

Reservoir Characteristics Of Coalbeds

Coalbeds are complex reservoirs. As previously discussed, methane is retained in coal in four ways: within a dual porosity system (fracture and matrix), sorbed on or within the internal surfaces of the dual porosity network, and dissolved in groundwater within the coalbed. Furthermore, the quantity of retained methane varies with coal rank and moisture content, reservoir temperature and pressure, and the *in situ* hydrogeologic environment. Movement of fluids (methane and water) within coalbeds is through fracture permeability and diffusion (Giron et al., 1984).

Permeability in coal is due primarily to the cleat system and other fractures resulting from local structure. Since face cleat is generally the best developed and most continuous fracture, relatively better permeability should be expected parallel to it. Tests by the Bureau of Mines (McCulloch et al., 1974) have shown that the difference in production is 2.5–10 times as great in holes drilled perpendicular to face cleat as compared to those drilled parallel to face cleat. This would indicate that vertical boreholes tend to drain in an elliptically shaped area in a coalbed with the major axis of the ellipse parallel to the face cleat and the short axis parallel to the butt cleat.

Conceptually, the production of methane from coalbeds can be considered a two-stage process, possibly involving two-phase flow. At each stage in the process, methane and naturally occurring water may compete for the fractures and pore volumes. Therefore, it may be necessary to dewater a reservoir first to promote methane production.

Initial first-stage production, often at substantial volume rates, results from the flow of fluids contained within the fracture system. Depending on cleat density and any other local fracture network, initial production rates may show a rapid decline. The second stage of fluid movement involves the diffusion of methane molecules from the molecular surfaces and microporosity into the fracture network. This stage is influenced by coal moisture.

Historically, production tests of coalbed reservoirs have been of relatively short duration and have predominantly tested only the fluids stored in the fracture network close to the borehole. Generally, either rapid production decline during the short-duration testing or the requirement to dewater the coalbed have conspired to label the rates of flow of coalbed gas as being uneconomical. However, with patience and testing of longer duration to involve second-stage fluid movement, it is possible that many coalbed reservoirs eventually may prove to be economical, although this might not have been apparent from the preliminary flow and pump tests.

GEOLOGIC CHARACTERIZATION

The evaluation of the coalbed methane resource potential in a given area uses classic assessment techniques applied to more conventional resources. These techniques usually include four tasks: (1) establishment of the regional geologic framework, (2) collection and evaluation of all existing data; (3) identification of gaps in the existing data and collection of the data necessary to eliminate or reduce these gaps; and (4) integration of the results of the first three tasks to assess the potential magnitude of the resource.

Tasks 1, 2, and 4 are fundamental and widely applied to resources of all types. Therefore, the following description of these tasks is brief. The third task, the identification and collection of new data, is discussed in more detail and includes both standard techniques and techniques unique to the evaluation of coalbed methane.

Regional Geologic Framework

Establishing the regional geologic framework of potential coalbed reservoirs requires an understanding of both the depositional environment and the tectonic and structural history. Understanding the depositional environment is critical to interpreting both the areal extent and stratigraphic distribution of coalbeds. For example, coals in the eastern United States were deposited during the Pennsylvanian Period along a northeast-trending elongate basin extending from the present location of Alabama to Pennsylvania. Clastic sediments were derived principally from a slowly eroding landmass lying to the east and southeast. Most coalbeds are relatively thin because they were deposited on a broad area of low relief during a long period of relatively stable tectonism with slow subsidence. Few coalbeds are thicker than 10 ft (3 m); most are <6 ft (1.8 m) in thickness. In many areas, the beds are very continuous and can be correlated for tens or even hundreds of miles. The coals are of relatively high rank, ranging from high-volatile bituminous on the western side of the basin to anthracite on the eastern side.

In contrast, coal deposits in the Rocky Mountain region were laid down in deltaic environments along the western

shore of an inland seaway that existed from Late Cretaceous to early Tertiary time. During most of Late Cretaceous time, a high, orogenically active Cordilleran mountain range—existing to the west of and generally paralleling the seaway—contributed abundant clastic sediment to the subsiding trough area along the western shoreline of the seaway. Land areas east of the seaway were relatively stable, had low relief, and contributed little sediment to the shallow eastern shelf. With a few exceptions, most individual coalbeds in the Rocky Mountain region are lenticular and can be traced for only short distances. Much of the western coal resource, however, is in beds of thicknesses >20 ft (6 m); some beds range in thickness from 50 to >200 ft (15 to >60 m). Because of their substantially younger age, coals in the western United States generally have lower ranks than coals in the east.

The tectonic and structural history of an area in relationship to coalbeds is important to an understanding of coal rank characteristics and in situ reservoir properties. Generally, depth of burial, and the thermal regimes associated with it, are the primary factors that establish existing coal rank. In some areas, however, coal rank established by depth of burial is increased by local and regional thermal anomalies. For example, Choate and Rightmire (1982) have suggested that there is an important relationship between the emplacement of the igneous complexes of the San Juan Mountains and adjacent areas, and the elevated coal rank and greater methane concentration in southeastern Piceance basin, northern San Juan basin, and Raton basin. This relationship is shown on the heat flow map of Figure 6.

In addition to effects on coal rank, the tectonic and structural history in an area may impact the reservoir characteristics of coalbeds. For example, local folding and faulting has the potential for significantly improving the in situ fracture network.

Evaluation of Existing Data

The existing data applicable to coal resource evaluation include the literature addressing regional and local coal assessments, geophysical log libraries, and core/cuttings collections. The literature on coal resources is generally limited to areas of basins where coalbeds crop out or are at shallow depths, and where it has been economical in the past to mine them. This literature, however, constitutes an extensive data base regarding the stratigraphic distribution of coalbeds and coal zones within each formation and the physical description of coals including rank, ash content, and heating values.

The evaluation of coal resources that exist below minable depths generally requires review of available geophysical logs. Detailed analysis of these logs can provide data on the depth, stratigraphic distribution, and thickness of individual coalbeds and coal zones. When selecting the wells to be studied, one must keep several things in mind. First, the highest quality logs and the most extensive log suites for interpretation of coalbeds normally are from those wells that have been drilled most recently. Ideally, a density log would be available. Also, in many wells a complete suite of logs has been run only across the conventional oil and gas reservoir zones of interest, which in many cases do not include the

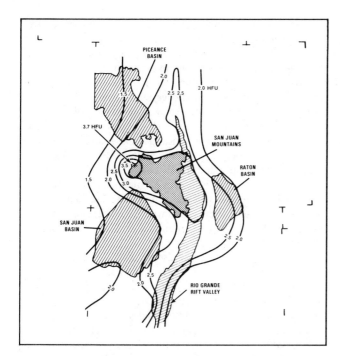

Figure 6. Coal-bearing basins and San Juan Mountains relative to the Rio Grande Rift Valley and generalized heat-flow line. From Choate and Rightmire (1982).

coal intervals. Commonly, a gamma-ray log may be the only log run the entire depth of the hole.

Geophysical logs provide data on the quantity and distribution of the coal present, but they do not provide any information on coal rank. A possible source of material that can be analyzed to establish coal rank distribution within a basin may exist in libraries of drilling chip samples. In selecting well cuttings for coal rank determination from sample libraries, the depth intervals containing coalbeds are identified from geophysical logs. The selected samples from those intervals then are scanned under a binocular microscope and several coal chips are removed for vitrinite reflectance (R_o) measurements. The R_o values then can be used to determine coal rank. This data, however, should be used with caution since a variety of external factors, such as heat, may have changed the true R_o values of the selected samples.

Collection of New Data

The collection of new data involves (1) obtaining information on the gas content of coal samples by direct measurement, (2) determining coal rank and chemical characteristics by R_o measurements and by ultimate and proximate analyses, and (3) establishing the stratigraphic distribution of coalbeds by running the proper suite of geophysical logs.

Sample Collection

Two types of samples can be collected during drilling operations for measurement cores or cuttings. Cores are the

most desirable samples that can be obtained for the direct measurement of coalbed methane content. Such samples, however, are expensive to obtain, and because most coals are highly friable, core recovery often is very poor. It is not uncommon in coring coal to experience recoveries of only 10–30%. If core recovery percentage is low, then it is difficult to determine the true gas content of that coalbed because the samples recovered are not representative of the highest quality coal present. The coal core recovered will be from the most competent portion of the coalbed, that is, the "dirtiest" zone, which is high in silt and clay content and thus where gas content will be relatively low. Techniques for improving core recovery include using either a rubber sleeve or a plastic liner in the core barrel. Such techniques have been shown to substantially improve core recovery.

An alternate low-cost method of collecting coal samples for methane gas content measurement is simply to collect cuttings off the shale shaker during normal drilling operations. Collecting cuttings in this way, however, can have several drawbacks. The most important is that, in some cases, the cuttings are too small for practical evaluation of the coalbed methane content. To ensure the recovery of usable cuttings, drilling bits that normally produce large cuttings should be selected for drilling through the major coal zones. A technique that has been successful is to drill the coal zone with an undersized bit and then to ream the hole to the designed diameter. Commonly, the reaming process produces very large coal cuttings. A second drawback in using cuttings, is that such samples are contaminated by cuttings from other noncoalbed interbeds and by mud and water that are encountered during drilling. This contamination may account for as much as 60% of the bulk sample weight. Therefore, laboratory measurements must be made of the contaminating water and noncoal cuttings. The amount of water contamination can be measured by air-drying the samples. The percentages of the coal and noncoal fractions then can be determined by density separation.

Measurement of Gas Content

The estimated and measured gas content of a coal sample is composed of three parts: (1) gas lost by the coal sample from the time it has been cut by the drilling bit until it reaches the surface and is placed in a canister, (2) gas that is directly measured as it desorbs from the sample, and (3) residual gas that remains in the sample after desorption ceases. Once a core or cuttings sample reaches the surface, it is washed in fresh water to remove drilling mud and then sealed in a desorption canister as quickly as possible. Gas desorption will be at the maximum rate when the coal sample is first placed in the canister, and it will then decrease exponentially with time. In order to estimate the amount of gas lost during a sample's trip uphole, it is necessary to establish an initial desorption rate for a fixed period of time. The plot of this data can then be extrapolated back to time zero to estimate the quantity of gas lost. Usually, initial desorption readings are made every 15 min for a period of 2 hr to establish the initial desorption curve. A convention established by the U.S. Bureau of Mines is to consider time zero that time at which the sample is halfway up the hole. A

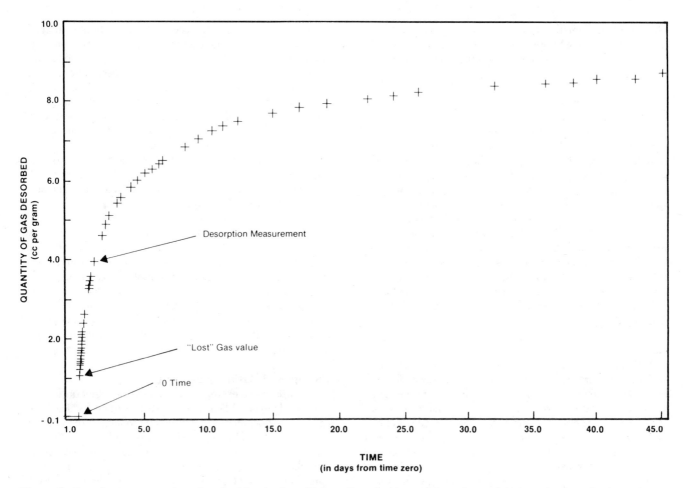

Figure 7. Cumulative curve of gas desorbed for the first 45 days, Sample 12 cored from the 14-ft (4.3-m) thick coalbed just above the Rollins Sandstone, depth 4,800 ft (1,440 m), Teton Energy well, Walck 28-2, Piceance basin, Colorado. After Choate et al. (1981).

detailed description of sample desorption and associated calculations is given by Diamond and Levine (1981). An example of a desorption curve for a coal sample is given in Figure 7.

In most wells sampled by TRW, ambient temperature and pressure readings were made each time a gas measurement was performed. This allows gas volume values to be corrected to standard temperature and pressure. For temperature measurements, a thermister thermometer was used so that a direct contact measurement of the canister wall temperature could be obtained rather than just ambient air temperature. In the field, we measured pressure by use of a high-quality survey aneroid barometer. Prior to going into the field and after return to the laboratory, we checked the aneroid barometer against a mercury barometer to correct for any instrument drift.

Measurement of Vitrinite Reflectance

As previously mentioned, measurement of vitrinite reflectance R_o of coal samples can provide a means of accurately determining coal rank. Measurements of R_o can be performed on very small amounts of coal; drill cuttings admirably serve this purpose. A major caution is that the

cuttings must not have had their reflectance characteristics changed, either from heating of the sample during drying or from chemicals that may have been placed on the samples. Normal drilling muds, however, do not seem to have any detrimental effect on cutting samples. Figure 4 shows the relationships existing between R_o values, coal rank, and gas content.

Analysis of Chemical Properties

In most coalbed methane assessment programs conducted to date, sample evaluation has usually involved performing several types of chemical analyses, including two groups of tests standard in the coal mining industry called "proximate" and "ultimate" analyses. Proximate analysis is the measurement of the percentage of volatile matter, fixed carbon, ash, and moisture. In addition, usually the heating value of the coal is determined and is measured in Btu's per pound. Ultimate analysis is the measurement of the percentage of moisture, hydrogen, carbon, nitrogen, oxygen, sulphur, and ash. Several of these measurements are of particular value in the evaluation of coalbed methane potential of an area. For example, the percentage of the volatile matter and fixed carbon can be related to the quantity of methane that has been generated during coal metamorphism. In addition, the percentage of ash is a good measure of the natural contamination that can be expected in a coalbed.

Geophysical Logging

Proper evaluation of the coalbed methane potential of a well requires obtaining a high-quality suite of geophysical logs. For detailed evaluation of the coal zones penetrated in a well, the following suite of logs have been found to be most useful (listed in relative order of importance): (1) compensated neutron/formation density, (2) gamma ray, (3) caliper, (4) dual induction resistivity, (5) self potential, and (6) sonic. This suite of logs will provide an excellent record not only of the coal zones but also of other potential gas reservoirs in the well. Other types of gas reservoirs commonly found in close association with gas-rich coalbeds include thick sandstones that commonly underlie (and in some cases overlie) major coal zones in the western United States, and tight gas-bearing sandstones that are either interbedded within coal zones or lie above or below them.

Resource Assessment Techniques

Assessments of coalbed methane potential on a basin or subbasin level probably can be most conveniently performed by construction of a series of basinwide maps that summarize all critical elements involved in the commercial recovery of coalbed methane gas. Such maps can then be used to define target areas where coalbed methane could be developed most economically. The first map to be constructed should be a structural contour map drawn on top of a rock unit that is closely associated with the coal deposits. Ideally, such a rock unit would have basinwide distribution. In many of the western U.S. intermontane basins that contain coal deposits, a major sandstone commonly underlies the basal coal zone within each principal coal-bearing formation. Examples include the Rollins–Trout Creek Sandstone that underlies the principal coal deposits in the Mesaverde Group in Piceance basin; the Pictured Cliffs Sandstone that underlies the coal-bearing Fruitland Formation in San Juan basin; and the Trinidad Sandstone that underlies the coal-bearing Vermejo Formation of Raton basin. Data for these structural contour maps can be obtained from geophysical well logs.

The second map to be constructed should be a depth-of-overburden map. Commonly, the reference plane used is the same as that used for the structural contour map, for example, the top of a sandstone unit underlying the base of a principal coal zone. This map is extremely useful because it identifies those areas lying within certain depth ranges to which an explorationist may want to confine drilling.

The next maps to be constructed should be isopach maps of coal thickness. Again, data for such maps would come from detailed interpretation of geophysical logs for wells distributed throughout the area of investigation. Such isopach maps can be either for an entire basin or for a subunit of a basin. The isopach maps can show the thickness of a single coalbed or coal zone or, more likely, the total coal thickness within a single formation. Such maps are useful in providing data regarding the total quantity of coal present within a basin, subbasin, or individual property.

The next series of maps to be constructed should either directly or indirectly indicate the gas concentration present throughout an area. If sufficient test data exist—such as the gas concentration measured in samples collected from wells—then a map can be constructed on which gas concentrations are contoured for the basin. A second technique is to construct a map showing R_o values for an area and then to contour these data points. A third technique is to plot one or more of the types of data obtained from proximate chemical analyses, for example, the percentage of volatile matter, which is directly related to gas generation. If R_o or chemical analysis data are plotted, an additional map must be constructed on which these data have been converted to estimated gas concentration in the coal.

The final step is to construct a map that shows the total gas content of a region. This map is obtained by multiplying, on a point-by-point basis, the coal volume present in a region times the gas concentration in the coal. This final map will be a contour map showing total gas content of an area in units such as billions of cubic feet of gas per square mile. From this map, we can then calculate the total amount of gas present in a basin or subbasin or within the boundaries of an individual property.

Examples of Basin Assessment

We present two examples of basinwide assessments here. The San Juan and Black Warrior basins will be discussed on the basis of documented commercial production of natural gas from coalbeds in those basins. The San Juan basin is given as an example of a quantitative assessment because of the relative abundance of coal and coal gas concentration data. The Warrior basin is addressed on a qualitative basis because we have much information on gas production from specific areas but less regional data on coal and coalbed gas distribution.

San Juan Basin

A map of total coal thickness in the Fruitland Formation of San Juan basin has been produced by Fassett and Hinds (1971). A simplified version of their map is shown in Figure 8. Figure 9 is a contour map of the average coalbed methane content of all samples for 28 individual wells distributed across the basin. Gas concentration data from these wells was obtained from both core and cuttings. All values obtained from cuttings have been corrected for contamination and dilution, both from water coating the cuttings and from noncoal contaminants. Figure 10 is a quantitative map showing the total gas content in San Juan Basin as constructed using Figure 8 and 9. The contours on Figure 10 are in billions of cubic feet of gas per square mile. On the basis of this township-by-township evaluation of gas content, it was determined that San Juan basin contains a total of 31 trillion cu ft (tcf) of gas in place in Fruitland Formation coalbeds greater than 2 ft thick.

Black Warrior Basin

For Black Warrior basin (hereafter referred to as the Warrior basin) only limited data on the distribution of deeper, potentially gassy coals are available. Most of the coal data are restricted to areas where the coalbeds are shallow

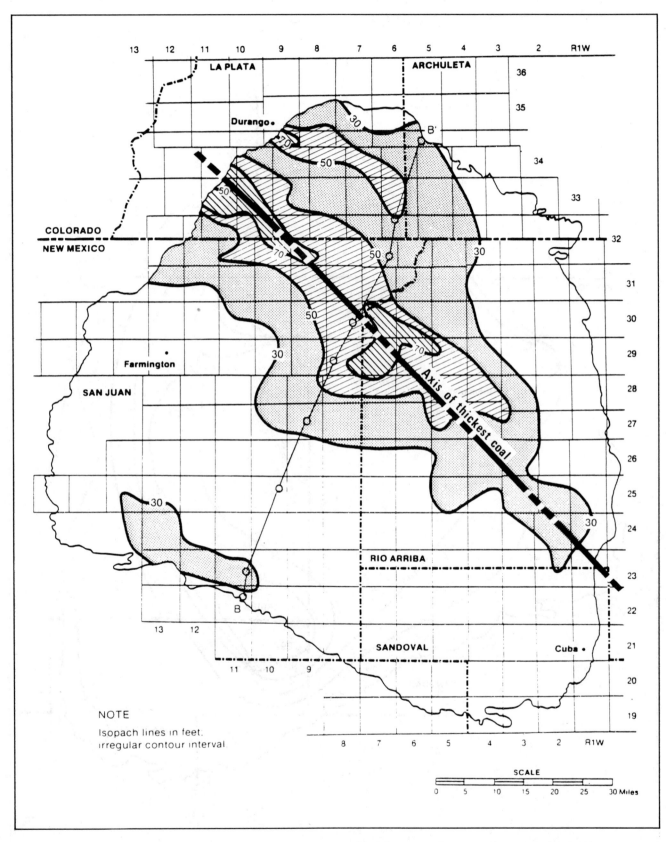

Figure 8. Simplified isopach map of total coal thickness in the Fruitland Formation for areas where total coal thickness exceeds 30 ft (9 m). Contours were selectively picked to define areas of thickest coal accumulation. From Choate et al. (1984); modified from Fassett and Hinds (1971).

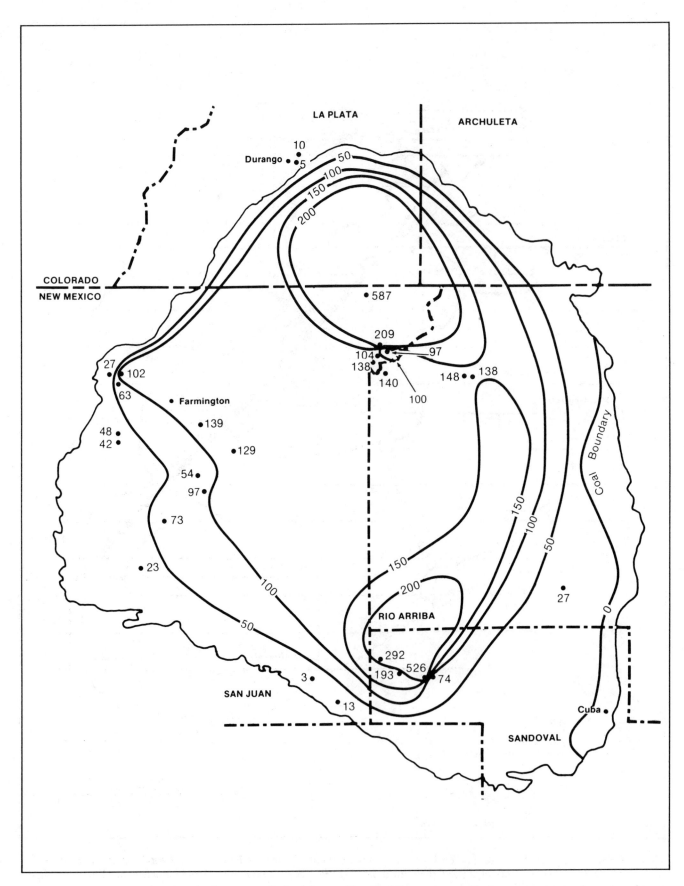

Figure 9. Contour map of averaged values per well of total gas in Fruitland Formation coalbeds; contour intervals are in cubic feet per ton. Contours are drawn to reflect stratigraphy, structure, and coal rank. From Choate et al. (1984).

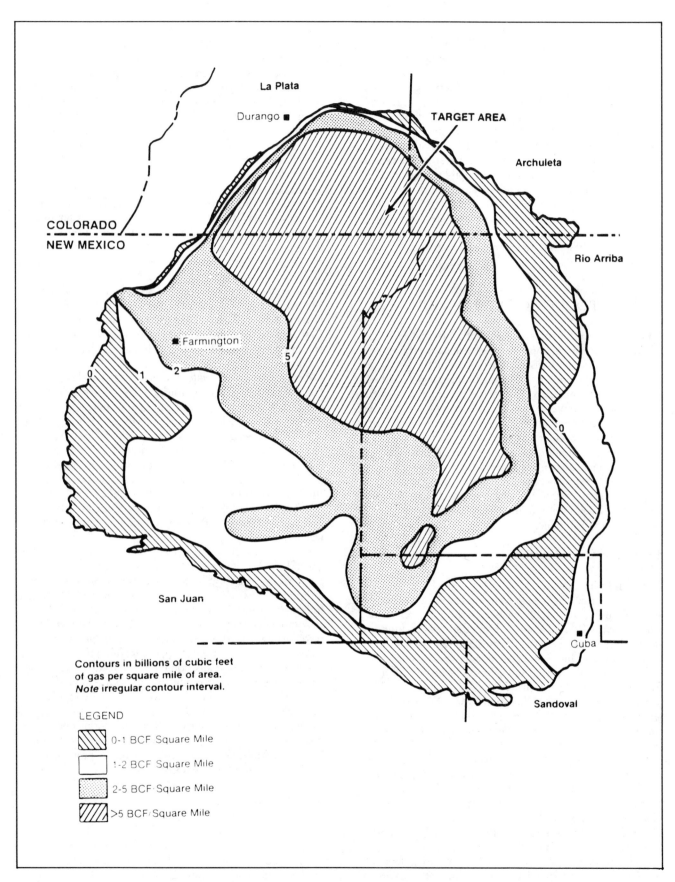

Figure 10. San Juan basin coalbed methane target area, contoured in billions of cubic feet per square mile. From Choate et al. (1984).

enough to be mined. The coalbeds of particular interest are contained within strata of the Pottsville Formation of Pennsylvanian age. Figure 11 shows the geographic distribution of stratigraphic intervals containing coals with the highest potential for methane production. Those coals in the Pratt, Mary Lee, and Black Creek coal groups are found in stratigraphic intervals identified in Figure 11 as intervals D, C, and B, respectively (Metzger, 1965). Cumulative coal thickness of > 20 ft (6 m) thick is normal; individual beds range from 2 to 4 ft (0.6 to 0.8 m). Gas contents in excess of 550 cu ft/ton have been reported (Diamond and Levine, 1981). On the basis of a qualitative assessment of the Warrior basin, it was estimated that approximately 10 tcf of methane is present in the coals of the Warrior basin (Hewitt, 1984). Figure 12 shows the area within the Warrior basin considered to have the highest potential for coalbed methane production.

PRODUCTION EVALUATION

Once the coal thickness and gas content have been determined by coring, borehole geophysical logging, and desorption testing, it is essential to determine the producibility of the reservoir. Production testing of coalbed gas reservoirs utilizes conventional test procedures. However, because of the complex nature of coalbed reservoirs and the potential for the reservoir to be water saturated, much longer testing periods are required. Experience has shown that long-term production testing of about 6 to 12 months is required to adequately involve the entire reservoir and to determine long-term dewatering requirements and gas production characteristics. We have selected the following production case histories to illustrate characteristics of coalbed methane production.

Production Case Histories

San Juan Basin

Gas production has been documented from coal-bearing intervals in the Fruitland Formation in the San Juan basin since the early 1950s. Figure 13 shows the areas of current production from the Fruitland Formation. Two wells representing San Juan basin coalbed methane production will be discussed in some detail below. The first well, Phillips Petroleum Well 6-17 San Juan Unit 32-7, produced for 29 years and was temporarily shut-in in 1982. The second of these wells is Amoco Production Cahn No. 1 Well, which produces more gas than any other single coalbed methane well identified to date.

Phillips Petroleum Well 6-17, San Juan 32-7 Unit. This well, completed by Phillips Petroleum Company on August 24, 1953, is the discovery well for the Los Pinos Fruitland South gas field and is located in the northeast corner of San Juan County, New Mexico, about 6 mi south of the Colorado/New Mexico state line (Sect. 17, T.31 N, R.7 W). The well was cased to a depth of 3,054 ft (920 m), then drilled with air and completed open-hole to a total depth of 3,240 ft (990 m). No stimulation treatment was necessary. A total of 1.1 billion cu ft (bcf) of gas has been produced from

this well through 1981. The annual production curve for the well (Figure 14) demonstrates the marked difference between gas production from a coalbed methane well and a conventional gas well. In a gas well that produces from a conventional reservoir, the normal production curve shows a steady decline over a period of several years. In contrast, the annual production in Phillips Petroleum Well 6-17, has steadily increased from an initial 27.7 million cu ft (mmcf) in 1953 to a maximum of 57.8 mmcf in 1974. Since 1974, there has been a tendency toward gradual decrease, to a 1981 production of 48.6 mcf. Over this long period (about three decades), there has been very little drop in pressure within the well. The initial shut-in pressure in 1953 was 1,504 psi. During a 43-day shut-in test in 1977, shut-in pressure had built back up to 1,472 psi, and was still rising at the end of the test period (Choate et al., 1984). The neutron log shows the presence of at least five coalbeds in the open-hole section of this well. Two beds (3 and 5 ft, or 0.9 and 1.5 m, thick) occur in the upper Fruitland coal zone, and three beds (4, 8, and 3 ft, or 1.2, 2.5, and 0.9 m, thick) are present in the lower coal zone. This equals a total coal thickness of at least 23 ft (7 m) in the 186-ft (56-m) section of open hole. No gas content determinations or isolated zone tests have been conducted on this well.

Amoco Production Company Cahn No. 1 Well. The Cahn No. 1 is in the Cedar Hill Fruitland Basal Coal gas field (formerly Mount Nebo field), San Juan County, 4 mi south of the Colorado/New Mexico state line (Sect. 33, T.32 N, R.10 W). The following information was obtained from public records at the New Mexico Oil Conservation Commission (Choate et al., 1984). Production casing was set at a depth of 2,795 ft (838 m), just above the basal Fruitland coalbed, and the well was drilled to a total depth of 2,812 ft (844 m), leaving a 17-ft (5-m) open-hole section. Gas contents in excess of 450 scf/ton have been reported for coals from this well (Amoco, 1984).

A 24-hr production test in May 1977 produced 350 mcf of gas and 239 bbl of water, with a flowing tubing pressure of 10 psi on a 0.375-in. (0.953-cm) choke. Shut-in measurements indicate a reservoir pressure of 1562 psi. All available records indicate that no stimulation was performed on the well.

Commercial gas production started in July 1979, and the records shown are through November 1984 (Figure 15). During this period, a total of approximately 450 mmcf of gas was produced. Gas production has been steadily increasing, while concurrent water production shows a general decline. Production for the Cahn No. 1 Well has required special treatment because the producing reservoir (the basal coalbed in the Fruitland Formation) contains substantial amounts of water, as it does in much of the basin. As can be seen from Figure 15, Amoco appears to have solved the problem of producing gas from wet coalbeds.

Black Warrior Basin

Wells in the Warrior basin are generally < 3,500 ft (1,050 m) in depth and penetrate numerous potentially gas-productive coalbeds. Two fields, the Oak Grove and Brookwood degasification fields (Figure 16), are currently producing commercial volumes of gas and delivering it to pipelines for transmission. We give a brief discussion here,

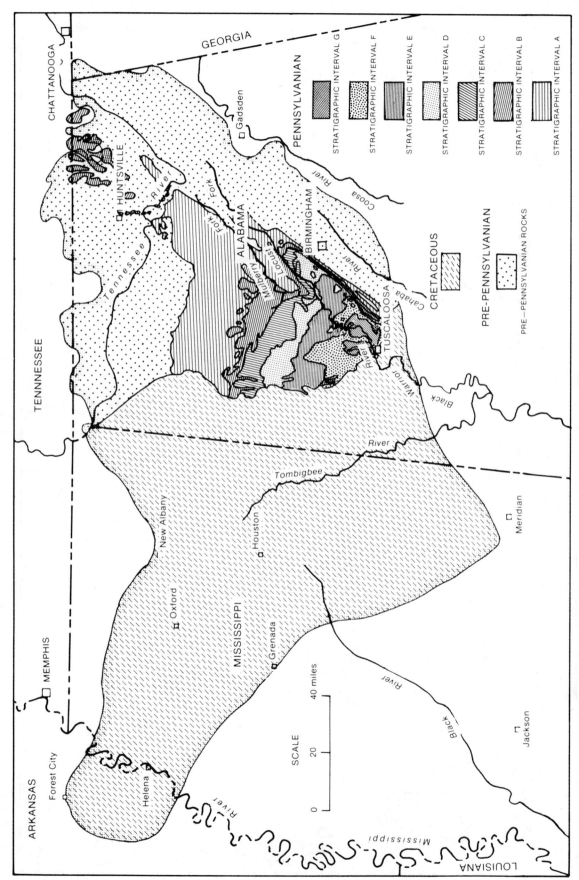

Figure 11. Outcrop map of the Pennsylvanian in the eastern Warrior basin. From Metzger (1965), and used with the permission of the Alabama Geological Survey.

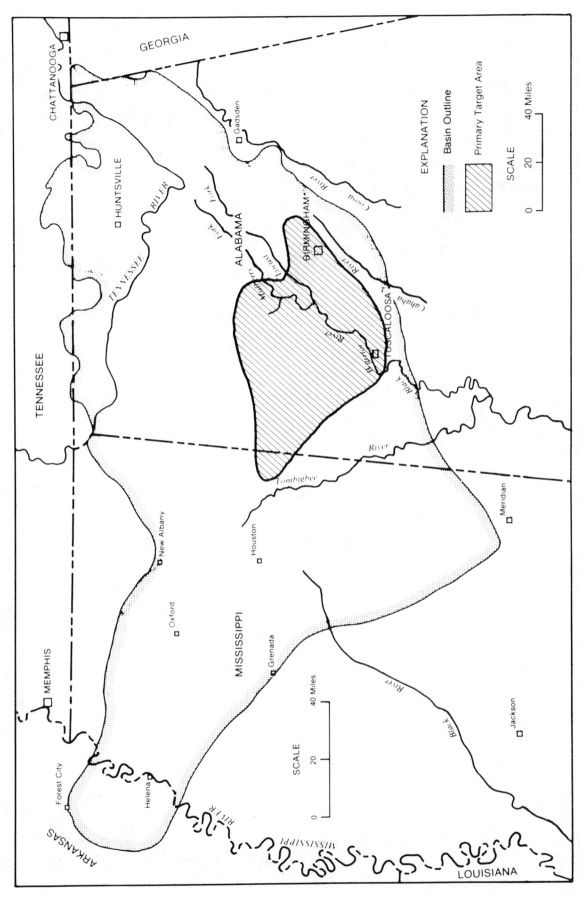

Figure 12. Primary target area of the Warrior basin. From Hewitt (1984).

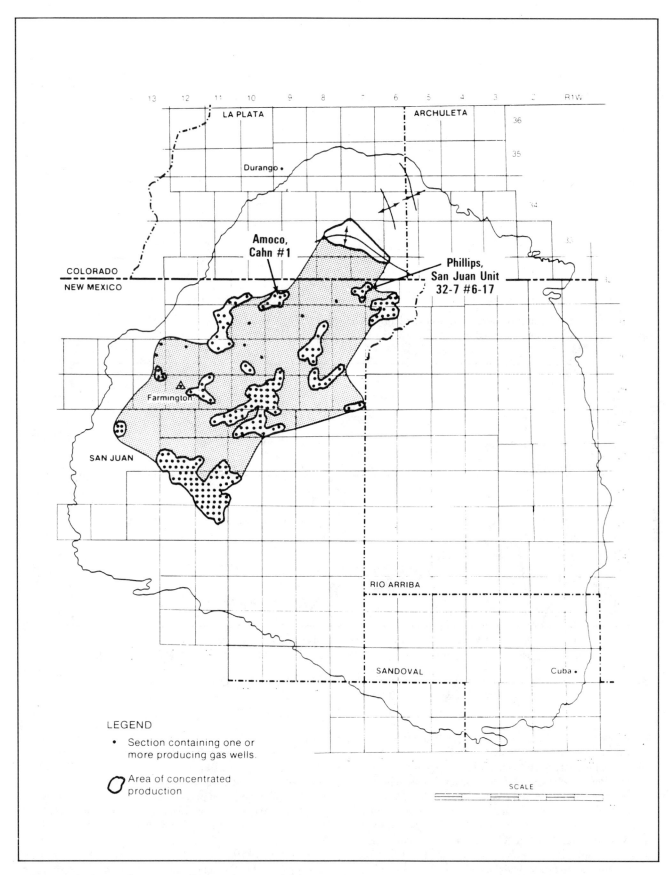

Figure 13. Areas of current production from Fruitland Formation and Pictured Cliffs Sandstone. From Choate et al. (1984).

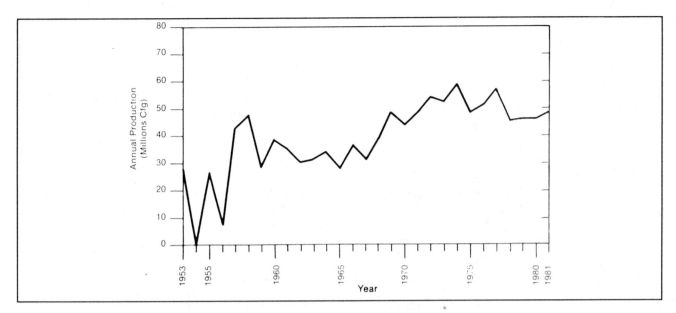

Figure 14. Annual production curve, Phillips Petroleum Well 6-17, San Juan 32-7 Unit, South Los Pinos Fruitland gas field, San Juan County, New Mexico. Modified from K. C. Bowman, *in* Fassett et al. (1978).

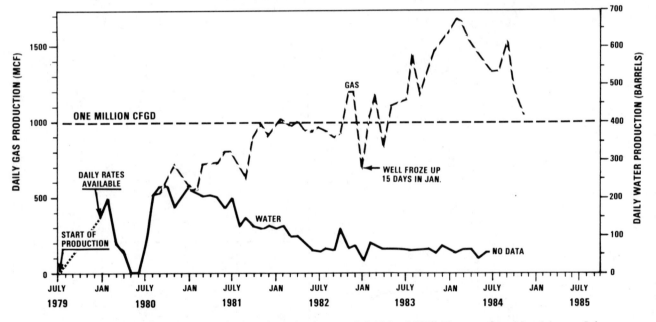

Figure 15. Daily rates of gas and water production from the Amoco Cahn No. 1 Well. Data are from New Mexico Oil Conservation Commission (July 1979 to November 1984).

but a more complete discussion of these fields and the surrounding area is given in Hewitt (1984).

Additional activity in the Warrior basin emphasizes multiple seam completions in wells in several areas. These wells are completed in several coalbeds in the Pratt, Mary Lee, and Black Creek coal groups to enhance the economical viability of the wells and to minimize the loss of gas resource.

The Gas Research Institute (GRI) is currently sponsoring a multizone completion research project in the vicinity of Oak Grove, Alabama, to prove that the technology works and to provide the industry with the means of improving the economics of these wells.

Oak Grove Field. The Oak Grove field began as a test of coalbed degasification in advance of mining. The seam being mined, the Mary Lee/Blue Creek coalbed, is 5.5 ft (1.6 m) thick and occurs at a depth of approximately 1,100 ft (330 m). Typical completion procedure for wells in this field is to drill to just above the Mary Lee/Blue Creek coalbed, set casing, drill through the coal and hydraulically fracture the coal.

At the end of 1983, 38 wells were on-line, with production ranging from 0 to 94 mcf per day, with an average of about 58 mcf per day per well. Early production from a typical Oak Grove well is shown in Figure 17. Total monthly production

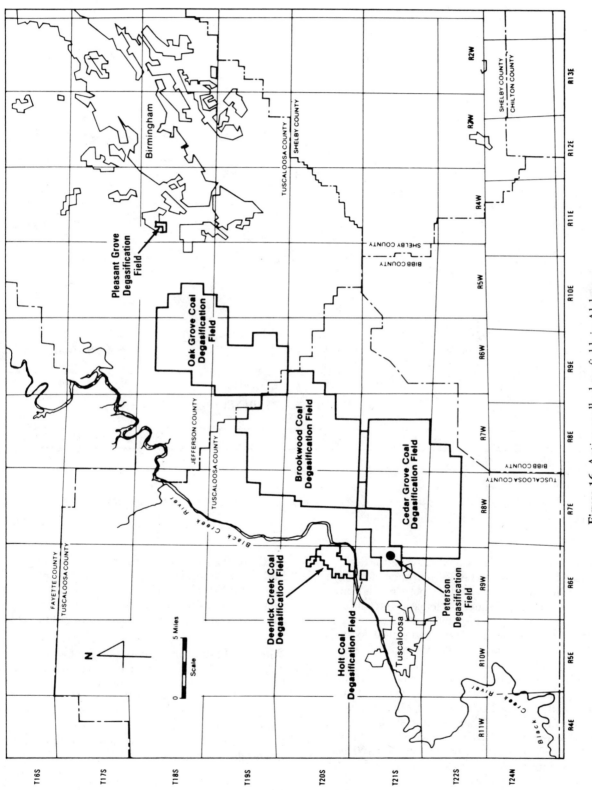

Figure 16. Active coalbed gas fields in Alabama.

WELL #5

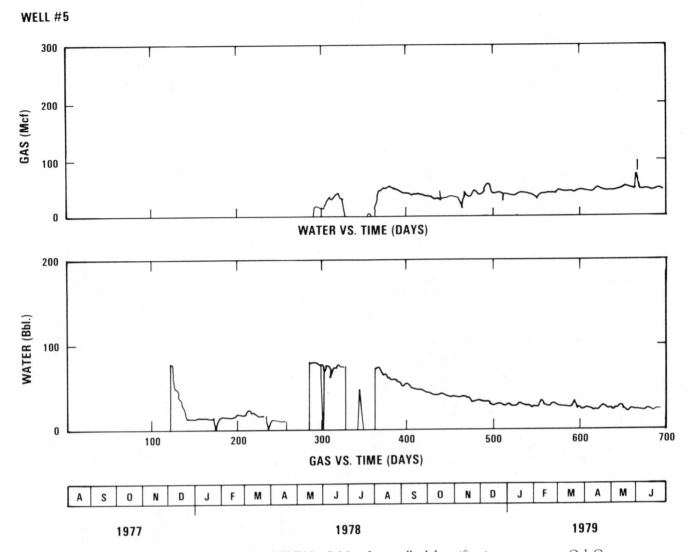

Figure 17. Daily production from foam-stimulated Well No. 5, Mary Lee coalbed degasification pattern near Oak Grove, Jefferson County, Alabama. After U.S. Steel (1980).

is about 64 mcf, and cumulative production for the field since December 1981 approaches 1.4 bcf. Water production, which initially may have exceeded 300 bbl per day for any individual well, usually declines to a stable maintenance level (20–50 bbl per day) after several months.

Brookwood Field. Brookwood degasification field, located in eastern Tuscaloosa County, Alabama, was initially established to provide degasification for Jim Walter's Resource mines in the area. Principal gas production is from a Mary Lee/Blue Creek coalbed that is 6–7 ft (1.8–2.1 m) thick at a depth of about 2,100 ft (630 m). Minor production has been obtained from the Pratt Group coalbeds. A total of 57 wells are currently producing gas ranging from 0.5 to >1,900 mcf per day, having an average rate of >160 mcf per day. Out of the 58 wells, 7 produce in excess of 540 mcf per day each and are draining areas influenced by mining operations. The remaining wells do not exceed 300 mcf per day.

The Brookwood wells are also primarily single-zone completions, being drilled to just above the target zone,

cased, and then drilled through the coal and completed. Initial water flows in some of the wells were reported to be in excess of 1,000 bbl per day. Figure 18 is a production plot for a representative Brookwood field well that has an average daily gas production of approximately 110 mcf per day.

Potential Testing and Production Problems

A number of technical problems remain to be addressed to facilitate the testing for and production of coalbed methane. Many of these focus on the utilization of conventional oil and gas technology to this unconventional resource. While it is true that conventional techniques are used, they commonly must be applied with finesse to minimize damage to a fragile reservoir system.

Testing

Because of the complexity of coalbed reservoirs and their potential to be water saturated, standard gas reservoir tests generally require much longer testing periods.

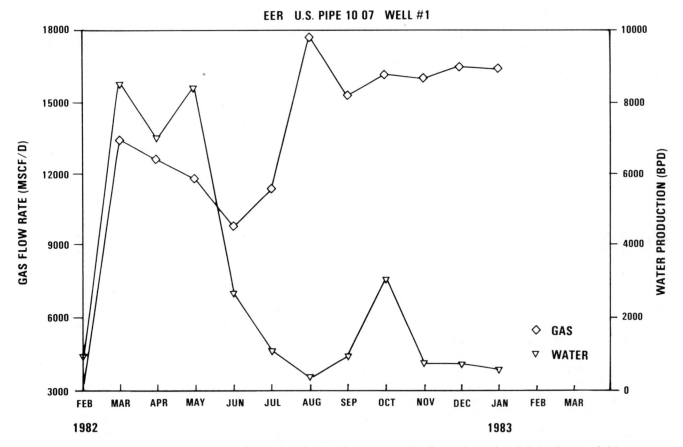

Figure 18. Monthly average gas and water production from a typical well, Brookwood coal degasification field.

Drill stem tests have the ability to provide some useful information if run for appropriate flow and shut-in periods. However, because of the very long shut-in periods required to adequately test coalbed reservoirs and the operational difficulties attendant to those long shut-in periods, standard drill stem tests are commonly not practical. Drill stem tests with short shut-in periods can be run, but the information generated will be generally qualitative in nature. Postcompletion testing of coalbed reservoirs might include pressure build-up or draw-down tests, injectivity tests, or both. A similar set of tests should be conducted before and after artificial stimulation.

Water

Water production and disposal can be a major problem in gas production from coalbeds. Where the coals are water saturated, some volume of water must be removed to provide the pressure differential leading to gas desorption and two-phase flow in the coalbed reservoir. Initial water flows may be as high as 1,000 bbl or more per day. When this is the case, stabilized water production over the life of the well may be close to 100 bbl per day. Normally, initial water production of 200–300 bbl per day may lead to maintenance levels of 20–50 bbl per day. The initial high rates of flow may last for 30–60 days or more, depending on the transmissitivity of the coalbeds and the possible interconnection with adjacent aquifer systems.

The chemistry of coalbed waters ranges from potable at relatively shallow depths to 30,000–50,000 mg/liter total dissolved solids; the waters may be either high in chloride or bicarbonate, depending on the stage of development of the hydrogeochemical cycle. Disposal of water with a high content of total dissolved solids may pose a potential environmental problem in some areas and may inhibit resource development.

Fines

Coal fines are often generated naturally during the flow of gas and/or water to a well bore along natural or artificially induced fractures. Production of fines may be mainly associated with actively generating, overpressured coals (F. F. Meissner, 1985, personal communication). Fines may be generated during stimulation and continue to form by spalling as the coals dewater and as gas and water flow to the well bore. In the stimulation of coals, although extremely fine coal particles can be carried in the gas phase if flow velocity is high enough, most of the particles are carried in the liquid phase. Since the water flows through the sand pack in a fracture-stimulated well, the potential for plugging of this pack by fines is increased.

Geochemical Considerations

Several geochemical considerations need to be addressed when completing and producing gas and water from coalbeds. The first is the effect of the established pressure differential on the potential outgassing of dissolved gases from the water, which has the possible side effect of mineral precipitation. Coalbed waters are commonly high in dissolved CO_2, and any outgassing of CO_2 could lead to carbonate precipitation in the form of either calcium carbonate or iron carbonate (siderite), depending on the chemistry of the water.

The introduction of dissolved oxygen in the fracturing fluid into reducing iron-rich environments could lead to the precipitation of a ferric oxyhydroxide that could cause plugging of the formation.

The potential of introducing colonial bacteria into the well with injected fluids could cause bacterial plugging when fracturing fluids contain organic substances to enhance proppant carrying capacity. This problem can be minimized by adding bacteriacides to the injected fluids.

CONCLUSIONS

Coalbed methane is a major natural gas resource of the United States that is presently being commerically produced in at least two regions of the country—the San Juan basin of Colorado and New Mexico and the Warrior basin of Alabama. As known conventional gas resources are depleted, coalbed methane will become increasingly more economical and will provide an important portion of U.S. gas production in future years.

Some of the properties and characteristics of coalbed methane, as now understood, are listed below.

1. Coalbeds are both source rocks and reservoirs for major accumulations of natural gas (methane) in the United States and other coal-bearing provinces of the world.
2. Coal rank and methane generation are dependent on the temperature regime a coal has been subjected to; the greater the temperature, the higher the coal rank and amount of gas generated.
3. Methane is retained in coalbeds as molecules sorbed on the internal surfaces or within the molecular structure of the coal, as gas in matrix porosity, as free gas in open fractures, and as gas dissolved in groundwater within the coal.
4. Sample analysis to determine coal rank (vitrinite reflectance and/or proximate analysis) provides a measure of the total amount of methane generated by the coal. Sample desorption provides a direct measure of the methane retained in the coal.
5. For many high-rank coals, the gas generated exceeds the retention capacity of the coalbed. Excess gas generated is free to migrate to, and charge, adjacent noncoalbed potential reservoirs.
6. The depositional, tectonic and structural history of a coal-bearing basin determines the coal rank stratigraphic distribution of potential coalbed reservoirs. Local structure may enhance the *in situ* reservoir characteristics of the coalbeds.
7. The existing data base on coal resources is extensive but is generally limited to basin margins where the coal is shallow and amenable to mining. Data are limited on the quantity, quality, and distribution of coals in basin interiors, especially at depths greater than 3,000 ft (900 m). Data on directly measured gas concentrations of deeper coals are also limited.
8. Because of the gas retention characteristics of coalbeds, gas migration to a wellbore is slow in coalbeds as compared to sandstone reservoirs. Conventional short-duration production testing techniques may not provide production data useful in projecting the long-term economics of a coalbed methane well.
9. The most successful means of determining the production characteristics and economic potential of coalbed methane resources is the detailed evaluation of drilling and completion techniques and costs and the long-term production records of successful coalbed methane wells.
10. At least several hundred wells have been completed in coalbeds or coal zones in the United States to date and are now commercially producing coalbed methane gas. The majority of these well completions are in the Fruitland Formation of northern San Juan basin, Colorado and New Mexico, and the Pottsville Formation of Warrior basin, Alabama.

ACKNOWLEDGMENTS

For critical review of the manuscript, we thank B. E. Law, USGS, Denver, Colorado; F. F. Meissner, Bird Oil Corporation, Denver, Colorado; and D. K. Murray, Consulting Geologist, Golden, Colorado. In addition, we thank K. C. Bowman, President, C.O.G. Resources, Inc., for his years of contributions to the understanding of this resource. We also thank TRW Exploration and Production for support in preparation of the paper.

REFERENCES CITED

Amoco, 1984, Exhibits to testimony before the N.W. Oil & Gas Conservation Commission, Case No. 7898, June 8, 1984.

Choate, R., and C. T. Rightmire, 1982, Influence of the San Juan mountain geothermal anomaly and other Tertiary igneous events on the coalbed methane potential in the Piceance, San Juan, and Raton basins, Colorado and New Mexico: SP/DOE 10805, Proceedings, Unconventional Gas Recovery Symposium, May 16–18, 1982, Pittsburgh, Pennsylvania, Society of Petroleum Engineers/U.S. Department of Energy, p, 151–164.

Choate, R., D. Jurich, and G. J. Saulnier, Jr., 1981, Geologic overview, coal deposits, and the potential for methane recovery from coalbeds, Piceance basin, Colorado: TRW Energy Engineering Division, McLean, Virginia and Lakewood, Colorado, for DOE/METC, Morgantown, West Virginia.

Choate, R., J. Lent, and C. T. Rightmire, 1984, Upper Cretaceous geology, coal, and the potential for methane recovery from coalbeds in San Juan basin, Colorado and New Mexico, *in* C. T. Rightmire, G. E. Eddy, and J. N. Kirr, eds., Coalbed Methane Resources of the United States: AAPG Studies in Geology Series, n. 17, p. 185–222.

Diamond, W. P., and J. R. Levine, 1981, Direct method determinations of the gas content of coals, procedures and results: U.S. Bureau of Mines, Report of Investigations 8515, 36 p.

Fassett, J. E., and J. S. Hinds, 1971, Geology and fuel resources of the Fruitland Formation and Kirtland Shale of the San Juan basin, New Mexico and Colorado: U.S. Geological Survey Professional Paper 676, 76 p.

Gan, H., S. P. Nandi, and P. L. Walker, Jr., 1972, Natural of porosity in American coals: Fuel, v. 51, p. 272–277.

Giron, A., A. M. Pavone, and F. C. Schwerer, 1984, Mathematical models for production of methane and water from coal seams: Quarterly Review of Methane from Coal Seams Technology, Gas Research Institute, v. 1, n. 4, p. 19–34.

Hewitt, J. L., 1984, Geologic overview, coal, and coalbed methane resources of the Warrior basin—Alabama and Mississippi: in C. T. Rightmire, G. E. Eddy, and J. N. Kirr, eds., Coalbed Methane Resources of the United States: AAPG Studies in Geology Series, n. 17, p. 73–104.

Hunt, John M., 1979, Petroleum Geochemistry and Geology: San Francisco, W. H. Freeman and Company, 617 p.

Juntgen, V. H., and J. Karweil, 1966, Formation and storage of gases in bituminous coals, Part I—Gas formation: Erdol and Kohle, v. 4, p. 251–258.

Kim, A. G., 1978, Methane drainage from coalbeds: research and utilization in methane gas from coalbeds development, production, and utilization: USDOE Morgantown Energy Research Center, SP-78/1, p. 13–17.

McCulloch, C. M., M. Deul, and P. W. Jeran, 1974, Cleat in bituminous coalbeds: U.S. Bureau of Mines Report of Investigations 7910, 15 p.

Meissner, F. F., 1984, Cretaceous and lower Tertiary coals as sources for gas accumulations in the Rocky Mountain area: Source Rocks of the Rocky Mountain Region, 1984 Guidebook, Rocky Mountain Association of Geologists, p. 401–431.

Metzger, W. J., 1965, Pennsylvanian stratigraphy of the Warrior basin, Alabama: Geological Survey of Alabama Circular 30, 80 p.

Nandi, S. P., and P. L. Walker, 1971, Adsorption of dyes from aqueous solution by coals, chars, and active carbons: Fuel, v. 50, p. 345–366.

Potential Gas Committee and Potential Gas Agency, 1981, Potential supply of natural gas in the United States (as of December 31, 1980): Golden, Colorado, Potential Gas Agency, Colorado School of Mines, 119 p.

Rightmire, C. T., 1984, Coalbed methane resource, in C. T. Rightmire, G. E. Eddy, and J. N. Kirr, eds., Coalbed methane resources of the United States: AAPG Studies in Geology Series, n. 17, p. 1–13.

Sato, T., 1981, Methane recovery from coalbeds: surface and physical properties of western United States coals: Masters thesis, University of New Mexico, Albuquerque, 78 p.

Tissot, B., B. Durand, J. Espitalie, and A. Combaz, 1974, Influence of nature and diagenesis of organic matter in formation of petroleum: AAPG Bulletin, v. 58, n. 3, p. 499–506.

U.S. Steel, 1980, Monthly Technical Progress Report: U.S. DOE Contract No. ET-75-C-01-9027, Morgantown Energy Technology Center, Morgantown, West Virginia.

Annotated Bibliography of Methodology for Assessment of Undiscovered Oil and Gas Resources

Ronald R. Charpentier and Jannette S. Wesley
U.S. Geological Survey
Denver, Colorado

This paper presents a bibliography of articles on methods of estimating amounts of undiscovered oil and gas resources. The articles are categorized according to which of ten estimation methods they emphasize.

SCOPE OF THE BIBLIOGRAPHY

This is a bibliography of appraisal methodology for estimating conventional and unconventional undiscovered oil and gas resources. Only those papers presenting some insight into quantitative appraisal methods are included. Those papers that merely mention, but do not explain, an assessment method are excluded as are those that discuss only nonquantitative aspects of resource appraisal. Also excluded are papers presenting only appraisal methods for nonhydrocarbon resources.

Methods for assessing the amount of reserves in discovered fields are not included here except for those methods that are used for calculating inferred reserves for the aggregate of fields in a region by extrapolation of past rates of reserve growth. These methods are similar to appraisal of undiscovered resources in their use of relatively uncertain, non-field-specific data, and thus papers about the latter methods are included in the bibliography, while those concerning field-specific petroleum engineering calculations are not.

Supply modeling, which is appraisal of what will be discovered over a particular interval of time as opposed to what is ultimately available for discovery, is not treated here. Although an assessment of undiscovered resources can be part of a supply model, supply models must also take into account additional economic factors. The emphasis on economics rather than geology qualifies supply modeling as a separate topic, and thus papers concerning it are not included here.

APPRAISAL METHODS

Methods of assessing undiscovered hydrocarbons are divided here into ten categories. The boundaries of these categories are not sharp and there is some overlap. Thus, a single method can have aspects of more than one category, or for a single study, the results of a method in one category can be used as input for a method in another category.

The main differences among the ten categories lie in what types of data are used. The data used in resource appraisal fall into three main types. The first type, geologic data, is a broad group of descriptive data ranging from areas of sedimentary basins to distributions of effective porosity in the pools of a particular play. The second type, deposit size data, includes measures of field or pool sizes. The third type, historical data, includes information on the history of hydrocarbon exploration and discovery. Some methods use data of only one type; some use data of two or all three types.

Suggested general references on oil and gas appraisal methodology are AAPG Studies in Geology Number 1[1], Adelman et al. (1983), Masters (1984)[2], and White and Gehman (1979).

Areal Yields

Areal yield methods use basic geologic data, usually just the areas of sedimentary rock in a region to be assessed. More detailed geologic data can be used, however, to aid in choosing appropriate analogs. These methods consist of the multiplication of the prospective area to be assessed by some yield of hydrocarbon per unit area calculated from a better explored analog area. Differences among methods are mainly in how the prospective area is calculated. These methods are most appropriate for large areas—basin size or larger. Unit regional values, when calculated specifically for hydrocarbon

---

[1] Haun, J. D., ed., 1975, Methods of estimating the volume of undiscovered oil and gas resources: AAPG Studies in Geology, n. 1, 206 p.; individual papers are cited in this bibliography.

[2] Masters, C. D., ed., 1984, Petroleum resource assessment: Ottawa, International Union of Geological Sciences Publication n. 17, 157 p.; individual papers are cited in this bibliography.

resources, fall into this category. Some of the major areal yield references are Hendricks (1965, 1975) and Weeks (1965, 1966).

Volumetric Yields

Volumetric yield methods are very similar to areal yield methods except that yields are calculated per unit volume of rock rather than per unit surface area. The methods use geologic data only slightly more detailed than that used for areal yield methods. Major volumetric yield references include Jones (1975), Mallory (1975), and Ulmishek and Harrison (1984).

Play Analysis

Play analysis is a more detailed form of volumetric yield analysis. Whereas areal yield and volumetric yield methods primarily use geologic data, play analysis is performed at a scale detailed enough to use both geologic and deposit size data. The methods consist of generating and then combining estimates of the number of undiscovered deposits and estimates of the sizes of undiscovered deposits. Often the estimates of sizes of undiscovered deposits are calculated using geologic data, such as structure sizes, reservoir thicknesses, and porosities, in a volumetric equation. The deposit size estimates may also come directly from analog deposit size data or from a combination of deposit size and historical data by way of a discovery model. Because of the more detailed scale of play analysis, risk is more likely to be assessed separately than it would be for areal or volumetric yield studies. Prospect analysis is a subset of play analysis; it is the special case where the number of prospects in a given play is determined to be one. Baker et al. (1984), Bird (1984), Canada Department of Energy, Mines, and Resources (1977), Lee and Wang (1983a, b), Miller (1981, 1982a), Procter and Taylor (1984), and L. P. White (1981) are some of the main references.

Field Size Analysis

One way to use deposit size data is to assume a distribution (e.g., lognormal or Zipf's Law) and then use the sizes of discovered deposits to calculate the expected sizes of undiscovered deposits. These methods have also been used to compare sizes of discovered deposits to deposit size distributions generated by a play analysis. Major references include: Coustau (1979), Folinsbee (1977), Harbaugh and Ducastaing (1981), Howarth et al. (1980), Kaufman (1962), Klemme (1983, 1984), McCrossan (1969), Riesz (1978), and Schuenemeyer and Drew (1983).

Historical Extrapolation

Historical extrapolation methods primarily use historical data. They relate discovery data to either time or some measure of exploratory effort. Individual methods mainly vary according to what time/effort measure is used and to what mathematical function is fit to the data. Drew (1975a, b), Hubbert (1962, 1967, 1969, 1974), Moore (1966a, c), Root and Drew (1979), J. T. Ryan (1973a, b), and Zapp (1962) all used historical extrapolation methods.

Discovery Modeling

Discovery modeling combines deposit size data with historical data and sometimes with geologic data as well. Generally it consists of a quantification of the relationship between deposit size and the order or sequence in which discoveries are made. More broadly, the category includes quantitative models of how deposits are discovered. Important examples include Arps and Roberts (1958), Barouch and Kaufman (1976a, b), Drew et al. (1980), Kaufman et al. (1975), Menard and Sharman (1975), and Root and Schuenemeyer (1980).

Reserve Growth

Calculation of inferred reserves on a field-by-field basis is an engineering problem and such methods are not included here. Inferred reserves, however, are sometimes calculated as an aggregate for large areas using historical trends in reserve growth. Such methods are represented in the bibliography by, for example, Arrington (1960), Hubbert (1967), Mast and Dingler (1975), and Root (1981).

Material Balance

The material balance approach (or volume genetic approach, as it is called in the Russian literature) begins with an assessment, based on geochemical data, of the amount of hydrocarbon generated. This amount is discounted by migration and trapping efficiencies on the basis of geologic analogs. Some of the major references on the material balance approach are Bishop et al. (1983), Conybeare (1965), Kontorovich (1950, 1976, 1984), and Neruchev (1964).

Direct Assessment

Direct assessment methods synthesize the results of other methods and/or modify the results in accordance with other data not taken into account by the primary methods. Delphi conferences are the most common direct assessment methods. Although traditionally called "direct subjective assessment methods," they are actually indirect because of their use of other primary methods for input. They are also not necessarily more subjective than other methods; subjectivity is usually just less obvious in the assumptions. Some examples of direct assessment procedures are given in Dolton et al. (1981) and Miller et al. (1975).

Probability Methods

Probability methods include statistical methods used to make further inferences based on resource assessments. These are mainly used to aggregate estimates for smaller areas into estimates for larger areas or to disaggregate estimates for larger areas into estimates for smaller areas. Also included are some basic probabilistic techniques, such as Monte Carlo simulation, which is also used in other methods. Included in our bibliography are a few selected papers that present such techniques, although the authors of these papers may not necessarily relate these techniques specifically to appraisal of undiscovered resources. The most important papers include Capen (1976), Crovelli (1981, 1983a, 1984a), Dolton et al. (1981), and Walstrom et al. (1967).

ANNOTATED BIBLIOGRAPHY

AAPG Strategic Committee on Public Affairs, 1976, The assessment of undiscovered petroleum resources: AAPG Background Paper n. 8, 5 p.
Volumetric yields

Abramovich, M. V., 1960, Estimate of reserves of prospective areas in folded oil and gas regions: Petroleum Geology, v. 4, n. 6-A, p. 315–318.
Play analysis

Adams, T. D., and M. A. Kirkby, 1975, Estimate of world gas reserves: Proceedings of the Ninth World Petroleum Congress, v.3, p. 3–9.
Reserve growth

Adelman, M. A., J. C. Houghton, G. Kaufman, and M. B. Zimmerman, 1983, Energy resources in an uncertain future: coal, gas, oil, and uranium supply forecasting: Cambridge, Massachusetts, Ballinger Publishing Co., 434 p.
Direct assessment, discovery modeling, historical extrapolation, play analysis, and volumetric yields

Arbatov, A. A., and A. V. Kondadov, 1979, Oil and gas occurrences in small intermontane basins: International Geology Review, v. 21, n. 3, p. 368–372.
Volumetric yields

Arps, J. J., and T. G. Roberts, 1958, Economics of drilling for Cretaceous oil on east flank of Denver–Julesburg Basin: AAPG Bulletin, v. 42, n. 11, p. 2549–2566.
Discovery modeling

Arps, J. J., M. Mortada, and A. E. Smith, 1971, Relationship between proved reserves and exploratory effort: Journal of Petroleum Technology, v. 23, n. 6, p. 671–675.
Historical extrapolation, reserve growth, and volumetric yields

Arrington, J. R., 1960, Predicting the size of crude reserves is key to evaluating exploration programs: Oil and Gas Journal, v. 58, n. 9, p. 130–134.
Reserve growth

——— , 1966, Estimation of future reserve revision in current fields, in Economics and the Petroleum Geologist Symposium Transactions: Midland, Texas, West Texas Geological Society Publication 66-53, p. 16–30.
Reserve growth

Ashton, P. R., 1981, Estimating potential reserves in Southeast Asian Neogene reefs, in Proceedings of the Seminar on Assessment of Undiscovered Oil and Gas, Kuala Lumpur, Malaysia: United Nations ESCAP CCOP Technical Publication n. 10, p. 244–259.
Play analysis

Attanasi, E. D., and L. J. Drew, 1977, Field expectations and the determinants of wildcat drilling: Southern Economic Journal, v. 44, n. 1, p. 53–67.
Discovery modeling

Attanasi, E. D., and D. H. Root, 1981, Petroleum potential of Latin America and Africa from a global perspective: Oil and Gas Journal, v. 79, n. 44, p. 187–205.
Historical extrapolation

Attanasi, E. D., and J. L. Haynes, 1983a, Economics and the appraisal of conventional oil and gas resources in the western Gulf of Mexico, in Proceedings of the Hydrocarbon Economics Evaluation Symposium, Society of Petroleum Engineers, n. 11297, p. 83–94.
Discovery modeling

——— , 1983b, Future supply of oil and gas from the Gulf of Mexico: U.S. Geological Survey Professional Paper 1294, 21 p.
Discovery modeling

Attanasi, E. D., T. M. Garland, J. H. Wood, W. D. Dietzman, and J. N. Hicks, 1979, Economics and resource appraisal—the case of the Permian Basin: Eighth Symposium of the Society of Petroleum Engineers of the American Institute of Mining, Metallurgical, and Petroleum Engineers on Petroleum Economics and Evaluation, n. 7738, p. 227–234. [Reprinted in 1981, Journal of Petroleum Technology, v. 33, n. 4, p. 603–616.]
Discovery modeling

Attanasi, E. D., L. J. Drew, and J. H. Schuenemeyer, 1980, Petroleum-resource appraisal and discovery rate forecasting in partially explored regions—an application to supply modeling: U.S. Geological Survey Professional Paper 1138-C, 20 p.
Discovery modeling

Attanasi, E. D., L. J. Drew, and D. H. Root, 1981, Physical variables and the petroleum discovery process, in J. B. Ramsey, ed., The economics of exploration for energy resources: Greenwich, Connecticut, JAI Press, p. 3–18.
Discovery modeling

Atwater, G. I., 1956, Future of Louisiana offshore oil province: AAPG Bulletin, v. 40, n. 11, p. 2624–2634.
Play analysis

Baecher, G. B., 1975, Subjective sampling approaches to resource estimation, in M. Grenon, ed., Proceedings of the First IIASA Conference on Energy Resources, Laxenburg, p. 251–274. [Reprinted in M. Grenon, ed., 1979, Methods and models for assessing energy resources, Proceedings of the First IIASA Conference on Energy Resources, Laxenburg, Austria: Oxford, Pergamon Press, p. 186–209.]
Discovery modeling

Baker, R. A., H. M. Gehman, W. R. James, and D. A. White, 1984, Geologic field number and size assessments of oil and gas plays: AAPG Bulletin, v. 68, n. 4, p. 426–437.
Field size analysis and play analysis

Bakirov, A., and Sc. G. Ovanessov, 1971, Scientific principles of calculation of potential oil and gas resources in connection with their evaluation in estimating different prospective oil and gas provinces: Proceedings of the Eighth World Petroleum Congress, v. 2, p. 315–322.
Areal yields, historical extrapolation, material balance, play analysis, and volumetric yields

Barouch, E., and G. M. Kaufman, 1975a, Predicting undiscovered oil and gas in a play using a stochastic model of discovery, in J. C. Davis, J. H. Doveton, and J. W. Harbaugh, convenors, Probability methods in oil exploration, AAPG Research Symposium, Stanford, California: Kansas Geological Survey, 7 p.
Discovery modeling

——— , 1975b, A probabilistic model of oil and gas discovery, in M. Grenon, ed., Proceedings of the First IIASA Conference on Energy Resources, Laxenburg, Austria, p. 311–324. [Reprinted in M. Grenon, ed., 1979, Methods and models for assessing energy resources, Proceedings of the First IIASA Conference on Energy Resources, Laxenburg, Austria: Oxford, Pergamon Press, p. 248–260.]
Discovery modeling and field size analysis

——— , 1976a, Oil and gas discovery modeled as sampling proportional to random size: Alfred P. Sloan School of Management Working Paper 888-76: Cambridge, Massachusetts Institute of Technology, 64 p.
Discovery modeling

——— , 1976b, Probabilistic modelling of oil and gas discovery, in F. S. Roberts, ed., Mathematics and models: Philadelphia, Society for Industrial and Applied Mathematics, p. 133–152.
Discovery modeling

——— , 1977, Estimation of undiscovered oil and gas, in Mathematical Aspects of Production and Distribution of Energy, Proceedings of the Symposium in Applied Mathematics of the American Mathematical Society, San Antonio, Texas: Providence, American Mathematical Society, v. 21, p. 77–91.
Discovery modeling

——— , 1978, The interface between geostatistical modeling of oil and gas discovery and economics: Journal of the International Association for Mathematical Geology, v. 10, n. 5, p. 611–627.
Discovery modeling and play analysis

Barss, D. L., 1978, The significance of petroleum resource estimates and their relation to exploration: Bulletin of Canadian Petroleum Geology, v. 26, n. 2, p. 275–291.
Direct assessment, historical extrapolation, play analysis, and volumetric yields

——— , 1980, Conventional petroleum resource estimates: methods of assessment and their implication for planning and policy issues, in G. D. Hobson, ed., Developments in petroleum geology—2: London, Applied Science Publishers, p. 299–338.
Field size analysis, historical extrapolation, play analysis, reserve growth, and volumetric yields

Baxter, G. G., S. M. Cargill, A. H. Chidester, P. E. Hart, G. M. Kaufman, and F. Urquidi-Barrau, 1978, Workshop on the Delphi method: Journal of the International Association for Mathematical Geology, v. 10, n. 5, p. 581–587.
Direct assessment

Beckie, K. N., 1975, A probabilistic assessment of Alberta's undiscovered petroleum, in J. C. Davis, J. H. Doveton, and J. W. Harbaugh, convenors, 1975, Probability methods in Oil Exploration, AAPG Research Symposium, Stanford, California: Kansas Geological Survey, p. 7–11.
Historical extrapolation

Beebe, B. W., R. J. Murdy, and E. A. Rassinier, 1975, Potential Gas Committee and undiscovered supplies of natural gas in United States, in J. D. Haun, ed., Methods of estimating the volume of undiscovered oil and gas resources: AAPG Studies in Geology, n. 1, p. 90–96.
Volumetric yields

Berg, R. R., J. C. Calhoun, Jr., and R. L. Whiting, 1974, Prognosis for expanded U.S. production of crude oil: Science, v. 184, n. 4134, p. 331–336.
Historical extrapolation

Beskow, K. M., and H. C. Ronnevik, 1981, Norwegian North Sea—a case history, in Proceedings of the Seminar on Assessment of Undiscovered Oil and Gas, Kuala Lumpur, Malaysia: United Nations ESCAP CCOP Technical Publication n. 10, p. 260–262.
Play analysis

Bilibin, V., 1936, Methods of estimating underground oil reserves: 17th International Geological Congress, Moscow, 1937: Leningrad [Preprint published by the Congress], 24 p.
Play analysis

Bird, K. J., 1984, A comparison of the play-analysis technique as applied in hydrocarbon resource assessments of the National Petroleum Reserve in Alaska and of the Arctic National Wildlife Refuge: U.S. Geological Survey Open-File Report 84-78, 18 p. [Reprinted in C. D. Masters, ed., Petroleum resource assessment: Ottawa, International Union of Geological Sciences Publication n. 17, p. 63–79.]
Play analysis

Bishop, R. S., H. M. Gehman, Jr., and A. Young, 1983, Concepts for estimating hydrocarbon accumulation and dispersion: AAPG Bulletin, v. 67, n. 3, p. 337–348. [Reprinted in G. Demaison and R. J. Murris, eds., 1984, Petroleum geochemistry and basin evaluation: AAPG Memoir, n. 35, p. 41–52.]
Material balance

Bleie, J., K. A. Oppeboen, and E. Nysaether, 1982, The hydrocarbon potential of the Northern Norwegian Shelf in the light of recent drilling: The geologic framework and hydrocarbon potential of basins in the northern seas, Offshore Northern Seas Conference and Exhibition, Stavanger, Norway, 1982, v. 3, p. E/5-1–E/5-39.
Volumetric yields

Bloomfield, P., K. S. Deffeys, B. Silverman, G. S. Watson, Y. Benjamini, and R. A. Stine, 1980, Volume and area of oil fields and their impact on order of discovery: Princeton University, Department of Statistics and Geology, 53 p.
Discovery modeling

Bois, C., H. Cousteau, A. Perrodon, and G. Pommier, 1980, Methodes d'estimation des reserves ultimes: Proceedings of the Tenth World Petroleum Congress, v. 2, p. 279–289.
Field size analysis, historical extrapolation, material balance, and play analysis

Borg, I. Y., 1975, Appraisal of current methods of evaluating crude oil resources: Livermore, California, Lawrence Livermore Laboratory Report UCRL-51848, 31 p.
Historical extrapolation, play analysis, and volumetric yields

Bradley, P. G., 1971, Exploration models and petroleum production economics, in M. A. Adelman, P. G. Bradley and C. A. Norman, Alaskan oil: costs and supply: New York, Praeger Publishers, p. 94–122.
Discovery modeling, field size analysis, and reserve growth

Brashear, J. P., F. Morra, C. Everett, F. H. Murphy, W. Hery, and R. Ciliano, 1982, A prospect specific simulation model of oil and gas exploration in the outer continental shelf: methodology, in S. I. Gass, ed., Proceedings of Oil and Gas Supply Modeling: Washington, National Bureau of Standards Special Publication 631, p. 688–738.
Play analysis

Brooks, J. R. V., 1981, Current status of UK exploration and estimation of undiscovered hydrocarbon on the UK continental shelf, in Proceedings of the Seminar on Assessment of Undiscovered Oil and Gas, Kuala Lumpur, Malaysia: United Nations ESCAP CCOP Technical Publication n. 10, p. 203–217.
Play analysis

Buyalov, N. I., and Ye. V. Zakharov, 1964, Use of the volume method for estimating prognostic reserves of oil: Petroleum Geology, v. 8, n. 7, p. 372–375.
Play analysis

Buyalov, N. I., N. I., V. G. Vasil'yev, N. D. Elin, N. S. Yerofeyev, M. S. L'vov, A. I. Kleshchev, N. M. Kudryashova, and V. L. Soklov, 1961, Method of estimating reserves of natural gas and oil: Petroleum Geology, v. 5, n. 1, p. 11–15.
Play analysis

Buyalov, N. I., N. S. Erofeev, N. A. Kalinen, A. I. Kleschev, N. M. Kudryashova, M. S. L'vov, S. N. Simakov, and V. G. Vasil'ev, 1964, Metodika otsenki prognoznykh zapasov nefti i gaza [Quantitative evaluation of predicted reserves of oil and

gas]: New York, Consultants bureau Translation, 69 p.
Areal yields, field size analysis, material balance, play analysis, and volumetric yields

Canada Department of Energy, Mines, and Resources, 1973, Energy reserves and potential resources, *in* An energy policy for Canada—phase 1: Ottawa, The Minister of Energy, Mines, and Resources, v. 2, p. 31–98.
Direct assessment, play analysis, and volumetric yields

——— , 1977, Oil and natural gas resources of Canada, 1976: Ottawa, Energy, Mines and Resources Canada Report EP 77-1, 76 p.
Play analysis, probability methods, volumetric yields

Capen, E. C., 1976, The difficulty of assessing uncertainty: Journal of Petroleum Technology, v. 28, p. 843–850.
Probability methods

Cargill, S. M., R. F. Meyer, D. D. Picklyk, and F. Urquidi, 1977, Summary of resource assessment methods resulting from the International Geological Correlation Programme Project 98: Journal of the International Association for Mathematical Geology, v. 9, n. 3, p. 211–220.
Areal yields, direct assessment, play analysis, and volumetric yields

Charpentier, R. R., W. de Witt, Jr., G. E. Claypool, L. D. Harris, R. F. Mast, J. D. Megeath, J. B. Roen, J. W. Schmoker, 1982, Estimates of unconventional natural-gas resources of the Devonian shale of the Appalachian Basin: U.S. Geological Survey Open-File Report 82-474, 43 p.
Play analysis

Chaube, A. N., and D. N. Avasthi, 1981, Methodology and results of prognostic resource assessment of Indian basins, *in* Proceedings of the Seminar on Assessment of Undiscovered Oil and Gas, Kuala Lumpur, Malaysia: United Nations ESCAP CCOP Technical Publication n. 10, p. 218–230.
Areal yields, material balance, play analysis

Cherniavsky, E. A., 1980, Long-range oil and gas forecasting methodologies: literature survey: Upton, New York, Brookhaven National Laboratory Report n. 51216, 22 p.
Historical extrapolation

Cherskii, N. V., and V. P. Tsarev, 1977 [1978], Estimating reserves in light of exploration for and extraction of natural gas from world ocean floor sediments: Soviet Geology and Geophysics [Geologiya i Geofizika], v. 18, n. 5, p. 21–31.
Material balance

Clark, A. L., 1981, Introduction to the process of resource assessment, *in* Proceedings of the Seminar on Assessment of Undiscovered Oil and Gas, Kuala Lumpur, Malaysia: United Nations ESCAP CCOP Technical Publication n. 10, p. 107–112.
Areal yields, discovery modeling, field size analysis, historical extrapolation, material balance, and volumetric yields

Claypool, G. E., A. H. Love, and E. K. Maughan, 1978, Organic geochemistry, incipient metamorphism, and oil generation in black shale members of Phosphoria Formation, western interior United States: AAPG Bulletin, v. 62, n. 1, p. 98–120. [Reprinted *in* G. Demaison and R. J. Murris, eds., 1984, Petroleum geochemistry and basin evaluation: AAPG Memoir 35, p. 139–158.]
Material balance

Combs, E. J., 1971, Summary of future petroleum potential of Region 8, Michigan Basin, *in* I. H. Cram, ed., Future petroleum provinces of the United States—their geology and potential: AAPG Memoir 15 v. 2, p. 1121–1164.
Areal yields and volumetric yields

Conybeare, C. E. B., 1965, Hydrocarbon-generation potential and hydrocarbon-yield capacity of sedimentary basins: Bulletin of

Canadian Petroleum Geology, v. 13, n. 4, p. 509–528.
Material balance and volumetric yields

Cook-Clark, Jennifer, 1983, World finding-rate studies—crude oil: U.S. Geological Survey Open-File Report 83-715, 257 p.
Historical extrapolation

Coustau, H., 1979, Logique de distribution des tailles des champs dans les bassins [Field size distribution in basins]: Petrole et techniques, n. 262, p. 23–30.
Field size analysis

——— , 1981, Habitat of hydrocarbons and field size distribution: a first step towards ultimate reserve assessment, *in* Proceedings of the Seminar on Assessment of Undiscovered Oil and Gas, Kuala Lumpur, Malaysia: United Nations ESCAP CCOP Technical Publication n. 10, p. 180–194.
Field size analysis

Crandall, K. H., 1975, Estimating petroleum resources, *in* AGI's White House papers on earth science: Geotimes, v. 20, n. 9, p. 24.
Volumetric yields

Crovelli, R. A., 1981, Probabilistic methodology for oil and gas resource appraisal: U.S. Geological Survey Open-File Report 81-1151, 77 p.
Probability methods

——— , 1983a, Probabilistic methodology for petroleum resource appraisal of wilderness lands, *in* B. M. Miller, ed., Petroleum potential of wilderness lands in the Western United States: U.S. Geological Survey Circular 902 A-P, p. 1–5.
Probability methods

——— , 1983b, Procedures for petroleum resource assessment used by the USGS statistical and probabilistic methodology: U.S. Geological Survey Open-File Report 83-402, 21 p.
Probability methods

——— , 1984a, Procedures for petroleum resource assessment used by the U.S. Geological Survey—statistical and probabilistic methodology, *in* C. D. Masters, ed., Petroleum resource assessment: Ottawa, International Union of Geological Sciences Publication n. 17, p. 23–38.
Discovery modeling and probability methods

——— , 1984b, U.S. Geological Survey probabilistic methodology for oil and gas resource appraisal of the United States: Journal of the International Association for Mathematical Geology, v. 16, n. 8, p. 797–808.
Probability methods

Davis, J. C., and J. W. Harbaugh, 1980, Oil and gas in offshore tracts: inexactness of resource estimates prior to drilling: Science, v. 209, n. 4460, p. 1047–1048.
Play analysis

——— , 1981a, A method for rapid evaluation of oil and gas prospects in OCS regions: Lawrence, Kansas, Kansas Geological Survey, prepared for the U.S. Geological Survey Contract n. 14-08-001-18785, 62 p. [Available for viewing in Reston, Va., at the U.S. Geological Survey Library.]
Play analysis

——— , 1981b, A simulation model for oil exploration policy on federal lands of the U.S. outer continental shelves, *in* J. B. Ramsey, ed., The economics of exploration for energy resources: Greenwich, Connecticut, JAI Press, p. 19–50.
Direct assessment and play analysis

——— , 1983, Statistical appraisal of seismic prospects in Louisiana and Texas outer continental shelf: AAPG Bulletin, v. 67, n. 3, p. 349–358.
Play analysis

Davis, W., 1958, A study of the future productive capacity and probable reserves of the U.S.: Oil and Gas Journal, v. 56, n. 8, p. 105–119.
Historical extrapolation

——, 1965, the enigma of oil and gas finding costs, *in* Third Symposium on petroleum Economics and Evaluation, Dallas: Society of Petroleum Engineers of the American Institute of Mining, Metallurgical and Petroleum Engineers, Dallas Section, p. 19–27.
Reserve growth

DeGolyer, E., 1951, On the estimation of undiscovered oil reserves: Journal of Petroleum Technology, v. 3, n. 1, p. 9–10.
Areal yields and volumetric yields

Dix, S. M., 1977, The petroleum figures, *in* Energy, a critical decision for the United States economy: Grand Rapids, Michigan, Energy Education Publishers, p. 63–103.
Historical extrapolation

Dolton, G. L., 1984, Basin assessment methods and approaches in the U.S. Geological Survey, *in* C. D. Masters, ed., Petroleum resource assessment: Ottawa, International Union of Geological Sciences Publication n. 17, p. 4–23.
Areal yields, direct assessment, historical extrapolation, material balance, play analysis, and volumetric yields

Dolton, G. L., G. L., A. B. Coury, S. E. Frezon, K. Robinson, K. L. Varnes, J. M. Wonder and R. W. Allen, 1979, Estimates of Undiscovered oil and gas, Permian Basin, West Texas and Southeast New Mexico: U.S. Geological Survey Open-File Report 79-838, 114 p.
Areal yields, direct assessment, field size analysis, historical extrapolation, probability methods, and volumetric yields

Dolton, G. L., K. H. Carlson, R. R. Charpentier, A. B. Coury, R. A. Crovelli, S. E. Frezon, A. S. Khan, J. H. Lister, R. H. McMullin, R. S. Pike, R. B. Powers, E. W. Scott, and K. L. Varnes, 1981, Estimates of undiscovered recoverable conventional resources of oil and gas in the United States: U.S. Geological Survey Circular 860, 87 p.
Areal yields, direct assessment, historical extrapolation, material balance, play analysis, probability methods, reserve growth, and volumetric yields

Drew, L. J., 1966, Grid drilling exploration and its application to the search for petroleum: Ph.D. thesis, Pennsylvania State University, University Park, 141 p.
Discovery modeling

——, 1967, Grid-drilling exploration and its application to the search for petroleum: Economic Geology, v. 62, n. 5, p. 698–710.
Discovery modeling

——, 1972, Spatial distribution of the probability of occurrence and the value of petroleum: Kansas, an example: Journal of the International Association for Mathematical Geology, v. 4, n. 2, p. 155–171.
Areal yields

——, 1974, Estimation of petroleum exploration success and the effects of resource base exhaustion via a simulation model: U.S. Geological Survey Bulletin 1328, 25 p.
Discovery modeling

——, 1975a, Analysis of the rate of wildcat drilling and deposit discovery: Journal of the International Association for Mathematical Geology, v. 7, n. 5/6, p. 395–414.
Historical extrapolation

——, 1975b, Analysis of the rate of wildcat drilling and petroleum deposit recovery, *in* J. C. Davis, J. H. Doveton, and J. W. Harbaugh, convenors, 1975, Probability methods in Oil Exploration, AAPG Research Symposium, Stanford, California: Kansas Geological Survey, p. 12–15.
Historical extrapolation

Drew, L. J., and D. H. Root, 1980, Data requirements for forecasting the year of world peak petroleum production, *in* D. A. Gardiner and T. Truett, compiler/ed., Proceedings of the Department of Energy Statistical Symposium, Gatlinburg, Tennessee, p. 227–230.
Historical extrapolation

——, 1982, Statistical estimate of tomorrow's offshore oil and gas fields: Ocean Industry, v. 17, n. 5, p. 54–58, 66.
Discovery modeling

Drew, L. J., E. D. Attanasi, and D. H. Root, 1977, Importance of physical parameters in petroleum supply models, *in* J. H. DeYoung, Jr., ed., Mineral policies in transition: Proceedings of the Third Mineral Economics Symposium, Washington, D.C.: American Association of Mining, Metallurgical, and Petroleum Engineers, p. 52–69. [Reprinted *in* 1979, Materials and Society, v. 3, n. 2, p. 163–174].
Discovery modeling

Drew, L. J., J. H. Schuenemeyer, and D. H. Root, 1977, Statistical history of petroleum exploration in Denver Basin (abs.): AAPG Bulletin, v. 61, n. 5, p. 782.
Discovery modeling

——, 1978, The use of a discovery process model based on the concept of area of influence of a drill hole to predict discovery rates in the Denver basin, *in* J. D. Pruit and P. E. Coffin, eds., Energy Resources of the Denver Basin: Denver, Rocky Mountain Association of Geologists, p. 31–34.
Discovery modeling

Drew, L. J., D. H. Root, and W. J. Bawiec, 1979, Estimating future rates of petroleum discovery in the Permian Basin: Eighth Symposium of the Society of Petroleum Engineers of the American Institute of Mining, Metallurgical, and Petroleum Engineers on Hydrocarbon Economics and Evaluation, n. 7722, p. 101–106.
Discovery modeling and field size analysis

Drew, L. J., J. H. Schuenemeyer, and W. J. Bawiec, 1979, Petroleum exhaustion maps of the Cretaceous "D–J" sandstone stratigraphic interval of the Denver Basin: U.S. Geological Survey Miscellaneous Investigations Series Map I-1138, scale 1:200,000, 4 sheets, 7 p.
Discovery modeling and field size analysis

Drew, L. J., J. H. Schuenemeyer, and D. H. Root, 1980, Petroleum-resource appraisal and discovery rate forecasting in partially explored regions—an application to the Denver Basin: U.S. Geological Survey Professional Paper 1138-A, p. A1–A11.
Discovery modeling and field size analysis

Drew, L. J., J. H. Schuenemeyer, and W. J. Bawiec, 1982, Estimation of the future rates of oil and gas discoveries in the western Gulf of Mexico: U.S. Geological Survey Professional Paper 1252, 26 p.
Discovery modeling, field size analysis, and historical extrapolation

Drew, L. J., G. C. Grender, and R. M. Turner, 1983, Atlas of discovery rate profiles showing oil and gas discovery rates by geological province in the United States: U.S. Geological Survey Open-File Report 83-75, 269 p.
Historical extrapolation

Ducastaing, M. and J. W. Harbaugh, 1980, Forecasting future oil field sizes through statistical analysis of historical changes in oil field populations, *in* S. I. Gass, ed., Oil and Gas Supply

Modeling Proceedings: Washington, D.C., National Bureau of Standards Special Publication 631, p. 200–256.
Field size analysis

Dvali, M. F., and T. P. Dmitrieva, 1976, Ob'emno-statisticheskii metod poscheta prognoznykh napasov nefti i gaza [Volumetric-statistical method calculates reserve forecasts for oil and gas]: Leningrad, Nedra, Trudy, vyp. 363, 110 p.
Volumetric yields

Edgar, N. T., and K. C. Bayer, 1979, Assessing oil and gas resources on the U.S. continental margin: Oceanus, v. 22, n. 3, p. 12–22.
Volumetric yields

Elliott, M. A., and H. R. Linden, 1968, A new analysis of U.S. natural gas supplies: Journal of Petroleum Technology, v. 20, Feb., p. 135–141.
Historical extrapolation

Fitzpatrick, A., B. Hitchon, and J. R. McGregor, 1973, Long-term growth of the oil industry in the United States: Journal of the International Association for Mathematical Geology, v. 5, n. 3, p. 237–267.
Historical extrapolation

Flawn, P. T., 1967, Concepts of resources: their effects on exploration and United States mineral policy, *in* Exploration and economics of the petroleum industry, v. 5: Houston, Gulf Publishing Co., p. 5–24.
Historical extrapolation and volumetric yields

Folinsbee, R. E., 1977, World's view—from Alph to Zipf: Geological Society of America Bulletin, v. 88, n. 7, p. 897–907.
Field size analysis

Forman, D. J., and A. L. Hinde, 1985, Improved statistical method for assessment of undiscovered petroleum resources: AAPG Bulletin, v. 69, n. 1, p. 106–118.
Discovery modeling

Gangwar, A., H. C. Kent, J. H. Schuenemeyer, J. Guzman, and S. Snow, 1983, Econometric and resource modeling methodology for projections of cost of development of U.S. natural gas potential: Society of Petroleum Engineers of the American Institute of Mining, Metallurgical, and Petroleum Engineers Hydrocarbon Economics and Evaluation Symposium, SPE 11296, p. 69–81.
Discovery modeling

Garland, T. M., M. Carrales, Jr., and J. S. Conway, 1974, Assessment of U.S. petroleum supply with varying drilling efforts: U.S. Bureau of Mines Information Circular 8634, 36 p.
Reserve growth

Garrett, R. W., Jr., G. R. Marsh, R. A. Baker, H. M. Gehman, and D. A. White, 1974, Assessing regional oil and gas potential, *in* K. H. Crandall and J. W. Harbaugh, convenors, Methods of estimating the volume of undiscovered oil and gas resources, Proceedings of the AAPG Research Symposium: Stanford, California, Stanford University, p. 311–344.
Historical extrapolation, probability methods, reserve growth, and volumetric yields.

Gehman, H. M., R. A. Baker, and D. A. White, 1975, Prospect risk analysis, *in* J. C. Davis, J. H. Doveton, and J. W. Harbaugh, convenors, Probability methods in oil exploration, AAPG Research Symposium, Stanford, California: Kansas Geological Survey, p. 16–20.
Play analysis

——— , 1981, Assessment methodology—an industry viewpoint, *in* Proceedings of the Seminar on Assessment of Undiscovered

Oil and Gas, Kuala Lumpur, Malaysia: United Nations ESCAP CCOP Technical Publication n. 10, p. 113–121.
Play analysis

Gess, G., and C. Bois, 1977, Study of petroleum zones: a contribution to the appraisal of hydrocarbon resources, *in* R. F. Meyer, ed., The future supply of nature-made petroleum and gas: First UNITAR Conference on Energy and the Future, Laxenburg, Austria, Technical Reports, p. 155–178.
Areal yields

Gill, D., and J. C. Griffiths, 1984, Areal value assessment of the mineral resources endowment of Israel: Journal of the International Association for Mathematical Geology, v. 16, n. 1, p. 37–89.
Areal yields

Gillette, R., 1974, Oil and gas resources—did U.S.G.S. gush too high?: Science, v. 185, n. 4146, p. 127–130.
Historical extrapolation, play analysis, and volumetric yields

Goff, J. C., 1983, Hydrocarbon generation and migration from Jurassic source rocks in the East Shetland Basin and Viking Graben of the Northern north Sea: Journal of the Geological Society of London, v. 140, pt. 3, p. 445–474. [Reprinted *in* G. Demaison and R. J. Murris, eds., 1984, Petroleum geochemistry and basin evaluation: AAPG Memoir 35, p. 139–158.]
Material balance

Golovin, L., 1970, Two mathematical models for oil and gas disposition: Cambridge, Sloan School of Management, Massachusetts Institute of Technology, M.S. dissertation, 65 p.
Discovery modeling

Gotautas, V. A., 1963, Quantitative analysis of prospect to determine whether it is drillable: AAPG Bulletin, v. 47, n. 10, p. 1794–1812.
Play analysis

Grenon, M., 1975, Methods for assessing petroleum resources, *in* M. Grenon, ed., Proceedings of the First IIASA Conference on Energy Resources, Laxenburg, Austria, p. 129–181. [Reprinted *in* M. Grenon, ed., 1979, Methods and models for assessing energy resources, Proceedings of the First IIASA Conference on Energy Resources, Laxenburg, Austria: Oxford, Pergamon Press, p. 115–116.]
Areal yields, field size analysis, historical extrapolation, and play analysis

Griffiths, J. C., 1962, Frequency distribution of some natural resource materials: 23rd Technical Conference on Petroleum Production, University Park, Pennsylvania: Pennsylvania State University, Mineral Industries Experiment Station Circular 63, p. 173–198.
Discovery modeling

——— , 1966, Exploration for natural resources: Operations Research, v. 14, n. 2, p. 189–209.
Discovery modeling

——— , 1967, Mathematical exploration strategy and decision-making: Proceedings of the Seventh World Petroleum Congress, Mexico, v. 2, p. 599–604.
Discovery modeling

Griffiths, J. C., and L. J. Drew, 1964, Simulation of exploration programs for natural resources by models: Quarterly of the Colorado School of Mines, v. 59, n. 4, pt. A, p. 187–206.
Discovery modeling

——— , 1966, Grid spacing and success ratios in exploration for the natural resources, *in* Proceedings of the symposium and short course on computers and operations research in mineral

industries, Sixth International Symposium on Computers and Operations Research: Pennsylvania State University, Mineral Industries Experiment Station Special Publication 2-65, p. Q-1–Q-24.
Discovery modeling

Griffiths, J. C., D. W. Menzie, M. L. Labovitz, 1975, Exploration for and evaluation of natural resources, *in* J. C. Davis, J. H. Doveton, and J. W. Harbaugh, convenors, 1975, Probability Methods in Oil Exploration, AAPG Research Symposium, Stanford, California: Kansas Geological Survey, p. 21–25.
Discovery modeling

Griggs, D. G., and R. J. Jaske, 1975, Preliminary USGS oil and gas resource estimate for federal OCS lease sale #35 off southern California, *in* J. C. Davis, J. H. Doveton, and J. W. Harbaugh, convenors, 1975, Probability Methods in Oil Exploration, AAPG Research Symposium, Stanford, California: Kansas Geological Survey, p. 26–31.
Play analysis

Grossling, B. F., 1975, In search of a statistical probability model for petroleum-resource assessment: U.S. Geological Survey Circular 724, 18 p. [revised *in* M. Grenon, ed., 1979, Methods and models for assessing energy resources, Proceedings of the First IIASA Conference on Energy Resources, 1st, Laxenburg, Austria: 1975: Oxford, Pergamon Press, p. 143–172.]
Discovery modeling and historical extrapolation

——, 1976, Window on oil: a survey of world petroleum sources: London, The Financial Times, 140 p.
Areal yields

——, 1977, A critical survey of world petroleum opportunities, *in* Congressional Research Service, Project Interdependence, U.S. and world energy outlook through 1990: 95th Congress, First Session, Committee Print 95-33, p. 645–658.
Areal yields and historical extrapolation

Halbouty, M. T., and J. D. Moody, 1980, World ultimate reserves of crude oil, *in* Proceedings of the Tenth World Petroleum Congress, Bucharest, v. 2, p. 291–301.
Play analysis and volumetric yields

Halbouty, M. T., A. A. Meyerhoff, R. E. King, R. H. Dott, Sr., H. D. Klemme, and T. Shabad, 1970, World's giant oil and gas fields, geologic factors affecting their formation, and basin classification, *in* M. T. Halbouty, ed., Geology of giant petroleum fields: AAPG Memoir 14, p. 502–555.
Volumetric yields

Hall, C. A. S., and C. J. Cleveland, 1981, Petroleum drilling and production in the United States: yield per effort and net energy analysis: Science, v. 211, n. 4482, p. 576–679.
Historical extrapolation

Harbaugh, J. W., and M. Ducastaing, 1981, Historical changes in oil-field populations as a method of forecasting field sizes of undiscovered populations: a comparison of Kansas, Wyoming, and California: Kansas Geological Survey, Subsurface Geology Series 5, 56 p.
Field size analysis

Harbaugh, J. W., J. H. Doveton, and J. C. Davis, 1977, Probability methods in oil exploration: New York, John Wiley, 269 p.
Areal yields, discovery modeling, field size analysis, and historical extrapolation

Harris, D. P., 1977, Quantitative methods for the appraisal of mineral resources: U.S. Department of Energy GJBX n. 14 (77), GJO-6344. [available from Grand Junction Office, U.S. Energy Research and Development Administration, prepared under contract nos. AT-05-1-16344 and E(05-1)-1665].
Historical extrapolation and volumetric yields

——, 1984, Mineral resources appraisal: mineral endowment, resources, and potential supply concepts, methods, and cases: Oxford, Claredon Press, 445 p.
Direct assessment, discovery modeling, and historical extrapolation

Haun, J. D., 1975a, Methods of estimating the volume of undiscovered oil and gas resources: AAPG Research Conference, *in* J. E. Haun, ed., Methods of estimating the volume of undiscovered oil and gas resources: AAPG Studies in Geology, n. 1, p. 1–7.
Areal yields, discovery modeling, historical extrapolation, and volumetric yields

——, 1975b, Statistical forecasting of exploration plays in the Muddy Formation of the Denver and Powder River Basins, *in* J. C. Davis, J. H. Doveton, and J. W. Harbaugh, convenors, 1975, Probability Methods in Oil Exploration, AAPG Research Symposium, Stanford, California: Kansas Geological Survey, 93 p.
Field size analysis

——, 1975c, What are ways of estimating unlocated oil, gas volumes?: Oil and Gas Journal, v. 73, n. 29, p. 94–96.
Areal yields, discovery modeling, historical extrapolation, and volumetric yields

Hedberg, H. D., 1975a, False precision in petroleum resource estimates, *in* Haun, J. D., ed., Methods of estimating the volume of undiscovered oil and gas resources: AAPG Studies in Geology, n. 1, p. 160.
Probability methods

——, 1975b, The volume-of-sediment fallacy in estimating petroleum resources, *in* J. D. Haun, ed., Methods of estimating the volume of undiscovered oil and gas resources: AAPG Studies in Geology, n. 1, p. 161.
Volumetric yields

Hendricks, T. A., 1965, Resources of oil, gas, and natural gas liquids in the United States and the world: U.S. Geological Survey Circular 522, 20 p.
Areal yields

——, 1975, Estimating resources of crude oil and natural gas in inadequately explored areas, *in* J. D. Haun, ed., Methods of estimating the volume of undiscovered oil and gas resources: AAPG Studies in Geology, n. 1, p. 19–22.
Areal yields

Herbert, J. H., 1982, A review and comparison of some commonly used methods of estimating petroleum resource availability: Energy Sources, v. 6, n. 4, p. 293–320.
Discovery modeling, historical extrapolation, and volumetric yields

——, 1983, A concise mathematical statement of the relationship between the Arps/Roberts and Barouch/Kaufmann models for estimating the petroleum resource base: Energy Sources, v. 7, n. 1, p. 33–42.
Discovery modeling

Hopkins, G. R., 1950, A projection of oil discovery 1949–1965: Journal of Petroleum Technology, June, sec. 1, p. 6–9, sec. 2, p. 6.
Historical extrapolation

Houpeurt, A. H., J. Groult, M. Mollier, C. L. Salle, P. R. Simandoux, and R. Thomere, 1975, Principe et methodes de calcul des reserves d'huile et de gaz, *in* Proceedings of the Ninth World Petroleum Congress: Tokyo, v. 3, p. 21–30.
Areal yields, historical extrapolation, material balance, and volumetric yields

Howarth, R. J., C. M. White, and G. S. Koch, Jr., 1980, On Zipf's Law applied to resource prediction: Transactions of the

Institution of Mining and Metallurgy, Applied Early Science, sec. b, v. 89, p. B182–B190.
Field size analysis

Hubbert, M. K., 1957, Nuclear energy and the fossil fuels: Drilling and production practice, Dallas, Texas, American Petroleum Institute, p. 7–25.
Historical extrapolation

——, 1959, Techniques of prediction with application to the petroleum industry: Shell Development Company Publication Preprint n. 204, 42 p.
Historical extrapolation

——, 1962, Energy resources: a report to the Committee on Natural Resources of the National Academy of Sciences—National Research Council: Washington, D.C., National Academy of Sciences—National Research Council Publication 1000-D, 141 p. [Reprinted, 1973, National Technical Information Service Report PB-222 401.]
Historical extrapolation

——, 1965, National Academy of Sciences report on energy resources: reply: AAPG Bulletin, v. 49, n. 10, p. 1720–1727.
Historical extrapolation

——, 1966a, M. King Hubbert's reply to J. M. Ryan: Journal of Petroleum Technology, v. 18, n. 3, p. 284–286.
Historical extrapolation

——, 1966b, History of petroleum geology and its bearing upon present and future exploration: AAPG Bulletin, v. 50, n. 12, p. 2504–2518.
Historical extrapolation

——, 1967, Degree of advancement of petroleum exploration in United States: AAPG Bulletin, v. 51, n. 11, p. 2207–2227.
Historical extrapolation, reserve growth, and volumetric yields

——, 1969, Energy resources, in National Research Council, Committee on Resources and Man, Resources and Man: a study and recommendations: San Francisco, W. H. Freeman and Co., p. 157–242.
Historical extrapolation

——, 1972, Estimation of oil and gas resources, in Proceedings of Workshop on Techniques of Mineral Resource Appraisal, Denver: U.S. Geological Survey, p. 16–50.
Historical extrapolation

——, 1973, Survey of world energy resources: Canadian Mining and Metallurgical Bulletin, v. 66, n. 735, p. 37–53.
Historical extrapolation, and volumetric yields

——, 1974, U.S. energy resources, a review as of 1972, in U.S. Senate Committee on Interior and Insular Affairs, U.S. energy resources, a review as of 1972, a background paper, pt. 1: 93rd Congress, 2nd Session, Committee Print, serial no. 93-40 (92-74), p. 1–201.
Historical extrapolation and volumetric yields

——, 1975, Ratio between recoverable oil per unit volume of sediments for future exploratory drilling to that of the past for the conterminous United States, in Mineral resources and the environment (Appendix to Sec. II), Report of Panel on Estimation of Mineral Reserves and Resources: Washington, D.C., National Academy of Sciences, p. 1–9 (p. 13–23).
Historical extrapolation and volumetric yields

——, 1979, Hubbert estimates from 1956 to 1974 of U.S. oil and gas, in M. Grenon, ed., Methods and models for assessing energy resources, Proceedings of the First IIASA Conference on Energy Resources, First, Laxenburg, Austria: Oxford, Pergamon Press, p. 370–383.
Historical extrapolation

——, 1982, Techniques of prediction as applied to the production of oil and gas, in S. I. Gass, ed., Oil and Gas Supply Modeling, Washington, 1980, Proceedings: National Bureau of Standards Special Publication 631, p. 16–141.
Historical extrapolation and volumetric yields

Hunt, J. M., 1962, Distribution of hydrocarbons in sedimentary rocks: Geochimica et Cosmochimica Acta, v. 22, n. 1, p. 37–49.
Volumetric yields

Ikoku, C. U., 1980, Decision analysis: how to make risk evaluations: World Oil, v. 191, n. 4, p. 71–74, 81, n. 5, p. 157, 158, 160, 162.
Play analysis

Ivanhoe, L. F., 1976, Evaluating prospective basins: Oil and Gas Journal, v. 74, n. 49, p. 154–155, n. 50, p. 108–110, n. 51, p. 82–84.
Field size analysis

Jeffries, F. S., 1975, Australian oil exploration—a great lottery: APEA Journal, v. 15, pt. 2, p 48–51.
Field size analysis, historical extrapolation and volumetric yields

Jones, D. A., N. A. Buck, and J. H. Kelsey, 1982, Choosing an optimum exploration strategy: World Oil, v. 195, n. 4, p. 71–86.
Discovery modeling

Jones, R. W., 1975, A quantitative geologic approach to prediction of petroleum resources, in J. D. Haun, ed., Methods of estimating the volume of undiscovered oil and gas resources: AAPG Studies in Geology, n. 1, p. 186–195.
Volumetric yields

Kaufman, G. M., 1962 [1963], Statistical decision and related techniques in oil and gas exploration: Ph.D. thesis, Harvard University, Cambridge, Massachusetts. [Reprinted, 1963, New York, Prentice Hall, 307 p.]
Field size analysis, discovery modeling, and reserve growth

——, 1965, Statistical analysis of the size distribution of oil and gas fields: Society of Petroleum Engineers of American Institute of Mining, Metallurgical, and Petroleum Engineers,, in Symposium on Petroleum Economics and Evaluation, Third Symposium on Petroleum Economics and Evaluation, Dallas, p. 109–124.
Field size analysis

——, 1974, Statistical methods for predicting the number and size distribution of undiscovered hydrocarbon deposits, in K. H. Crandall, and J. W. Harbaugh, convenors, Methods of estimating the volume of undiscovered oil and gas resources, Proceedings of the AAPG Research Symposium: Stanford, California, Stanford University, p. 247–310.
Discovery modeling and field size analysis

——, 1975, Models and methods for estimating oil and gas—what they do and do not do, in M. Grenon, ed., Proceedings of the First IIASA Conference on Energy Resources, Laxenburg, Austria, p. 237–249. [Reprinted in M. Grenon, ed., 1979, Methods and models for assessing energy resources, Proceedings of the First IIASA Conference on Energy Resources, Laxenburg, Austria: Oxford, Pergamon Press, p. 173–185.]
Historical extrapolation, direct assessment, discovery modeling, reserve growth, and volumetric yields

——, 1982, Issues past and present in modeling oil and gas supply, in S. I. Gass, ed., Oil and Gas Supply Modeling Proceedings: National Bureau of Standards Special Publication 631, p. 257–271.
Direct assessment, discovery modeling, and play analysis

Kaufman, G. M., and P. G. Bradley, 1973, Two stochastic models useful in petroleum exploration, in M. G. Pitcher, ed., Arctic Geology: AAPG Memoir 19, p. 633–637.
Discovery modeling

Kaufman, G. M., and J. W. Wang, 1980, Model mis-specification and the Princeton study of volume and area of oil fields and their impact on the order of discovery: Cambridge, Massachusetts Institute of Technology, Energy Laboratory Working Paper n. MIT-EL 80-003WP, 21 p.
Discovery modeling

Kaufman, G. M., Y. Balcer, and D. Kruyt, 1974, A probabilistic model of the oil and gas discovery process, in P. Benenson, H. Ruderman, D. Merrill, and J. Sathaye, eds., Proceedings of the Conference on Energy Modeling and Forecasting: Berkeley, University of California, Lawrence Berkeley Laboratory,: Springfield, Virginia, National Technical Information Service LBL 3635, p. 13–26.
Discovery modeling and field size analysis

————, 1975, A probabilistic model of oil and gas discovery, in J. D. Haun, ed., Methods of estimating the volume of undiscovered oil and gas resources: AAPG Studies in Geology, n. 1, p.113–142.
Discovery modeling and field size analysis

Kaufman, G. M., W. Runggaldier, and A. Livne, 1981, Predicting the time rate of supply from a petroleum play, in J. B. Ramsey, ed., The economics of exploration for energy resources: Greenwich, Connecticut, Jai Press, p. 69–102.
Discovery modeling

Kechek, G. A., 1965, Calculation of prognostic reserves of oil and gas fields: Petroleum Geology, v. 6, n. 3, p. 184–186.
Areal yields and play analysis

Kingston, G. A., M. David, R. F. Meyer, A. T. Ovenshine, S. Slamet, and J. J. Schanz, 1978, Workshop on volumetric estimation: Journal of the International Association for Mathematical Geology, v. 10, n. 5, p. 495–499.
Volumetric yields

Klemme, H. D., 1971, The giants and the supergiants: Oil and Gas Journal, v. 69, n. 9, p. 85–90, n. 10, p. 103–110, n. 11, p. 96–100.
Field size analysis

————, 1975a, Giant oil fields, related to their geologic setting: a possible guide to exploration: Bulletin of Canadian Petroleum Geology, v. 23, n. 1, p. 30–66.
Field size analysis and volumetric yields

————, 1975b, Geothermal gradients, heat flow, and hydrocarbon recovery, in A. G. Fischer and S. Judson, eds., Petroleum and global tectonics: Princeton, New Jersey, Princeton University Press, p. 251–304.
Volumetric yields

————, 1977, World oil and gas reserves from analysis of giant fields and petroleum basins (provinces), in R. F. Meyer, ed., The future supply of nature-made petroleum and gas, in Technical Reports of the First UNITAR Conference on Energy and the Future, Laxenburg, Austria: New York, Pergamon Press, p. 217–260.
Volumetric yields

————, 1980, Petroleum basins—classifications and characteristics: Journal of Petroleum Geology, v. 3, n. 2, p. 187–207.
Volumetric yields

————, 1983, Field size distribution related to basin characteristics: Oil and Gas Journal, v. 81, n. 52, p. 168–176.
Field size analysis

————, 1984, Field size distribution related to basin characteristics, in C. D. Masters, ed., Petroleum resource assessment: Ottawa, International Union of Geological Sciences Publication n. 17, p. 95–121.
Field size analysis

Kontorovich, A. E., 1950, Teoreticheskie osnovy obemno-geneticheskogo metoda otsenki potentsialnykh resursov nefti i gaza [Theoretical basis for the volume-genetic method of estimating potential resources of oil and gas]: Sibirskogo Nauchno-Issledovatel'skogo Institut Geologii, Geofiziki i Mineral'nogo Syr'ia Trudy, v. 95., 51 p.
Material balance

————, 1976, Geokhimicheskie metody kolichestvennogo prognoza neftegazonosnosti [Geochemical methods for the quantitative evaluation of petroleum potential]: Moskva, Sibirskii Nauchno-Issledovatel'skii Institut Geologii, Geofiziki i Mineral'nogo Syr'ia, Trudy, v. 229, 249 p.
Material balance

————, 1984, Geochemical methods for the quantitative evaluation of the petroleum potential of sedimentary basins, in G. Demaison and R. J. Murris, eds., Petroleum geochemistry and basin evaluation: AAPG Memoir 35, p. 79–109.
Material balance

Kontorovich, A. E., and V. I. Demin, 1977 [1979], Metod otsenki kolichestva i raspredeleniya po zapasam mestorozhdeniy nefti i gaza v krupnykh neftegazonosnykh basseynakh [A method of assessing the amount and distribution of oil and gas reserves in large oil and gas basins]: International Geology Review, v. 21, p. 361–367.
Field size analysis

Korotkov, S. T., 1959, Method of planning exploration in the Azov–Kuban oil–gas basin: Petroleum Geology, v. 3, n. 9A, p. 526–528.
Play analysis

Kunin, N. Ya., and T. B. Krasil'nikova, 1980, A statistical method of forecasting the number and size of petroleum-bearing anticline structures in platform regions: International Geology Review, v. 23, n. 12, p. 1443–1448.
Field size analysis

Lador, M., 1981, Historical oil discovery trend in Libya, in Proceesings of the Seminar on Assessment of Undiscovered Oil and Gas, Kuala Lumpur, Malaysia: United Nations ESCAP CCOP Technical Publication n. 10, p. 231–243.
Historical extrapolation

Landes, K. K., 1973, The estimation of undiscovered natural gas supplies: National Gas Survey, U.S. Federal Power Commission, v. 5, p. 237–256.
Areal yields, historical extrapolation, and volumetric yields

Lee, P. J., and P. C. C. Wang, 1983a, Probabilistic formulation of a method for the evaluation of petroleum resources: Journal of the International Association for Mathematical Geology, v. 15, n. 1, p. 163–181.
Field size analysis and play analysis

————, 1983b, Conditional analysis for petroleum resource evaluation: Journal of the International Association for Mathematical Geology, v. 15, n. 2, p. 349–361.
Field size analysis and play analysis

————, 1984, PRIMES: a petroleum resources information management and evaluation system: Oil and Gas Journal, v. 82, n. 40, p. 204–206.
Discovery modeling and play analysis

Levorsen, A. I., 1950, Estimates of undiscovered petroleum reserves, in Proceedings of Plenary Meetings, United Nations

Scientific Conference on the Conservation and utilization of Resources: Lake Success, New York, United Nations Department of Economic Affairs, v. 1, p. 94–99.
Volumetric yields

Lin, J. and J. Wang, 1982, Geologic model for basin study: Geologic Research Institute of Shengli Oil Field, China, 21 p.
Material balance

Lovejoy, W. F., and P. T. Homan, 1965, Methods of estimating reserves of crude oil, natural gas, and natural gas liquids: Washington, D.C., Resources for the Future, and Baltimore, John Hopkins Press, 163 p.
Historical extrapolation

Mackay, I. H., and F. K. North, 1975, Undiscovered oil reserves, *in* J. D. Haun, ed., Methods of estimating the volume of undiscovered oil and gas resources: AAPG Studies in Geology, n. 1, p. 76–86.
Historical extrapolation

Mallory, W. W., 1972, A statistical–stratigraphic method for computing undiscovered resources of petroleum and natural gas including an estimate for eastern Colorado, Workshop on Techniques of Mineral Resources Appraisal, Denver, March 23–24: U.S. Geological Survey, p. 10–16.
Volumetric yields

——— , 1975a, Accelerated National Oil and Gas Resource Appraisal (ANOGRE), *in* J. D. Haun, ed., Methods of estimating the volume of undiscovered oil and gas resources: AAPG Studies in Geology, n. 1, p. 23–30.
Volumetric yields

——— , 1975b, Synopsis of procedure: Accelerated National Oil and Gas Resource Evaluation, *in* Mineral resources and the environment (Appendix to Sec. II), Report of Panel on Estimation of Mineral Reserves and Resources: Washington, National Academy of Sciences, p. 1–5.
Volumetric yields

Marland, G., 1978, A random drilling model for placing limits on ultimately recoverable crude oil in the conterminous U.S.: Materials and Society, v. 2, n. 1-2, p. 5–14.
Discovery modeling

Marsh, G. R., 1971, How much oil are we really finding?: Oil and Gas Journal, v. 69, n. 14, p. 100–104.
Reserve growth

Martinez, A. R., 1961, Tecnicas de prediccion aplicables a la industria petrolera de Venezuela, *in* Third Congreso Geologico Venezolano, 3rd, Caracas, 1959, Memoria, T. 4: Venezuela, Ministerio de Minas e Hidrocarburos, Direccion de Geologia, Boletin de Geologia, Publication Especial n. 3, p. 1531–1554.
Historical extrapolation and volumetric yields

——— , 1963, Estimation of the magnitude and duration of petroleum resources: Proceedings of the Sixth World Petroleum Congress, Frankfurt, sec. 8, paper 17, p. 133–148.
Historical extrapolation and volumetric yields

——— , 1966, Estimation of petroleum resources: AAPG Bulletin, v. 50, n. 9, p. 2001–2008.
Areal yields

Mast, R. F., and J. Dingler, 1975, Estimates of inferred plus indicated reserves for the United States by states, *in* B. M. Miller, H. L. Thomsen, G. L. Dolton, A. B. Coury, T. A. Hendricks, F. E. Lennartz, R. B. Powers, E. G. Sable, and K. L. Varnes, Geological estimates of undiscovered recoverable oil and gas resources in the United States: U.S. Geological Survey Circular 725, p. 73–78.
Reserve growth

Mast, R. F., R. H. McMullin, K. J. Bird, and W. P. Brosge, 1980, Resource appraisal of undiscovered oil and gas resources in the William O. Douglas Arctic Wildlife Range: U.S. Geological Survey Open-File Report 80-916, 62 p.
Play analysis

Master, C. D., D. H. Root, and W. D. Dietzman, 1983, Distribution and quantitative assessment of world crude oil reserves and resources: Proceedings of the Eleventh World Petroleum Congress, London, (Preprint), p. 229–237.
Direct assessment

Mattick, R. E., W. J. Perry, Jr., E. Robbins, E. C. Rhodehamel, E. G. A. Weed, D. J. Taylor, H. L. Krivoy, K. C. Bayer, J. A. Lees, and C. P. Clifford, 1975, Sediments, structural framework, petroleum potential, environmental conditions, and operational considerations of the United States Mid-Atlantic Outer Continental Shelf: U.S. Geological Survey Open-File Report 75-61, 143 p.
Areal yields, historical extrapolation and volumetric yields

Maximov, S. P., and S. A. Vinnikovski, 1983, Development of methods for the quantitative evaluation of petroleum potential in the U.S.S.R.: Journal of Petroleum Geology, v. 5, n. 3, p. 309–314.
Areal yields, material balance, and volumetric yields

Mayer, L. S., R. A. Stine, B. W. Silverman, D. M. Snyder, S. L. Zeger, D. J. Venzon, and A. B. Bruce, 1980, The use of field size distribution in resource estimation: Princeton, New Jersey, Princeton University, Department of Statistics and Geology, [Informal technical report], 39 p.
Discovery modeling and field size analysis

McCrossan, R. G., 1969, An analysis of size frequency distribution of oil and gas reserves of Western Canada: Canadian Journal of Earth Sciences, v. 6, n. 2, p. 201–211.
Field size analysis

McCrossan, R. G., and J. W. Porter, 1973, The geology and petroleum potential of the Canadian sedimentary basins—a synthesis, *in* R. G. McCrossan, ed., The future petroleum provinces of Canada—their geology and potential: Canadian Society of Petroleum Geologists Memoir 1, p. 589–720.
Historical extrapolation, play analysis, and volumetric yields

McCulloh, T. H., 1973, Oil and gas, *in* D. A. Brobst and W. P. Pratt, eds., United States mineral resources: U.S. Geological Survey Professional Paper 820, p. 477–496.
Areal yields, historical extrapolation, and volumetric yields

McDowell, A. N., 1975, What are the problems in estimating the oil potential of a basin?: Oil and Gas Journal, v. 73, n. 23, p. 85–90.
Material balance

McKelvey, V. E., 1968, Contradictions in energy resource estimates, *in* L. B. Holmes, ed., Proceedings of the Seventh Energy Gas Dynamics Symposium: Evanston, Northwestern University Press, p. 18–26.
Historical extrapolation and volumetric yields

——— , 1972, Mineral resource estimates and public policy: American Scientist, v. 60, n. 1, p. 32–40. [Reprinted *in* D. A. Brost and W. P. Pratt, eds., 1973, United States mineral resources: U.S. Geological Survey Professional Paper 820, p. 9–19.]
Field size analysis, historical extrapolation, and volumetric yields

——— , 1984, Undiscovered oil and gas resources: procedures and problems of estimation: Proceedings of the 27th International Geological Congress, v. 13, p. 333–352.
Areal yields, direct assessment, field size analysis, historical extrapolation, material balance, play analysis, and volumetric yields

McKelvey, V. E., F. H. Wang, S. P. Schweinfurth, and W. C. Overstreet, 1969, Potential mineral resources of the United States outer continental shelf, in Nossaman, Waters, Scott, Krueger and Riordan, Study of outer continental shelf lands of the United States: National Technical Information Service PB 188 717, v. IV, p. 5-A-1–5-A-117.
Volumetric yields

Megill, R. E., 1958, How much does it cost to find oil?: Oil and Gas Journal, v. 56, n. 19, p. 189, 192, 196, 198.
Historical extrapolation

———, 1981, An explorationist's approach to prospect evaluation, in Proceedings of the Seminar on Assessment of Undiscovered Oil and Gas, Kuala Lumpur, Malaysia: United Nationas ESCAP CCOP Technical Publication n. 10, p. 263–275.
Play analysis and probability methods

———, 1984, An introduction to risk analysis, 2nd Ed.: Tulsa, Oklahoma, PennWell Books, 274 p.
Direct assessment, historical extrapolation, play analysis, probability methods, and volumetric yields

Meisner, J., and F. Demirmen, 1981, The creaming method: a Bayesian procedure to forecast future oil and gas discoveries in mature exploration provinces: Journal of Royal Statistical Society A, v. 114, n. 1, p. 131.
Discovery modeling

Menard, H. W., 1981, Toward a rational strategy for oil exploration: Scientific American, v. 244, n. 1, p. 55–65.
Discovery modeling

Menard, H. W., and G. Sharman, 1975, Scientific uses of random drilling models: Science, v. 190, n. 4212, p. 337–343.
Discovery modeling

Menzie, D. W., M. L. Labovitz, and J. C. Griffiths, 1977, Evaluation of mineral resources and the unit regional valve concept, in R. V. Ramani, ed., Application of Computer Methods in the Mineral Industry, 14th Symposium Proceedings, University Park, Pennsylvania State University: New York, Society of Mining Engineers of AIME, p. 322–338.
Areal yields

Meyer, R. F., 1977, Petroleum resource data systems: Journal of the International Association for Mathematical Geology, v. 9, n. 3, p. 281–299.
Direct assessment and volumetric yields

———, 1978a, A look at natural gas resources: Oil and Gas Journal, v. 76, n. 19, p. 334, 336, 338, 341, 342, 344.
Reserve growth

———, 1978b, The volumetric method for petroleum resource estimation: Journal of the International Association for Mathematical Geology, v. 10, n. 5, p. 501–518.
Volumetric yields

Meyer, R. F., P. A. Fulton, L. J. Drew, D. R. Root, and G. Grender, 1983, The resource potential of small oil and gas fields, in R. F. Meyer and J. C. Olson, eds., International Conference on The Future of Small Energy Resources, Los Angeles: New York, Unitar and McGraw Hill, p. 9–29.
Discovery modeling and field size analysis

Meyer, R. F., P. A. Fulton, and W. D. Dietzman, 1984, A preliminary estimate of world heavy crude oil and bitumen resources, in R. F. Meyer, J. C. Wynn, and J. C. Olson, eds., Second International Conference on the Future of Heavy Crude and Tar Sands, Caracas, Venezuela: New York, Unitar and McGraw Hill, p. 97–158.
Historical extrapolation

Miller, B. M., 1977, Probabilistic and computer methodologies used by the U.S. Geological Survey for geological estimates of undiscovered oil and gas resources in the United States, in R. V. Ramani, ed., Application of Computer Methods in the Mineral Industry, 14th Symposium Proceedings, University Park, Pennsylvania State University: New York, Society of Mining Engineers of AIME, p. 419–430.
Direct assessment and probability methods

———, 1979, The evolution in the development of the petroleum resource appraisal procedures in the U.S. Geological Survey and a summary of current assessments for the United States: Society of Petroleum Engineers of the American Institute of Mining, Metallurgical, and Petroleum Engineers Paper SPE 7720, p. 79–87.
Direct assessment, field size analysis, historical extrapolation, play analysis, and volumetric yields

———, 1981, Methods of estimating potential hydrocarbon resources by the U.S. Geological Survey: case studies in resource assessment in the National Petroleum Reserve in Alaska and the William O. Douglas Arctic Wildlife Range, in Exploration and Economics of the Petroleum Industry: New York, Matthew Bender & Co., v. 19, p. 57–96.
Play analysis

———, 1982a, Application of exploration play analysis techniques to the assessment of conventional petroleum resources by the USGS: Journal of Petroleum Technology, v. 34, n. 1, p. 55–64.
Play analysis

———, 1982b, The evolution in the development of petroleum resource appraisal procedures in the U.S. Geological Survey, in S. I. Gass, ed., Oil and Gas Supply Modeling, Proceedings: National Bureau of Standards Special Publication 631, p. 171–199.
Historical extrapolation, play analysis, and volumetric yields

———, 1983, Petroleum resource assessments of the wilderness lands in the western United States, in B. M. Miller, ed., Petroleum potential of wilderness lands in the Western United States: U.S. Geological Survey Circular 902 A-P, p. A1-A10.
Probability methods

———, 1984, Oil, gas potential of west U.S. wilderness lands: Oil and Gas Journal, v. 82, n. 18, p. 83.
Probability methods

Miller, B. M., and H. L. Thomsen, 1976, Geological estimates of undiscovered oil and gas resources in the United States, in R. E. Jantzen, ed., Tomorrow's oil from today's provinces: AAPG Miscellaneous Publication 24, p. 1–18.
Direct assessment and probability methods

Miller, B. M., and H. L. Thomsen, G. L. Dolton, A. B. Coury, T. A. Hendricks, F. E. Lennartz, R. B. Powers, E. G. Sable, and K. L. Varnes, 1975, Geological estimates of undiscovered recoverable oil and gas resources in the United States: U.S. Geological Survey Circular 725, 78 p.
Areal yields, direct assessment, historical extrapolation, probability methods, reserve growth, and volumetric yields

Momper, J. A., 1979, Domestic oil reserves forecasting method, regional potential assessment: Oil and Gas Journal, v. 77, n. 33, p. 144–149.
Field size analysis and historical extrapolation

———, 1984, Petroleum resource assessments derived from geochemistry studies (abs.), in C. D. Masters, ed., Petroleum resource assessment: Ottawa, International Union of Geological Sciences Publication n. 17, p. 122.
Material balance

Momper, J. A., and J. A. Williams, 1984, Geochemical exploration in the Powder River Basin, *in* G. Demaison and R. J. Murris, eds., Petroleum geochemistry and basin evaluation: AAPG Memoir 35, p. 181–191.
Material balance

Moody, J. D., and R. W. Esser, 1975, An estimate of world's recoverable crude oil resources, *in* Proceedings of the Ninth World Petroleum Congress, Tokyo, v. 3, p. 11–20.
Play analysis

Moody, J. D., and R. E. Geiger, 1975, Petroleum resources: how much oil and where?: Technology review, v. 77, n. 5, p. 39–45.
Historical extrapolation, play analysis, and volumetric yields

Moody, J. D., J. W. Mooney, and J. Spivak, 1970, Giant oil fields of North America, *in* M. T. Halbouty, ed., Geology of giant petroleum fields: AAPG Memoir 14, p. 8–17.
Field size analysis and historical extrapolation

Moore, C. L., 1962, Method for evaluating U.S. crude oil resources and projecting domestic crude oil availability: U.S. Department of the Interior, Office of Oil and Gas, 112 p.
Historical extrapolation

———, 1966a, Analyses and projections of the historic patterns of U.S. domestic supply of crude oil, natural gas, and natural gas liquids: U.S. Department of the Interior, Office of Oil and Gas, 83 p.
Historical extrapolation

———, 1966b, C. L. Moore's reply to J. M. Ryan: Journal of Petroleum Technology, v. 18, n. 3, p. 286–287.
Historical extrapolation

———, 1966c, Projections of U.S. petroleum supply to 1980: U.S. Department of Interior, Office of Oil and Gas, 42 p.
Historical extrapolation

———, 1966d, Ultimate domestic petroleum discoveries and discovery rates, *in* Economics and the Petroleum Geologists, Symposium Transactions: Midland, Texas, West Texas Geological Society Publication n. 66-53, p. 118–135.
Historical extrapolation and volumetric yields

———, 1971, Analysis and projection of historic patterns of U.S. crude oil and natural gas, *in* I. H. Cram, ed., Future petroleum provinces of the United States—their geology and potential: AAPG Memoir 15, v. 1, p. 50–54.
Historical extrapolation

Nakayama, K., and D. C. Van Siclen, 1981, Simulation model for petroleum exploration: AAPG Bulletin, v. 65, n. 7, p. 1230–1255.
Material balance

Nalivkin, V. D., M. D. Belonin, V. S. Lazarev, S. G. Neruchev, and G. P. Sverchkov, 1976, Kriterii i metody kolichestvennoy otsenki neftegazonosnosti slaboizuchennykh kruppnykh territoriy [Criteria and methods of quantitative assessment of petroelum prospects in poorly studied large territories]: International geological Review, v. 18, n. 11, p. 1259–1268. [in English]
Material balance and volumetric yields

Nassichuk, W. W., 1983, Petroleum potential in arctic North America and Greenland: Cold Regions Science and Technology, v. 7, special issue, p. 51–88.
Play analysis

National Petroleum Council, Committee on Arctic Oil and Gas Resource, 1981, Working papers of the Resource Assessment Task Group of the National Petroleum Council's Committee on Arctic Oil and Gas Resources: Washington, National Petroleum Council, 79 p.
Direct assessment

National Petroleum Council, Committee on Possible Future Petroleum Provinces of the U.S., 1970, Future petroleum provinces of the United States—a summary: Washington, National Petroleum Council, 138 p.
Historical extrapolation and volumetric yields

Nederlof, M. H., 1980, The use of the habitat of oil model in exploration, prospect appraisal, *in* Proceedings of the Tenth World Petroleum Congress, Bucharest: London, Heyden, v. 2, p. 13–21.
Material balance

———, 1981, Calibrated computer simulation as a tool for exploration prospect assessment, *in* Proceedings of the Seminar on Assessment of Undiscovered Oil and Gas, Kuala Lumpur, Malaysia: United Nations ESCAP CCOP Technical Publication n. 10, p. 122–138.
Play analysis, probability methods

Nehring, R., 1978, Giant oil fields and world oil resources: Santa Monica, California, Rand Corporation Report R-2284-CIA, 162 p.
Direct assessment, field size analysis, and historical extrapolation

———, 1981, The discovery of significant oil and gas fields in the United States: Santa Monica, California, Rand Corporation Report USGS/DOE, RAND R-2654/1, 236 p.
Direct assessment and historical extrapolation

Neruchev, S. G., 1964, Possibilities of estimating prognostic reserves of oil on a genetic basis: Petroleum Geology, v. 8, n. 7, p. 368–372.
Material balance

North, F. K., 1973, A sane look at U.S. gas resources: National gas survey, U.S. Federal Power Commission, v. 5, p. 113–156.
Historical extrapolation and volumetric yields

Oil and Gas Journal, 1969, Vast Delaware–Val Verde reserve seen: Oil and Gas Journal, v. 67, n. 16, p. 44.
Areal yields

Ovasenov, G. P., and A. D. Nadezhkin, 1962, Method of calculating prognostic reserves of oil and gas: Petroleum Geology, v. 6, n. 4, p. 230–233.
Play analysis

Parker, J. M., compiler, 1977, American Association of Petroleum Geologists Petroleum Resources Estimation Project pilot study—Rocky Mountain Area: Tulsa, Oklahoma, AAPG, 61 p.
Play analysis and probability methods

Pelto, C. R., 1973, Forecasting ultimate oil recovery: Society of Petroleum Engineers of the American Institute of Mining, Metallurgical, and Petroleum Engineers, Symposium on Petroleum Economics and Evaluation, Dallas, SPE 4261, p. 45–52.
Historical extrapolation and reserve growth

Perrodon, A., 1972, Provinces petroliers: approaches statistique et geologique de quelques types d'habitat de l'huile et du gaz: Mineral Fuels, sec. 5, 24th International Geological Congress, Proceedings of Section Reports, p. 176–186.
Areal yields

Porter, J. W., and R. G. McCrossan, 1975, Basin consanguinity in petroleum resource estimation, *in* J. D. Haun, ed., Methods of estimating the volume of undiscovered oil and gas resources: AAPG Studies in Geology, n. 1, p. 50–75.
Field size analysis and play analysis

Potential Gas Committee, 1967, Potential supply of natural gas in the United States as of December 31, 1966: Golden, Colorado, Potential Gas Agency, Colorado School of Mines, 38 p.
Areal yields and volumetric yields

———, 1969, Potential supply of natural gas in the United States (as of December 31, 1968): Golden, Colorado, Potential Gas Agency, Mineral Resources Institute, Colorado School of Mines Foundation, 39 p.
Volumetric yields

———, 1971, Potential supply of natural gas in the Unite States (as of December 31, 1970): Golden, Colorado, Potential Gas Agency, Mineral Resources Institute, Colorado School of Mines Foundation, 41 p.
Volumetric yields

———, 1973, Potential supply of natural gas in the United States (as of December 31, 1972): Golden, Colorado, Potential Gas Agency, Mineral Resources Institute, Colorado School of Mines Foundation, 48 p.
Volumetric yields

———, 1977a, A comparison of estimates of ultimately recoverable quantities of natural gas in the United States: Golden, Colorado, Potential Gas Agency, Mineral Resources Institute, Colorado School of Mines Foundation, Gas Resources Studies n. 1, 27 p.
Direct assessment, historical extrapolation, play analysis, and volumetric yields

———, 1977b, Potential supply of natural gas in the United States (as of December 31, 1976): Golden, Colorado, Potential Gas Agency, Mineral Resources Institute, Colorado School of Mines Foundation, 45 p.
Volumetric yields

———, 1979, Potential supply of natural gas in the United States (as of December 31, 1978): Golden, Colorado, Potential Gas Agency, Mineral Resources Institute, Colorado School of Mines Foundation, 75 p.
Volumetric yields

———, 1981, Potential supply of natural gas in the United States (as of December 31, 1980): Golden, Colorado, Potential Gas Agency, Mineral Resources Institute, Colorado School of Mines Foundation, 119 p.
Play analysis and volumetric yields

Potential Gas Committee, Committee on Definitions and Procedures, 1984, Definitions and procedures for estimation of potential gas resources: Golden, Colorado, Potential Gas Agency, Gas Resources Studies, Colorado School of Mines, n. 2, 16 p.
Play analysis and volumetric yields

Pratt, W. E., 1937, Discovery rates in oil finding: AAPG Bulletin, v. 21, n. 6, p. 697–705.
Historical extrapolation

———, 1951, On the stimulation of undiscovered oil reserves: Journal of Petroleum Technology, v. 3, n. 4, p. 9–10.
Volumetric yields

Procter, R. M., and G. C. Taylor, 1984, Evaluation of oil and gas potential of an offshore west coast Canada play an example of Geological Survey of Canada methodology, in C. D. Masters, ed., Petroleum resource assessment: Ottawa, International Union of Geological Sciences Publication n. 17, p. 39–62.
Play analysis

Qahwash, A. A., and R. A. Akkad, 1978, Spatial distribution of the occurrence probability of oil fields: The Arabian Journal for Science and Engineering, special issue, p. 1–4.
Discovery modeling

Rapoport, L. A., and G. C. Grender, 1983, Discovery function for oil or gas by depth zones throughout the U.S. basins (abs.): AAPG Bulletin, v. 67, n. 3, p. 538.
Discovery modeling

Reznik, V. S., 1981, Metod veroiatnostnoi otsenki resursov nefti i gaza sedimentatsionnykh basseinov [Probabilistic method of assessing oil and gas reserves in sedimentary basins]: Geologizya Nefti i Gaza, n. 4, p. 24–29. [Reprinted in 1982, International Geology Review, v. 24, n. 7, p. 797–802.]
Volumetric yields

Riesz, E. J., 1978, Can rank-size "laws" be used for undiscovered petroleum and mineral assessments?: BMR Journal of Australian Geology and Geophysics, v. 3, n. 3, p. 253–256.
Field size analysis

Roadifer, R. E., 1975, A probability approach to estimate volumes of undiscovered oil and gas, in M. Grenon, ed., Proceedings of the First IIASA conference on Energy Resources, Laxenburg, Austria, p. 333–343. [Reprinted in M. Grenon, ed., 1979, Methods and models for assessing energy resources, Proceedings of the First IIASA Conference on Energy Resources, Laxenburg, Austria: Oxford, Pergamon Press, p. 268–278.]
Play analysis

Root, D. H., 1981, Estimation of inferred plus indicated reserves for the United States, in G. L. Dolton, K. H. Carlson, R. R. Charpentier, A. B. Coury, R. A. Crovelli, S. E. Frezon, A. S. Khan, J. H. Lister, R. H. McMullin, R. S. Pike, R. B. Powers, E. W. Scott, and K. L. Varnes, Estimates of undiscovered recoverable conventional resources of oil and gas in the United States: U.S. Geological Survey Circular 860, p 83–87.
Reserve growth

———, 1982, Historical growth of estimates of oil- and gas-field size, in S. I. Gass, ed., Oil and Gas Supply Modeling Proceedings: National Bureau of Standards Special Publication 631, p. 350–368.
Reserve growth

Root, D. H., and L. J. Drew, 1979, The pattern of petroleum discovery rates: American Scientist, v. 67, n. 6, p. 648–652.
Field size analysis and historical extrapolation

———, 1984, Practical solutions to problems in the application of statistical analysis to oil and gas resource appraisal illustrated by case studies, in C. D. Masters, ed., Petroleum resource assessment: Ottawa, International Union of Geological Sciences Publication n. 17, p. 123–139.
Historical extrapolation

Root, D. H., and J. H. Schuenemeyer, 1980, Petroleum-resource appraisal and discovery rate forecasting in partially explored regions—mathematical foundations: U.S. Geological Survey Professional Paper 1138-B, 9 p.
Discovery modeling

Rose, P. R., 1975, Procedures for assessing U.S. petroleum resources and utilization of result in M. Grenon, ed., Proceedings of the First IIASA Conference on Energy Resources, Laxenburg, Austria, p. 291–309. [Reprinted in M. Grenon, ed., 1979, Methods and models for assessing energy resources, Proceedings of the First IIASA Conference on Energy Resources, Laxenburg, Austria: Oxford, Pergamon Press, p. 229–247.]
Direct assessment, historical extrapolation, and volumetric yields

Roy, K. J., 1974, Quantitative assessment of resource potential using Monte Carlo simulation, in T. Gordon and W. W. Hutchinson, eds., Computer use in projects of the Geological Survey of Canada: Geological Survey of Canada Paper 76-60, p. 81.
Play analysis

———, 1975, Hydrocarbon assessment using subjective probability and Monte Carlo methods, in M. Grenon, ed., Proceedings of the First IIASA Conference on Energy

Resources, Laxenburg, Austria, p. 345–359. [Reprinted in M. Grenon, ed., 1979, Methods and models for assessing energy resources, Proceedings of the First IIASA Conference on Energy Resources, Laxenburg, Austria: Oxford, Pergamon Press, p. 279–290.]
Play analysis

Roy, K. J., R. M. Procter, and R. G. McCrossan, 1975, Hydrocarbon assessment using subjective probability, *in* J. C. Davis, J. H. Doveton, and J. W. Harbaugh, convenors, Probability Methods in Oil Exploration, AAPG Research Symposium, Stanford, California: Kansas Geological Survey, p. 56–60.
Play analysis and volumetric yields

Rozanov, Y. A., 1975, Hypothetical probabilistic prototype of an undiscovered resources model, *in* M. Grenon, ed., Proceedings of the First IIASA Conference on Energy Resources, Laxenburg, Austria, p. 323–331. [Reprinted *in* M. Grenon, ed., 1979, Methods and models for assessing energy resources, Proceedings of the First IIASA Conference on Energy Resources, Laxenburg, Austria: Oxford, Pergamon Press, p. 261–267.]
Areal yields

Ryan, J. M., 1965, National Academy of Sciences report on energy resources: discussion of limitations of logistic projections: AAPG Bulletin, v. 49, n. 10, p. 1713–1720.
Historical extrapolation

———, 1966, Limitations of statistical methods for predicting petroleum and natural gas reserves and availability: Journal of Petroleum Technology, v. 18, n. 3, p. 281–284.
Historical extrapolation and volumetric yields

Ryan, J. T., 1973a, An analysis of crude-oil discovery rate in Alberta: Bulletin of Canadian Petroleum Geology, v. 21, n. 2, p. 219–235.
Historical extrapolation

———, 1973b, An estimate of the conventional crude-oil potential in Alberta: Bulletin of Canadian Petroleum Geology, v. 21, n. 2, p. 236–246.
Historical extrapolation and reserve growth

Ryockborst, H., 1980, Determining probable maximum size of oil, gas pools, oil-in-place volumes: Oil and Gas Journal, v. 78, n. 6, p. 150–154.
Field size analysis

Schagen, I. P., 1980, A stochastic model for the occurrence of oilfields and its application to some North Sea data: Applied Statistics, v. 29, n. 3, p. 282–291.
Field size analysis

Schuenemeyer, J. H., 1980, An analysis of a statistical model to predict future oil, *in* D. A. Gardiner and Tykey Truett, compiler/ed., Proceedings of the Department of Energy Statistical Symposium, Gatlinburg, Tennessee, p. 223–225.
Discovery modeling

Schuenemeyer, J. H., and L. J. Drew, 1983, A procedure to estimate the parent population of the size of oil and gas fields as revealed by a study of economic truncation: Journal of the International Association for Mathematical Geology, v. 15, n. 1, p. 145–161.
Field size analysis

Schuenemeyer, J. H., and D. H. Root, 1977, Computational aspects of a probabilistic oil discovery model: Proceedings of the Statistical Computing Section, Annual Meeting of the American Statistical Association, Washington, D.C., p. 347–351.
Discovery modeling

Schuenemeyer, J. H., L. J. Drew, and W. Bawiec, 1979, Predicting future oil using three-dimensional discovery-process model (abs.): AAPG Bulletin, v. 63, n. 3, p. 522–523.
Discovery modeling

Schuenemeyer, J. H., W. J. Bawiec, and L. J. Drew, 1980, Computational methods for a three-dimensional model of the petroleum-discovery process: Computers and Geosciences, v. 6, n. 4, p. 323–360.
Discovery modeling

Schultz, P. R., 1952, What is the future of petroleum discovery: Oil and Gas Journal, v. 51, n. 12, p. 258–259, 295–300.
Historical extrapolation

Seigneurin, A., 1975, Probabilistic evaluation technique, *in* M. Grenon, ed., Proceedings of the First IIASA Conference on Energy Resources, Laxenburg, Austria, p. 373–381. [Reprinted *in* M. Grenon, ed., 1979, Methods and models for assessing energy resources, Proceedings of the First IIASA Conference on Energy Resources, Laxenburg, Austria: Oxford, Pergamon Press, p. 302–310.]
Play analysis

Semenovich, V. V., N. I. Buyalov, V. N. Kramarenko, A. E. Kontorovich, Yu. Ya. Kuznetsov, S. P. Maksimov, M. Sh. Modelevsky, and I. I. Nesterov, 1977, Methods used in the U.S.S.R. for estimating potential petroleum resources, *in* R. F. Meyer, ed., The future supply of nature-made petroleum and gas, First UNITAR Conference on Energy and the Future, Laxenburg, Austria, Technical Reports: New York, Pergamon Press, p. 139–153.
Historical extrapolation, material balance, play analysis, and volumetric yields

Sheldon, R. P., 1976, Estimates of undiscovered petroleum resources—a perspective: U.S. Geological Survey Annual Report Fiscal Year 1975, p. 11–22. [Reprinted *in* R. F. Meyer, ed., 1977, The future supply of nature-made petroleum and gas technical reports, First UNITAR Conference on Energy and the Future, Laxenburg, Austria, Technical Reports: New York, Pergamon Press, p. 997–1023.]
Historical extrapolation and volumetric yields

Sickler, R. A., 1975, World petroleum resources, Part 1: method and models used to estimate world petroleum resources, *in* M. Grenon, ed., Proceedings of the First IIASA Conference on Energy Resources, Laxenburg, Austria, p. 183–206. [Reprinted *in* M. Grenon, ed., 1979, Methods and models for assessing energy resources, Proceedings of the First IIASA Conference on Energy Resources, Laxenburg, Austria: Oxford, Pergamon Press, p. 117–131.]
Areal yields, direct assessment, field size analysis, historical extrapolation, material balance, and volumetric yields

Singer, D. A., 1971, Multivariate statistical analysis of the unit regional value of mineral resources: Ph.D. thesis, Pennsylvania State University, University Park, 210 p.
Areal yields

Singer, D. A., and L. J. Drew, 1975, The area of influence of an exploratory hole, *in* J. C. Davis, J. H. Doveton, and J. W. Harbaugh, convenors, 1975, Probability Methods in Oil Exploration, AAPG Research Symposium, Stanford, California: Kansas Geological Survey, p. 61–65. [Reprinted in 1976, Economic Geology, v. 71, n. 3, p. 642–647.]
Historical extrapolation

Singer, S. F., 1975, Oil resource estimates: Science, v. 188, n. 1487, p. 401.
Historical extrapolation and volumetric yields

Sluijk, D., and M. H. Nederlof, 1984, Worldwide geological experience as a systematic basis for prospect appraisal, in G. Demaison and R. J. Murris, eds., Petroleum geochemistry and basin evaluation: AAPG Memoir 35, p. 15–26.
Material balance

Smith, M. B., 1968, Estimate reserves by using computer simulation method: Oil and Gas Journal, v. 66, n. 11, p. 80–84.
Play analysis and probability methods

Steinhart, J. S., and B. J. McKellar, 1981, Future availability of oil for the United States, in L. C. Ruedisili and M. W. Firebaugh, eds., Perspectives on energy: issues, ideas, and environmental dilemmas: New York, Oxford University Press, 156–186.
Field size analysis and historical extrapolation

Steinhart, J. S., and M. Bultman, 1983, How undiscovered oil is estimated: Oceanus, v. 26, n. 3, p. 40–45.
Historical extrapolation

Stoian, E., 1965, Fundamentals and applications of the Monte Carlo method: Journal of Canadian Petroleum Technology, v. 4, n. 4, p. 120–129.
Probability methods

Suardy, A., F. X. Suyanto, and N. Hariadi, 1981, Assessment of undiscovered recoverable hydrocarbon resources in Indonesia, in Proceedings of the Seminar on Assessment of Undiscovered Oil and Gas, Seminar, Kuala Lumpur, Malaysia: United Nations ESCAP CCOP Technical Publication n. 10, p. 164–169.
Volumetric yields

Surkov, V. S., 1977, Evaluation methods of hydrocarbon potential reserves in east Siberia, in R. F. Meyer, ed., The future supply of nature-made petroleum and gas, First UNITAR Conference on Energy and the Future, Laxenburg, Austria: New York, Pergamon Press, p. 381–388.
Material balance and volumetric yields

Swetland, P. J., and D. M. Demshur, 1981, Geochemistry and basin assessment, in Proceedings of the Seminar on Assessment of Undiscovered Oil and Gas, Kuala Lumpur, Malaysia: United Nations ESCAP CCOP Technical Publication n. 10, p. 62–65.
Material balance

Tanner, W. F., 1978, Future oil: what is the outlook?: World Oil, v. 187, n. 6, p. 123, 124, 128, 133, 136, 140.
Historical extrapolation

Tissot, B., 1973, Vers l'evaluation quantiative du petrole forme dans les bassins sedimentaires: Revue de l'Association Francais des Technicieus du Petrole, n. 222, p. 27–31.
Material balance

Tissot, B., and J. Espitalie, 1975, L'evolution thermique de la matiere organique des sediments: applications d'une simulation mathematique: Revue de l'Institut Francais du Petrole, v. 30, p. 743–777.
Material balance

Tissot, B., and D. H. Welte, 1978, Petroleum formation and occurrence: Berlin, Springer Verlag, 538 p.
Material balance

Trask, P. D., 1936, Proportion of organic matter converted into oil in Santa Fe Springs Field, California: AAPG Bulletin, v. 20, n. 3, p. 245–257.
Material balance

Uhler, R. S., and R. G. Bradley, 1970, A stochastic model for determining the economic prospects of petroleum exploration over large regions: Journal of the American Statistical Association, v. 65, n. 330, p. 623–630.
Areal yields and field size analysis

Ulmishek, G., 1982, Petroleum geology and resource assessment of the Timan-Pechora Basin, U.S.S.R., and the adjacent Barents-Northern Kara Shelf: Argonne, Illinois, Argonne National Laboratory Report ANL/EES-TM-199, 197 p.
Volumetric yields

Ulmishek, G., and W. Harrison, 1981, Petroleum geology and resource assessment of the Middle Caspian Basin, U.S.S.R., with special emphasis on the Uzen Field: Argonne, Illinois, Argonne National Laboratory Report ANL/ES-116, 147 p.
Material balance and volumetric yields

—— , 1984, Quantitative methods for assessment of petroleum resources of poorly known basins, in C. D. Masters, ed., Petroleum resource assessment: Ottawa, International Union of Geological Sciences Publication n. 17, p. 80–94.
Volumetric yields

Uman, M. F., W. R. James, and H. R. Tomlinson, 1979, Oil and gas in offshore tracts: estimates before and after drilling: Science, v. 205, n. 4405, p. 489–491.
Play analysis

—— , 1980, (Reply to Davis and Harbaugh): Science, v. 209, n. 4460, p. 1048.
Play analysis

Ungerer, P., F. Bessis, P. Y. Chenet, B. Durand, E. Nogaret, A. Chiarelli, J. L. Oudin, and J. F. Perrin, 1984, Geological and geochemical models in oil exploration; principles and practical examples, in G. Demaison and R. J. Murris, eds., Petroleum geochemistry and basin evaluation: AAPG Memoir 35, p. 53–77.
Material balance

U.S. Department of Energy and U.S. Geological Survey, 1979, Report on the petroleum resources of the Federal Republic of Nigeria: U.S. Department of Energy Report DOE/IA-0008, 137 p.
Field size analysis, historical extrapolation, and volumetric yields

U.S. Department of the Interior, Office of Minerals Policy and Research Analysis, 1980, Final report of the 105(b) Economic and policy analysis: Alternative overall procedures for the exploration, development, production, transportation and distribution of the petroleum resources of the National Petroleum Reserve in Alaska (NPRA): U.S. Department of the Interior, 145 p.
Play analysis

U.S. Geological Survey, 1953, Potential oil and gas reserves of the continental shelf off the coasts of Louisiana, Texas, and California: Statement prepared by the Fuels Branch, Geologic Division, U.S. Geological Survey, at request of the Committee on Interior and Insular Affairs, U.S. Senate, February 16, 1953, 11 p.
Areal yields

U.S. Interagency Oil and Gas Supply Project, 1980, Future supply of oil and gas from the Permian Basin of west Texas and southeastern New Mexico: U.S. Geological Survey Circular 828, 57 p.
Areal yields, discovery modeling, field size analysis, historical extrapolation, reserve growth, and volumetric yields

Uri, N. D., 1978, A re-examination of undiscovered oil resources in the United States: U.S. Energy Information Administration Technical Memorandum TM/ES/79-03, 20 p.
Historical extrapolation

———, 1979, A hybrid approach to the estimation of undiscovered oil resources in the United States: Energy, v. 4, n. 6, p. 1079–1085.
Historical extrapolation

Uspenskiy, V. A., O. A. Radchenko, and N. F. Yarovaya, 1980, Quantitative determination of the oil-source potential of rocks: International Geology Review, v. 22, n. 6, p. 680–684.
Material balance

Vinkovetsky, Y., and V. Rokhlin, 1982, Quantitative evaluation of the contribution of geologic knowledge in exploration for petroleum, *in* G. DeMarsily and D. F. Merriam, Predictive Geology: Oxford and New York, Pergamon Press, p. 171–190.
Discovery modeling

Walstrom, J. E., T. D. Mueller, and R. C. McFarlane, 1967, Evaluating uncertainty in engineering calculations: Journal of Petroleum Technology, v. 19, n. 12, p. 155–1603. [Reprinted from the Society of Petroleum Engineers of the American Institute of Mining, Metallurgical and Petroleum Engineers Paper SPE 1928.]
Probability methods

Waples, D. W., 1979, Simple method for oil source bed evaluation: AAPG Bulletin, v. 63, n. 2, p. 239–248.
Material balance

———, 1980, Time and temperature in petroleum formation: application of Lopatin's method to petroleum exploration: AAPG Bulletin, v. 64, n. 6, p. 916–926.
Material balance

Weeks, L. G., 1950, Concerning estimates of potential oil reserves: AAPG Bulletin, v. 34, n. 10, p. 1947–1953.
Areal yields and volumetric yields

———, 1958, Fuel reserves of the future: AAPG Bulletin, v. 42, n. 2, p. 431–438.
Areal yields and volumetric yields

———, 1965, World offshore petroleum resources: AAPG Bulletin, v. 49, n. 10, p. 1680–1693.
Areal yields

———, 1966, Assessment of the world's offshore petroleum resources and exploration review, *in* Exploration and Economics of the Petroleum Industry: New York, Matthew Bender & Co., v. 4, p. 115–148.
Areal yields

———, 1975, Potential petroleum resources—classification, estimation, and status, *in* J. D. Haun, ed., Methods of estimating the volume of undiscovered oil and gas resources: AAPG Studies in Geology, n. 1, p. 31–49.
Areal yields, historical extrapolation, and volumetric yields

Welte, D. H., and M. A. Yukler, 1980, Evolution of sedimentary basins from the standpoint of petroleum origin and accumulation—an approach for a quantitative basin study: Organic Geochemistry, v. 2, n. 1, p. 1–8.
Material balance

———, 1981, Petroleum origin and accumulation in basin evolution—a quantitative model: AAPG Bulletin, v. 65, n. 8, p. 1387–1396. [Reprinted in G. Demaison and R. J. Murris, eds., 1984, Petroleum geochemistry and basin evaluation: AAPG Memoir 35, p. 27–39.]
Material balance

Welte, D. H., and M. A. Yukler, M. Radke, and D. Leythaeuser, 1981, Application of organic geochemistry and quantitative analysis to petroleum origin and accumulation—an approach for a quantitative basin study, *in* G. Atkinson, and J. J.

Zuckerman, eds., Origin and chemistry of petroleum: Oxford and New york, Pergamon Press, p. 67–88.
Material balance

West, J., 1974, U.S. oil policy riddle: how much left to find?: Oil and Gas Journal, v. 72, n. 37, p. 25–28.
Historical extrapolation and volumetric yields

White, D. A., 1980, Assessing oil and gas plays in facies-cycle wedges: AAPG Bulletin, v. 64, n. 8, p. 1158–1178.
Areal yields, historical extrapolation, play analysis, and volumetric yields

White, D. A., and T. A. Fitzgerald, 1976, Random drilling: Science, v. 192, n. 4236, p. 206–207.
Discovery modeling

White, D. A., and H. M. Gehman, 1979, Methods of estimating oil and gas resources: AAPG Bulletin, v. 63, n. 12, p. 2183–2192.
Areal yields, direct assessment, historical extrapolation, material balance, play analysis, probability methods, and volumetric yields

White, D. A., R. W. Garrett, Jr., G. R. Marsh, R. A. Baker, and H. M. Gehman, 1975a, Three methods assess regional oil, gas potential: Oil and Gas Journal, v. 73, n. 34, p. 140–142, n. 35, p. 140–143.
Historical extrapolation, probability methods, reserve growth, and volumetric yields

———, 1975b, Assessing regional oil and gas potential, *in* J. D. Haun, ed., Methods of estimating the volume of undiscovered oil and gas resources: AAPG Studies in Geology, n. 1, p. 143–159.
Historical extrapolation, probability methods, reserve growth, and volumetric yields

White, L. P., 1981, A play approach to hydrocarbon resource assessment and evaluation, *in* J. B. Ramsey, ed., The economics of exploration for energy resources: Greenwich, Connecticut, Jai Press, p. 51–68.
Play analysis and volumetric yields

Willums, J. O., 1981, A hybrid model for assessing regional petroleum potential with a case study of China's continental shelf, *in* Proceedings of the Seminar on Assessment of Undiscovered Oil and Gas, Kuala Lumpur, Malaysia: United Nations ESCAP CCOP Technical Publication n. 10, p. 139–148.
Discovery modeling, field size analysis, historical extrapolation, and volumetric yields

Wiorkowski, J. J., 1981, Estimating volumes of remaining fossil fuel resources: a critical review: Journal of the American Statistical Association, v. 76, n. 375, p. 534–548.
Discovery modeling, historical extrapolation, reserve growth, and volumetric yields

Zapp, A. D., 1962, Future petroleum producing capacity of the United States: U.S. Geological Survey Bulletin 1142-H, 36 p.
Historical extrapolation

Zhdanov, M. A., 1962, Prognostic reserves of oil and gas and problems of methods of estimating them: Petroleum Geology, v. 6, n. 3, p. 177–184.
Areal yields, play analysis, and volumetric yields

Zielinski, R. E., and R. D. McIver, 1982, Resource and exploration assessment of the oil and gas potential in the Devonian gas shales of the Appalachian Basin: Monsanto Research Corp. for the U.S. Department of Energy, Morgantown Energy Technology Center, Contract no. DE-AC04-76-OP00053, Report n. MLM-MU-82-61-0002, DOE/DP/0053-1125, 365 p.
Material balance

Index

A reference is indexed according to its important, or "key" words.

Three columns are to the left of a keyword entry. The first column, a letter entry, represents the AAPG book series from which the reference originated. In this case, ST stands for Studies in Geology Series. Every five years, AAPG will merge all its indexes together, and the letters ST will differentiate this reference from those of the AAPG Memoir Series (ME) or from the AAPG Bulletin (B).

The following number is the series number. In this case, 21 represents a reference from Studies in Geology Series 21. The third column lists the page number of this volume on which the reference can be found.

† = titles   * = authors